Collective Spatial Cognition

This book integrates the science of spatial cognition and the science of team cognition to explore the social, psychological, and behavioral phenomenon of spatial cognition as it occurs in human collectives such as dyads and work teams.

It represents the culmination of a process of outlining and defining a growing field of research termed *Collective Spatial Cognition*. It engages contributions from an international and multi-disciplinary community of scholars, who have collaborated to provide a foundation for knowledge discovery regarding how groups of people of varying size acquire information and solve problems involving spatiality as a key component, leading to action that incorporates the spatial information and problem-solving collectively achieved. The collectives under study can be as small as dyads (teams of two) to large teams-of-teams who are working alongside each other to complete a mutual goal. The book lays the foundation for multidisciplinary and interdisciplinary work regarding *Collective Spatial Cognition* in the years to come, and this book documents that foundation.

This book will be of interest to those researching spatial, behavioral, cognitive, and information sciences in the fields of human geography, sociology, psychology, and computer science.

Kevin M. Curtin is Professor of Geography and Director of the Laboratory for Location Science at the University of Alabama. He received his B.A. and M.A. in urban and environmental geography from the University of Illinois at Chicago and his Ph.D. in geography from the University of California, Santa Barbara. He performs primary research in the field of geographic information science with specializations in location science, transportation and logistics, urban resource allocation, spatial statistics, and network geographic information systems (GIS). Application areas for his research include autonomous vehicle logistics, transportation geography, crime studies, health and nutrition, and geospatial intelligence, and he teaches extensively at both the undergraduate and graduate university levels. His research program has been supported by the Office for Naval Research, the National Science Foundation, and the Army Research Institute, among others.

Daniel R. Montello is Professor of Geography and Affiliated Professor of Psychological and Brain Sciences at the University of California, Santa Barbara. His educational background is in environmental, cognitive, and developmental psychology. His research is in the areas of spatial, environmental, and geographic perception, cognition, affect, and behavior. Dan has authored or co-authored over 100 articles and chapters, and co-authored or edited seven books, including the *Handbook of Behavioral and Cognitive Geography* (2018); *Space in Mind: Concepts for Spatial Learning and Education* (2014) with Karl Grossner and Don Janelle; and *An Introduction to Scientific Research Methods in Geography and Environmental Studies*, 2nd edition (2013) with Paul C. Sutton. He co-edits the academic journal *Spatial Cognition and Computation.*

Collective Spatial Cognition

A Research Agenda

Edited by Kevin M. Curtin
and Daniel R. Montello

LONDON AND NEW YORK

First published 2024
by Routledge
4 Park Square, Milton Park, Abingdon, Oxon OX14 4RN

and by Routledge
605 Third Avenue, New York, NY 10158

Routledge is an imprint of the Taylor & Francis Group, an informa business

British Library Cataloguing-in-Publication Data
A catalogue record for this book is available from the British Library

Library of Congress Cataloging-in-Publication Data
Names: Curtin, Kevin M., 1967– editor. | Montello, Daniel R., 1959– editor.
Title: Collective spatial cognition / edited by Kevin M. Curtin and Daniel R. Montello.
Description: Abingdon, Oxon ; New York, NY : Routledge, [2023] | Includes bibliographical references and index.
Identifiers: LCCN 2023020857 (print) | LCCN 2023020858 (ebook) | ISBN 9781032065427 (hardback) | ISBN 9781032065434 (paperback) | ISBN 9781003202738 (ebook)
Subjects: LCSH: Space perception. | Spatial ability.
Classification: LCC BF469 .C635 2023 (print) | LCC BF469 (ebook) | DDC 153.7/52—dc23/eng/20230722
LC record available at https://lccn.loc.gov/2023020857
LC ebook record available at https://lccn.loc.gov/2023020858

ISBN: 978-1-032-06542-7 (hbk)
ISBN: 978-1-032-06543-4 (pbk)
ISBN: 978-1-003-20273-8 (ebk)

DOI: 10.4324/9781003202738

Typeset in Times New Roman
by Apex CoVantage, LLC

Contents

Acknowledgments

This research activity was sponsored by the U.S. Army Research Institute for the Behavioral and Social Sciences (ARI) and accomplished under Grant Number W911NF-18–1 - 0273. The views, opinions, and/or findings contained in this volume are those of the editors and contributors and shall not be construed as an official department of the army position, policy, or decision, unless so designated by other documents. The editors would like to acknowledge the critical role Alec D. Barker played in the conception of the idea for the Collective Spatial Cognition research effort and for crafting the motivating material that led to the successful funding of the activities. We would also like to thank Penelope Mitchell and Crystal Bae for their supportive assistance with this project. We greatly appreciate the efforts of all of our contributing authors and others who participated in the Collective Spatial Cognition activities along the way.

Contributors

Jessica Andrews-Todd is Research Scientist at the Cognitive and Learning Sciences Group, Educational Testing Service. Her research has focused on collaborative learning and assessment of collaborative problem-solving and communication skills. Her research on learning examines how individuals acquire accurate and inaccurate information as a function of their collaborative experiences. In her assessment work, Dr. Andrews-Todd has developed and applied approaches for (1) identifying measurable components of complex constructs such as collaborative problem-solving and (2) reasoning up from learner behaviors in fine-grained log data from games and simulations to inferences about high-level proficiencies. Her research has been funded by the National Science Foundation, the Institute of Education Sciences, and the Gordon Commission/MacArthur Foundation.

Crystal Bae is Assistant Instructional Professor of GIScience at the Division of the Social Sciences, University of Chicago. She received her Ph.D. in geography with an emphasis on cognitive science from the University of California, Santa Barbara (UCSB), in 2020, and was a postdoctoral researcher at the Movement Data Science Lab, UCSB. Dr. Bae's research focuses on the spatial cognition of neighborhoods and regions, social interaction and decision-making during navigation, and geographic movement visualization. Her core research questions investigate the use of social and spatial cues to guide wayfinding behavior in real-world environments and assess the interpretation and use of geovisualizations and other representations of spatial information.

Michael R. Baumann is Professor at The University of Texas at San Antonio. He earned his Ph.D. in psychology from the Social, Organizational, and Individual Differences Division, University of Illinois at Urbana–Champaign. Dr. Baumann's oldest line of research involves problem-solving and decision-making in groups and teams, particularly issues of how member inputs are combined (e.g., recognizing and leveraging expertise, information sharing, information weighting, and transactive memory). More recently, he has begun examining related issues in multiteam systems. Dr. Baumann also researches performance in complex, dynamic, and high-stakes environments, including civilian firefighting and military settings.

Tad T. Brunyé received a Ph.D. in experimental cognitive psychology from Tufts University in 2007 and currently holds positions as Senior Cognitive Scientist at the U.S. Army DEVCOM Soldier Center, Scientific Manager at the Center for Applied Brain and Cognitive Sciences, and Visiting Associate Professor at Tufts University. His research interests include biosensing, predictive modeling of cognitive and physical performance, and technologies for performance optimization and enhancement. Dr. Brunyé holds over 170 publications in these domains, serves as an associate editor for three scholarly journals, and was honored with the Presidential Early Career Award for Scientists and Engineers (PECASE).

Dorothy R. Carter received her Ph.D. from the Georgia Institute of Technology, and is Associate Professor of Organizational Psychology at Michigan State University. Her research investigates the phenomena that enable leaders, teams, and larger systems to tackle complex challenges in a variety of organizational contexts, including the military, medicine, scientific research, corporations, and spaceflight. She has co-authored more than 40 articles and chapters on teamwork and leadership, many of which are published in top outlets in the organizational sciences. Her current research program is funded by the United States Army Research Institute, the National Science Foundation, and the National Aeronautics and Space Administration.

You (Lily) Cheng is a Ph.D. candidate with the Department of Cognitive Sciences, University of California, Irvine. She received a master's degree from the University of California, Irvine, a master's degree from the University of California, Santa Barbara, and a bachelor's degree from South China Normal University.

Elizabeth R. Chrastil is Associate Professor at the Department of Neurobiology and Behavior, University of California, Irvine. She was previously a faculty member with the Department of Geography, University of California, Santa Barbara. She received her Ph.D. from Brown University, with a master's degree from Tufts University and bachelor's degree from Washington University in St. Louis.

Thomas J. Cova is Professor of Geography at The University of Utah. His research interests are environmental hazards, transportation, and geographic information science with a particular focus on wildfire evacuation planning and management. He has a B.S. in computer and information science from the University of Oregon and an M.S. and Ph.D. in geography from the University of California, Santa Barbara.

Bianca Dalangin graduated from the University of California, Santa Barbara, with B.A. degrees in communication and psychology in 2019. She is a researcher at DCS Corp with a research background in behavioral science and a technical expertise in psychology research methodology. Her current research interests include individual differences in visual attention and learning performance as well as communication in heterogeneous teams in human–computer interactive spaces.

Frank A. Drews is Professor of Cognitive Psychology at The University Utah. He created the department's award-winning Human Factors Certificate program and served as its Director for 15 years. He also served as the Director of the Center for Human Factors in Patient Safety at Salt Lake City VA. Before joining The University of Utah in 2000, he received his Ph.D. in cognitive psychology from the Technical University of Berlin, Germany. Dr. Drews' research has been funded by federal agencies, such as NIOSH, NIH, NSF, NASA, CDC, the VA, and DARPA, as well as numerous industry sponsors.

Dario Esposito has a Ph.D. in environmental risk and territorial and building development from the Polytechnic University of Bari, Italy. He is currently a research fellow at the Polytechnic University of Bari, Italy, where he teaches environmental and territorial engineering. His research interests range from urban development modeling and planning to the design of decision support systems and also include complex adaptive systems analysis for disaster risk management. In his research, he develops computational approaches to assess qualitative aspects that influence the resilient and sustainable use of space and infrastructures in cities, principally related to human factors and knowledge. He is also Vice-Chair of the Technical Commission n. 465 Sustainable Cities and Communities for the European Committee for Standardization.

Benjamin T. Files has a Ph.D. in neuroscience from the University of Southern California, where he received the College Doctoral Fellowship and a Hearing and Communication Science Fellowship. He did postdoctoral work at DEVCOM ARL in Aberdeen Proving Ground, Maryland, on brain–computer interfaces for human/autonomy teaming. He is currently a researcher at DEVCOM ARL, Los Angeles, California, where he does use-inspired basic research on the effects of individual differences on training effectiveness, uncertainty visualization, and technology adaptation.

Steven D. Fleming is Professor of Geospatial Science and Intelligence with the Institute for Environmental and Spatial Analysis and the University of North Georgia, where he teaches courses in geospatial intelligence, military geography, and spatial science. From 2015 to 2022, he served as Professor at the University of Southern California's Spatial Sciences Institute and Institute for Creative Technologies. Prior to joining USC, he served on active duty for 30 years in the U.S. Army, fulfilling joint and NATO duties on multiple deployments in support of Operation Enduring Freedom. He culminated his military career at the United States Military Academy, where he was Academy Professor of Geospatial Information Science. His research specialties focus on applications of geospatial technologies for national defense (military operations, homeland security, and disaster management).

Aaron L. Gardony received his joint Ph.D. in cognitive science and psychology from Tufts University in 2016. His graduate work focused on basic and applied research topics in spatial cognition, including navigation, spatial memory, and mental rotation. Dr. Gardony held a postgraduate research fellowship sponsored by the Oak Ridge Institute of Science and Education (ORISE) and received numerous accolades, including the Outstanding Academic Scholarship Award. As Cognitive Scientist at the U.S. Army DEVCOM Soldier Center, his research examines spatial perception and cognition using both behavioral and neural approaches, as well as with emerging virtual and augmented reality technologies.

Ioannis Giannopoulos is Full Professor of Geoinformation at the Vienna University of Technology (TU Wien). Before coming to Vienna, Ioannis was Postdoctoral Researcher and Lecturer at ETH Zürich. In 2015, he successfully defended his doctoral thesis at ETH Zürich, which was honored with the ETH Culmann Award for outstanding dissertations. He further holds a B.Sc. and M.Sc. in computer science from Saarland University with a focus on artificial intelligence and human–computer interaction (HCI). His publications are mostly focusing on the areas of geographic information science, HCI, and cognitive sciences. More specifically, his research interests lie in urban computing, geoAI, mobile and remote eye tracking, and pedestrian navigation.

Gerald F. Goodwin is Senior Scientist (ST—Personnel Sciences) at the U.S. Army Research Institute for Behavioral and Social Sciences (ARI). He earned his Ph.D. in industrial/organizational psychology from Pennsylvania State University. He has previously served as Chief of Basic Research at ARI and led research programs on cross-cultural competence, leadership in multinational contexts, and team effectiveness. His core research focus areas include team socio-cognitive processes, team staffing, and personnel testing and measurement.

Qiliang He is a postdoctoral fellow at the Department of Psychology, Georgia Institute of Technology. He received his Ph.D. from Vanderbilt University with concentration in spatial cognition. He is interested in understanding the cognitive and neural mechanisms underlying memory formation, integration, and retrieval. Using functional magnetic resonance imaging (fMRI) and virtual reality (VR), his research mainly focuses on how environmental complexity and individual differences affect the acquisition and transfer of spatial knowledge. A second line of his research involves investigating how external-induced neural oscillations impact attention and memory, and whether these oscillations can be used to combat memory loss such as Alzheimer's disease in humans.

Dalit Hendel is a graduate student in the Data Analytics Program at Tufts University, with interests in machine learning and visualization of sustainability and environmental data. As a senior research coordinator at the Center for Applied Brain and Cognitive Sciences, she led participant recruitment and scheduling, experimental execution, and the processing and analysis of multivariate time-series data sets in the domains of spatial visualization and navigation.

Verlin B. Hinsz, Ph.D., is Dale Hogoboom Presidential Professor at North Dakota State University in Fargo, having served 40 years in various professor and administrative roles. His research and teaching efforts cover social, work, and organizational psychology. He has published more than 100 articles and chapters and has contributed to over 200 national and international research presentations. His research focuses on the thought processes and social influences that impact the judgments and decisions of groups and individuals, which ultimately affect their task performance. Prof. Hinsz is a fellow of ten professional organizations and has received numerous awards for his research and mentoring activities.

Erika K. Hussey is Program Manager in the Human Systems Portfolio at the Defense Innovation Unit (DIU) and a cognitive neuroscientist by training. She has published over 40 peer-reviewed scientific articles in the areas of cognitive function and neuroenhancement. Erika has worked for the U.S. Department of Defense since 2016 in various positions, leading projects focused on characterizing and addressing gaps in warfighter cognitive and physical performance. In her current role at DIU, Erika oversees a breadth of efforts that use commercial technologies, including biosensors, brain–computer interfaces, and extended reality technologies to optimize the health and performance of U.S. service members.

Toru Ishikawa is a professor at the Department of Information Networking for Innovation and Design, Toyo University, Tokyo, Japan. He has a Ph.D. in geography from the University of California, Santa Barbara. He specializes in cognitive-behavioral geography, geographic information science, and urban residential environments and planning, and his research interests include cognitive maps and mapping, wayfinding and navigation, spatial thinking in geoscience, and geospatial awareness and technology. He is an editorial board member of the Journal of Environmental Psychology, Spatial Cognition and Computation, Cognitive Research: Principles and Implications, and the Journal of Architectural and Planning Research, and the author of the book *Human Spatial Cognition and Experience: Mind in the World, World in the Mind* (Routledge, 2020).

Yehuda E. Kalay is Professor Emeritus at the Technion—Israel Institute of Technology, where he served as Dean of the Faculty of Architecture and Town Planning from 2010 through 2016. He is also Professor Emeritus at the University of California, Berkeley, where he served as Professor of Architecture for 18 years. Prior to his tenure at Berkeley, for ten years, he taught at the Department of Architecture, State University of New York at Buffalo. Kalay's research explores the implications and applications of digital technologies for architectural design, in particular as they pertain to the interactions between spaces, the people who inhabit them, and the activities they engage in healthcare settings. Prof. Kalay has published more than 120 scholarly papers and authored or edited nine books.

Peter Khooshabehadeh is a cognitive scientist and the regional lead of DEVCOM ARL West. He leads a group of interdisciplinary scientists and engineers who collaborate to operationalize science with trusted academic and industry partners. He earned a Ph.D. and an M.S. in psychological and brain sciences from the University of California, Santa Barbara, and a B.A. in cognitive science from the University of California, Berkeley.

Donald R. Kretz is Director of Psychological Sciences at Applied Research Associates (ARA). He earned a Ph.D. in cognitive neuroscience at The University of Texas at Dallas. A former U.S. Marine intelligence specialist, he conducts cognitive science research in the defense and intelligence domains and studies the cognitive factors in complex human judgment and decision-making. Prior to joining ARA, Dr. Kretz was Principal Investigator and two-time Technical Honors recipient at Raytheon, where he studied cognitive technologies, human profiling and targeting, and sociocultural semantic alignment. He also won the "Bobby R. Inman Award for Scholarship in Intelligence" from the Clements Center for National Security and Strauss Center for International Security and Law, and was featured on a BBC Horizon science documentary alongside Nobel Prize winner Daniel Kahneman.

Cynthia K. Maupin has a Ph.D. from the University of Georgia and is Assistant Professor of Management at the School of Business Administration, University of Mississippi. Her research investigates the phenomena that support effective teamwork and leadership processes within teams and multiteam systems operating in dynamic environments. Her research has been supported by the United States Army Research Institute, the Center for Collective Dynamics of Complex Systems, and the Ziskin Future of Work Award. She previously served as Doctoral Fellow and Senior Research Fellow for the U.S. Army Research Institute (ARI) for the Behavioral and Social Sciences.

Neil G. MacLaren is Postdoctoral Associate in Applied Mathematics at the State University of New York at Buffalo and Fellow of the Bernard M. and Ruth R. Bass Center for Leadership Studies at Binghamton University, State University of New York. He received his Ph.D. in management from Binghamton University. His research focuses on network science and communication, social structure, and performance behaviors in work groups and organizations. He served in the U.S. Marine Corps for 12 years, including tours as a company commander and a training advisor.

Penelope Mitchell is a Ph.D. candidate at the Department of Geography, Laboratory for Location Science, University of Alabama. She is a human geographer and a computational social scientist with research interests in complex spatial systems. Penelope integrates methods from geography, spatial statistics, geographic information systems, and operations research in an effort to make advancements in spatial decision support in health and human systems. Application areas include health and behavioral geography, transportation geography, and the influence of weather and climate on social systems.

Ashley H. Oiknine is currently a scientist with DCS Corporation, Playa Vista, California, supporting the Army Research Laboratory research efforts since 2016. She received her bachelor's degree in psychology and applied psychology with an emphasis on cognitive science at the University of California, Santa Barbara. Her current research involves understanding human performance in the context of human autonomy teaming and adaptation to future technologies. Additionally, much of her work has utilized mixed reality technologies and multi-modal physiology to evaluate individual and team performance.

Ernest S. Park is a social psychologist and associate professor at Grand Valley State University. He earned his undergraduate degree from the University of Virginia and his doctorate from Michigan State University. His research explores a variety of ways in which group contexts influence the motivation, cognition, affect, and behavior of group members. He is particularly interested in understanding how people communicate and coordinate to make group decisions and perform group actions. Ernest has published over 20 articles and chapters and has been funded by the National Science Foundation. He also serves as an associate editor for Group Dynamics and is on the editorial review board for *Small Group Research.*

Kimberly A. Pollard is a biologist at the Humans in Complex Systems Division, DEVCOM ARL-West, where she specializes in research on human–technology interaction. Her work examines the ways in which training, communication, individual differences, and task elements come together to affect performance in human–technology domains. Dr. Pollard earned her B.A. in biology from Rice University and her Ph.D. in biology from UCLA.

Tiffany R. Raber is a cognitive scientist at the Battlefield Information Systems Branch, DEVCOM ARL-West. She explores behavioral navigation and decision-making techniques, as well as information visualization to increase situational in cross-reality (XR) common operating environments. She earned her M.S. in biomedical visualization from the University of Illinois, Chicago, and is currently completing her Ph.D. in cognitive neuroscience at the University of California, Irvine.

David N. Rapp is Professor at the School of Education and Social Policy and the Department of Psychology and is Charles Deering McCormick Professor of Teaching Excellence at Northwestern University. His research examines language and memory, focusing on the cognitive mechanisms responsible for successful learning and knowledge failures, including the consequences of exposure to inaccurate information from diverse sources and discourse experiences (including fake news and erroneous ideas). His recent books include the co-edited volumes *Processing Inaccurate Information: Theoretical and Applied Perspectives from Cognitive Science and the Educational Sciences* and *The Handbook of Discourse Processes, second edition* from Routledge. Dr. Rapp's projects have been funded by agencies, including the National Science Foundation, the U.S. Department of Education, the National Institute on Aging, and Instagram. For his work, he has received the McKnight Land-Grant Professor Award from the University of Minnesota, and the Tom Trabasso Young Investigator Award from the Society for Text & Discourse. He is Fellow of the Association for Psychological Science. He has served as Editor for the journal *Discourse Processes* and Associate Editor for the *Journal of Experimental Psychology: Applied* and the *Journal of Educational Psychology*.

Megan Rondinelli is an undergraduate researcher at the Laboratory for Location Science, University of Alabama. She performs spatial analysis in the context of the integration of geographic information systems and operations research techniques for autonomous logistics operations. She has researched the use of agent-based simulations for the study of collective spatial cognition and the broad range of spatial analytic techniques that can be used to broaden the scope of research in this area.

Davide Schaumann is Assistant Professor at the Faculty of Architecture and Town Planning at the Technion—Israel Institute of Technology, where he directs the Intelligent Place Lab (IPL). He conducts research at the intersection of architecture, AI, and human behavior science to develop novel approaches for intelligent and adaptive built environments, which foster operational efficiency and people's well-being. This research agenda integrates Dr. Schaumann's interdisciplinary education in architecture (B.Sc. and M.Sc. degrees from the Politecnico di Milano and Ph.D. from Technion), computer science (Postdoc at Rutgers University), and business/entrepreneurship (Postdoc at Cornell Tech). Dr. Schaumann has published more than 40 papers in established scientific journals and conferences and won a prestigious postdoctoral fellowship and two best paper awards in leading conferences.

Anne M. Sinatra is Research Psychologist at the U.S. Army DEVCOM Soldier Center, Simulation and Training Technology Center (STTC), Orlando, FL, USA. Her research focuses on applying cognitive psychology and human factors principles to computer-based education and adaptive training to enhance learning. She is a member of the research team for the award-winning Generalized Intelligent Framework for Tutoring (GIFT) software. She is currently the lead editor of the *Design Recommendations for Intelligent Tutoring Systems* book series. Dr. Sinatra holds a Ph.D. in applied experimental and human factors psychology from the University of Central Florida.

Holly A. Taylor received her bachelor's degree in mathematics, with a minor in psychology from Dartmouth College in 1987. She earned a Ph.D. in cognitive psychology from Stanford University in 1992. She has been a faculty member at Tufts since 1994. Her research examines the mental representation of information, sometimes referred to as mental models or situation models. She is particularly interested in the domains of spatial cognition and comprehension. Her work focuses on how information sources influence mental models. In addition to basic research in this area, she is interested in applications to real-world information sources.

Introduction

1 With a Little Help From My Friends

An Overview of Collective Spatial Cognition

Daniel R. Montello, Kevin M. Curtin, Crystal Bae, and Penelope Mitchell

Introduction

Consider how humans perceive, learn, communicate, decide, and act in the context of spatial problems in the following scenarios.

(1) Three European friends have traveled to South Asia, hoping to visit several of the tourist cities none of them has visited before. One day, they set out on foot to explore the city of Bangkok, with the proviso that they must be able to find their way back to their hotel by 5:00 in the afternoon. They know almost nothing about the physical appearance and layout of the city, other than some basic assumptions about the layouts of cities derived from common features of cities they have experienced. Some of these assumptions will be wrong. Although the three share some interests in what they should visit, they have somewhat different interests as well. They have guidebooks and maps (paper and digital), and they can potentially interact with local residents, although none speaks Thai (many residents of Bangkok do speak English). What knowledge do the three bring to bear on the task of deciding where to visit and how to get there? How do the activity patterns of other pedestrians, many of whom are tourists, influence their decisions about where to explore, possibly unintentionally? How does geographic information influence their decisions? What social and cognitive processes play a part in their reasoning, learning, and behavior in these unfamiliar places? How do the personalities and abilities of each friend combine with interpersonal processes to produce a particular course of action?

(2) Or consider an Army platoon of 40 soldiers who are part of a larger task force newly deployed to a foreign country emerging from years of civil war. Their mission is to help provide stability and security as well as assist the host nation to implement and monitor the cease-fire agreement. The platoon is moving through an unfamiliar and densely urbanized section of a very large city that has sustained severe infrastructure damage after years of war. While the platoon did not expect their maps and overhead imagery to accurately depict everything about passable routes through the city, they quickly realize that these wayfinding aids are nearly useless. The city is a maze of dead ends, narrow passages, and improvised routes created by rubble. The platoon fans

DOI: 10.4324/9781003202738-2

out to find—collectively—the best way through the city, working together to test routes, report obstacles, and understand the urban landscape, all while avoiding hostile threats including improvised explosive devices. Unexpectedly, their headquarters calls and directs them to immediately move in a new direction toward where a hostile attack is in progress. During both phases of this operation, how should the platoon's leader assign movement objectives to subordinates while considering differences in their individual characteristics? How does the leader perceive risk and progress toward the objective in terms of distance, dispersion, speed, and other spatial considerations? How does the leader receive, process, and respond to the spatial representations communicated by the members of the platoon? How do the subordinate teams and squads adapt their movement behaviors cooperatively—in concert with the directions of their leaders—to ensure rapid and safe transit? What cognitive assistance could enhance the unit's wayfinding performance in this context?

(3) A third scenario finds a married couple putting together their new furniture after a shopping trip to the furniture superstore. As is typical for shopping trips to this store, the couple has been able to purchase a brand-new piece of furniture that is very functional, if modest in its style and durability, and all for a very reasonable price. However, that affordability results partly from the fact that the couple must assemble their furniture purchase themselves at home. Unfortunately, the couple has purchased one of the most complicated pieces of furniture to assemble that this store offers—the PAX wardrobe, which comes in huge seven-foot-tall boxes. It has over 150 parts, including sliding doors and numerous drawers and trays. The instructions estimate it will require some four hours to assemble and that assumes a relatively smooth and mistake-free process that many amateur builders of limited skill (and patience) will not experience. Successfully assembling this wardrobe is a highly challenging spatial and social task. Will the two spouses find a way to combine their respective spatial reasoning and mechanical skills, so they put the wardrobe together more quickly and accurately than either would alone or will they somehow interfere with each other's skills, causing the whole process to go more slowly and the finished product to be a little "off"? How will the spouses tacitly or explicitly adjust their respective roles in order to perform complementary tasks while building their wardrobe? How will the long-acquired expectations each spouse has of their own skills and abilities blend with the expectations they have of the skills and abilities of their spouse during this (possible) ordeal?

(4) In a fourth scenario, multiple teams of first responders respond to a fire emergency in a densely populated urban area. The teams are comprised of both generalists (e.g., firefighters and police officers) and specialists (e.g., communications technicians, search and rescue officers, medical technicians, and equipment specialists) who need to work together, within and across teams, to search a high-rise office building for several trapped individuals. Simultaneously, they must track and extinguish a large fire spreading rapidly and unpredictably in three dimensions. How does this system of teams spatially plan an

efficient mission while mitigating immediate risk? How do variables such as leadership, training, and cohesion influence shared awareness of this spatially dynamic situation? How do the dynamics of the situation (e.g., fire spread and partial collapse) change the spatial behavior of teams and team members? How do spatial and communication technologies affect outcomes? How could the overall response be enhanced through improving spatial communication within the group? How does the spatial cognition and behavior of the local population influence the response of teams to the emergency?

(5) As a final example, consider an international, ad hoc group of government, non-governmental, humanitarian, and commercial organizations coming together in the aftermath of a regional natural disaster—a massively destructive earthquake—to coordinate and execute response and relief efforts. The earthquake has affected a subcontinental region comprised of four neighboring countries with distinct differences in geographical, cultural, and infrastructural character. The responding organizations range widely in capacity (e.g., small civic organizations vs. military forces and major international charities) and specialty (e.g., planning, communication, urban search and rescue, disease prevention, and relief logistics). Personnel endure significant cognitive strain, pressure to perform, and, in some cases, physical or cognitive impairment. How does this highly heterogeneous system of teams perceive, communicate, and make decisions about tasks relative to the terrain? Do concepts of team spatial cognition scale up from teams of individuals at the microscale ($\leq$10) to the mesoscale (>10 and <1,000) to larger systems of different types of teams ($\geq$1,000)? How do systems of teams think about space, terrain, and other aspects of geography as the area of concern grows in size (local area to national territory to subcontinental region) and complexity (a variously developed composite of rural, urban, mountainous, littoral, wetland, and forested land)? How is the changing nature of the response—search and rescue, recovery, stability, and redevelopment—informed by the spatial knowledge obtained about the area?

In each of these examples, the common elements are a collective of people—as small as a pair and as large as hundreds—operating in a space about which they have limited or incomplete knowledge, and with goals that require them to learn something about the space and subsequently execute spatial behaviors in pursuit of their goal. These examples represent only a few of the highly motivating—in some cases even life-saving—applications that could benefit from greater knowledge of the components of—and outcomes from—collective cognition and action in space. The motivating premise for this volume is that the study of these groups as they collectively learn about and act in their space is a heretofore underdeveloped area of fundamental research. We seek to understand, define, and develop this research area, which we term "collective spatial cognition." Our goal is to lay the foundation for a long-term research program that will span disciplines and methodological approaches. We believe that the rapid developments in areas, such as movement analysis and mobile sensing, as well as the near ubiquity of wayfinding aids, will only increase the need for a greater understanding of how groups of people learn about and act in spaces.

This volume represents the outcome of a process through which an international and highly multidisciplinary group of scholars were brought together to accelerate research in the area of collective spatial cognition. Depending on how one counts them, at least ten academic disciplines were represented among this group: psychology, geography, cognitive science (including cognitive neuroscience), architecture, interior design, landscape architecture, environmental planning, education, computer science, information science and studies, and engineering (civil, environmental, and building). We certainly do not see this list as exhaustive of the disciplinary perspectives relevant to the study of collective spatial cognition—sociology, anthropology, management, and education come to mind readily. The chapters in this book were initiated by a Specialist Meeting that we hosted on April 17–19, 2019, in Santa Barbara, California. The meeting was supported by funding from the U.S. Army Research Institute for Behavioral and Social Science; the funding, however, supported basic-science research discussions that did not necessarily connect to military activities or situations.

The Three Pillars of Collective Spatial Cognition

The five scenarios above exemplify the topic of this volume—collective spatial cognition. Throughout the rest of this chapter and the other chapters of this volume, we and the other authors discuss this concept and its constituent components in considerable detail. Collective spatial cognition (CSC) refers to a wide range of phenomena in which people solve spatial problems in human collectives, from dyads (two-person groups) and small groups to multiteam systems to crowds. Spatial problems include a broad array of activities, including navigation and wayfinding, spatial knowledge acquisition, location and allocation, planning, design and construction, and spatial communication. They occur at a range of spatial scales, theoretically from the subatomic to the galactic, although the motivating challenge problems above demonstrate that there is significant interest in spatial problems that occur at scales where people can operate in that space based on what they have learned about it. People apprehend and interact with spatial entities directly, as when people choose which trail to take as they walk through a forest, or indirectly, via symbolic media such as maps, images, and verbal descriptions.

The concept of CSC is a hybrid of the three pillars of collective, spatial, and cognition. We can begin our exploration of the hybrid concept, in the first place, by considering each of these constituent pillars in turn. Starting with the noun pillar, "cognition" refers to structures and processes of knowledge and knowing, whether by human beings, nonhuman animals, or intelligent machines (Glass & Holyoak, 1986; Montello, 2008; Wilson & Keil, 1999). Traditional examples include sensation, perception, thinking, learning, memory, attention, imagination, conceptualization, language, and reasoning and problem-solving. Cognitive structures and processes are conscious or nonconscious (explicit or implicit). They are bottom-up, derived from incoming sensory information, or top-down, derived from prior beliefs and expectations; usually, they are based on a combination of the two. Importantly, cognition is functionally and experientially interwoven with affect (emotion), motivation, and behavior.

The brain provides a critical anatomical and physiological basis for cognition, and the last decades have seen amazing advances in understanding the brain as the organ of cognition (Gazzaniga, 2000; Ochsner & Kosslyn, 2014), including spatial cognition (Byrne et al., 2007; Maguire et al., 2000). Much of this progress has resulted from new technologies to study the brain of active, awake persons, the technology of functional magnetic resonance imaging (fMRI) being especially central. This important research continues; it is clearly one of the most significant frontiers of current scientific work in any domain. That said, it must be recognized that the mind is not the same thing as the brain, nor is it reducible to just the brain. The mind emerges from the brain (and the rest of the nervous system) operating within the physical body, which acts within a physical and sociocultural world. Research on the brain does not do away with the value of studying cognition at functional, semantic, experiential, computational, environmental, and social levels.

Our second pillar, the adjective "spatial," is a great example of a concept that everyone intuitively understands but is very hard to formally define (Grünbaum, 1973; Huggett & Hoefer, 2018). It is exceptionally challenging to define it without being circular, no doubt because of the ubiquity and omnipresence of space and spatiality, not only in reality but in experience and mind. Spatiality is the collection of all "extensional properties" of reality, stretching or taking up space, which is circular (the logical term "circular" is a spatial metaphor!). Montello and Raubal (2012) proposed that spatiality is the property of reality reflecting the fact that everything is not at one place, a little tongue in cheek (another spatial metaphor!) and unfortunately still circular. Alternatively, we can do a less circular job of describing—if not defining—space and spatiality by listing examples of spatial properties, such as size, location, distance, direction, shape, connectivity, containment, overlap, dimensionality, hierarchy, and so on. It is important to note that these properties range across the entire spectrum of geometric sophistication, from topological to fully metric, and can include spatial information of varying resolution, from very precise to very vague or approximate. They also include properties of any spatial dimensionality, from points to volumes, and beyond (e.g., the hyperspaces that exist in mathematics such as in multiple regression and linear programming for optimization).

Our third pillar is also an adjective, "collective." In simplest terms, we use collective to mean involving more than one person. Near synonyms are social, group, shared, crowd-sourced, and collaborative. The smallest collective in number is two people. There is no theoretical upper limit to how numerous a collective could be; in the case of human collectives, the constraint is the number of extant humans, currently in the billions. Important issues concerning collectives are whether they form accidentally or are planned and whether their members interact/influence one another with conscious intent or only unintentionally, perhaps without members being aware of the interaction or influence of the collective. In a narrow sense, the term collective might necessarily imply an intentional group, but we see value for our analysis in more broadly including de facto collectives, such as crowds of strangers. Finally, we focus mostly on collectives of humans in this book, but collectives can consist of other types of sentient entities, such as other animals or intelligent machines. Indeed, an increasingly important issue concerns how robots or avatars can play the role of group member in collectives of actual living humans. The nature of what constitutes a collective was one of the most intellectually challenging topics that came out of this process, with significantly differing expert opinions. We discuss this in more detail below in this chapter.

How the Pillars of Collective Spatial Cognition Support a Research Structure

From our own research, and that of the group of researchers assembled for this effort, there is no question that our pillars have been combined as pairs in fundamental research for many decades. Therefore, we continue our exploration of CSC by examining the pairwise integration of the pillars, considering extant conceptualizations and evidence that focus on the two compound concepts of spatial cognition and collective cognition. This will pave the way for our central interest in this book—to consider the integration of all three pillars simultaneously.

Spatial Cognition

Spatial cognition is cognition that is primarily about spatiality—that serves primarily to solve problems involving spatial properties as a central component, whether properties of places or environments, of objects or events, or of symbolic representations such as maps, language, or virtual realities (VRs) (Golledge, 2004; Montello & Raubal, 2012; Waller & Nadel, 2012). These are generally restricted to problems that centrally involve (potentially) explicit mental representations of space, not just perceptual-motor coordination with the external world. That is, deciding which hallway to walk down employs spatial cognition but moving through the hallway without bumping into the walls is merely locomotion, the latter undeniably a spatial problem but arguably not cognitive as such. Montello and Raubal point out that it is misleading and of little use to define spatial cognition simply as any cognition involving spatial properties or taking place in space, as all cognition presumably occurs in space, and most cognitive problems involve spatial information, at least implicitly. Reading is not generally understood to be spatial cognition, even though recognizing letters and words from orthography is clearly a spatial task, in part. Many tasks which we would not consider spatial cognition do have spatial-cognitive components, and most, if not all, spatial-cognitive tasks have non-spatial and non-cognitive components. In many cases, a task can usually be done in alternative ways that are more or less spatial and more or less cognitive.

Spatial cognition is central to many human activities and useful in solving many human problems (Allen, 2007; Montello & Raubal, 2012). It is both a fundamental component of human experience and functional and relevant in many situations. It is difficult to overstate the importance and even ubiquity of spatial cognition in mental and behavioral structures and processes. Its study helps us understand how spatial knowledge is acquired and changes over time; how spatial knowledge is stored in the mind and brain; how people decide where to go and how they find their way there; how people communicate spatial knowledge with words, gestures, maps, and other symbolic media; and how spatial knowledge and thinking are similar or different among individuals or groups.

A key concept in the study of spatial cognition is the "cognitive map." This metaphor was coined by Tolman in a 1948 paper on learning by rats to refer to internally (mentally) represented spatial models of the environment (to use modern terminology); the concept by other names was used before Tolman, however (Trowbridge, 1913, referred to "imaginary maps"). Cognitive or mental maps include what a person knows about landmarks, routes, and the general spatial layout of the environment. Cognitive maps also encode nonspatial attributes of places and objects, as well as emotional associations (Downs & Stea, 1973). Although it is an attractive metaphor, it is misleading to overextend the idea that a mental representation of the environment is like a cartographic "map" (Tversky, 2000). It is not a unitary integrated representation like a picture, but a set of discrete mental representations stored in memory that include landmarks, route segments, regions, and more. The component representations are often partially linked or associated as full or partial hierarchies, such as when a place is stored as being inside of a larger region (McNamara et al., 2008), but much of it is disconnected and mutually uncoordinated. Further, it is inappropriate to conceptualize spatial knowledge as encoding only purely metric information such as Euclidean geometry. Cognitive maps are derived from a rich and complex assortment of information sources acquired via different modalities and media, in different scales, formats, and perspectives. In fact, we believe that many misinterpretations of the cognitive-map metaphor also derive from inaccurately limited understanding of what cartographic maps really are and can be. For instance, cartographic maps need not be intentionally designed to depict accurate and comprehensive metric spatial information.

Our understanding of the multifarious, fragmented, disconnected, incomplete, nonuniform, partially incoherent, and uncoordinated nature of the cognitive map comes from many studies that have tied distortions in the way people answer spatial questions to the form in which spatial knowledge is stored and processed by human minds. These include distorted beliefs about a wide array of spatial properties, including location, distance, direction, size, order, containment, and more (Friedman & Brown, 2000; Golledge, 1987; McNamara et al., 2008; Tversky, 2000). Overviews of spatial cognition from various disciplinary perspectives can be found in Allen (2004), Bloom et al. (1996), Burgess (2014), Denis (2018), Ekstrom et al. (2018), Golledge (1999), Kitchin and Blades (2002), McNamara et al. (2008), Montello (2018), Newcombe and Huttenlocher (2000), Shah and Miyake (2005), Tversky (2000), and Waller and Nadel (2012).

Collective Cognition

Turning to the second pairwise integration of our pillars, collective cognition is the performance of cognitive tasks, such as thinking or problem-solving, by two or more individuals. As we discussed earlier, we broadly conceive that collectives may or may not be intentional. Of course, there are varying degrees and types of collective influence or interaction, which we touch on further below. But when one considers collective cognition in the broadest sense, one can argue that all cognition is collective (Ostrom, 1984; Rogoff, 1990; Wertsch, 1985). Even when thinking and reasoning in isolation, our cognition is influenced by cultural practices, by past instruction and observational learning that have involved others, and by the involvement of material artifacts or physical situations that were produced by other people's cognition and behavior. This book focuses on more overt and explicit examples of collective cognition, but it is important to remember in the end that collective cognition is important as a research domain because it illuminates cognition in virtually all contexts, to some nontrivial degree.

Research on group behavior and collective cognition has historically been very important in social psychology. Several thousand research books and articles have been published, going back to the late 19th century, that discuss concepts, theories, and empirical results regarding aspects of collective cognition (e.g., Asch, 1951; Brandstätter et al., 1982; Bruch & Feinberg, 2017; Ellis & Fisher, 1994; Kerr & Tindale, 2004; Levine & Moreland, 1990; McGrath & Kravitz, 1982; Steiner, 1964; Swap, 1984; Tindale et al., 2003). As Gerard and Miller declared, "[p]roblem solving behavior of groups is perhaps the oldest problem in experimental social psychology" (1967, p. 298). These publications include considerations of groups ranging from two (the dyad) to hundreds or even millions (as in collective problem-solving by a nation's citizens or humanity as a whole). Nearly any text in social psychology, for instance, will include several chapters on group psychology, particularly ways that groups influence individuals and individuals influence groups.

Research on group behavior and cognition has considered collections of individuals who know each other or are strangers, have varying levels of different personality traits, influence each other intentionally or unintentionally, and interact directly or only indirectly. It has looked at numerous situational contexts, types of tasks or problems, and means of interacting. Studies have focused on the nature of communication within groups, the effects of the spatial arrangement of group interactants, the emergence of leadership, and every variation of motivation from cooperative to competitive. In addition to experimental and nonexperimental studies involving the observation of actual behavior, a great number of mathematical and computational models of group decision-making have been developed and evaluated (Coleman, 1959; Levine & Moreland, 1990).

A major thrust of group cognition research has been to compare the performance of cognitive tasks by groups to performance by individuals; in some cases, group performance is treated simply as the sum of performance by individuals (Ellis & Fisher, 1994; Hall et al., 1963; McGrath & Kravitz, 1982; Tindale et al., 2003). In such a context, the average judgment over a group of people who are not collaborating is better than that of most of the individual people because extreme judgments balance each other out (famously described by Galton, 1907). When considered as collaborating entities, collectives are typically thought to make better decisions than individuals because collectives contain more skills and knowledge, and research supports the general superiority of group decisions. Group members interact and collaborate with one another, sharing ideas and knowledge, to widely different degrees. The degree of "social sharedness" is seen as critical to understanding group decision-making (Kameda et al., 2002), particularly the processes by which groups achieve consensus and arrive at a final decision.

However, collective problem-solving is not always superior to individual problem-solving, and collective problem-solving introduces additional challenges to communication and coordination. There has been interest in the ways that group processes can systematically lead to suboptimal decisions. Perhaps the most famous concept in this regard is that of "groupthink," the phenomenon whereby certain subtle and overt group processes (such as the tendency people often display to conform with others, e.g., Asch, 1951) can lead to the suppression of individual expression, leading individuals not to express what would otherwise be legitimate doubts about group decisions. This, in turn, leads to suboptimal—even tragically bad—group decisions. Researchers also consider the deindividuation that groups can inspire, which can produce a phenomenon known as "risky shift," wherein group decisions are riskier than the average of the decisions that would be made by the individual members. Similarly, group polarization is the tendency for groups to arrive at more extreme decisions than those of any of the group members. Some research simply suggests that there are conditions under which group problem-solving is neither better nor worse than individual, finding that gains from collective effort depend on situational and procedural contexts that differentially affect group motivation and the coordination of resources (e.g., Kerr & Tindale, 2004).

Behavioral scientists, especially organizational psychologists, have studied teams for many decades and produced sophisticated models of team mentality (Fiore et al., 2010; Salas et al., 2008, 2013). Task interdependence, communication, conflict, roles, and leadership are persistent themes in the team research literature (Bell & Kozlowski, 2002; Curseu et al., 2008; Fisher et al., 1998; Hinds & Bailey, 2000, 2003; Rico & Cohen, 2005; Senior, 1997; Yoo & Alavi, 2004; Zaccaro et al., 2001). Team cognition is the shared understanding of factors and information pertinent to a human collective's goal and of the communicative processes and relationships among the members (Beck & Keyton, 2012). Work teams are collectives of two or more individuals that interact socially and work interdependently, under constraints, to accomplish organizationally relevant tasks and goals (Kozlowski & Bell, 2003).

Adding additional complexity, a multiteam system (MTS) is a network of teams that interdependently achieve their own team-level tasks while striving to accomplish some overarching system-level goal(s) (Mathieu et al., 2002). Insofar as MTSs may be understood as teams of teams, they exhibit similarities with teams in the significance of their communications, culture, and leadership. MTSs organize individuals, according to roles and tasks, into teams that work interdependently (Zaccaro et al., 2012). These complex networks demand leadership that can iteratively make sense of multiple sets of requirements, establish objectives, and coordinate the achievement toward these goals (DeChurch & Marks, 2006). Culturally, MTSs rely upon normative processes and shared meaning rooted in systems of values rather than just rules or constraints (Zaccaro & DeChurch, 2012). MTSs organize teams according to their tasks and enable team interdependence so that cross-boundary work sharing occurs where needed to efficiently achieve system-level goals (Davidson & Hollenbeck, 2012). Thus, inter-organizational collaborative effort is the cornerstone of MTS communication, which, in turn, enables collaboration by creating a sense of shared meaning—and possibly shared cognition—illuminating task requirements and priorities (Keyton et al., 2012).

Bringing the Three Pillars Together: Collective Spatial Cognition

This brief review demonstrates that we know a great deal about group interaction in reasoning and problem-solving contexts and we also know somewhat less, if still quite a bit, about human spatial cognition. But scholarship on the combined concept of CSC has heretofore been quite sparse in the basic research literature. Stated concisely, our aim in this book is to combine the study of spatial cognition with the study of collective cognition: collective spatial cognition. We hope that this will help lay the groundwork for a new domain of basic and applied research activities; we also hope to facilitate better interdisciplinary communication and collaboration among researchers from the many disciplines interested in CSC.

The role of collectivity in spatial cognition is broad and impactful. Collectivity can influence what people perceive, what information they have, how they apply problem-solving strategies and heuristics, what expectations and motivations they bring to problems, and more. This suggests the great breadth—even ubiquity—of the influence of the social in spatial cognition. Just as we argued earlier that virtually all cognition is partly collective, in the broadest sense, so is all *spatial* cognition (Dalton et al., 2019; Hutchins, 1995; Xiao & Liu, 2007). For instance, the design and comprehension of navigation technologies and artifacts (maps, compasses, sextants, and satellite positioning systems) always reflect, in part, the past decisions and influences of other people, cultural traits, and social practices. Both built spaces and signage are typically designed and located by other people, sometimes following regulatory guidelines or widespread cultural practices, sometimes not (Hillier & Hanson, 1984; Weisman, 1981). Some of the spatial strategies that are regularly employed are ones we learned from instruction by others, in person, or from books. Even in the least social cases of a lone person, without a map, in a wilderness environment, spatial decisions are partly based on advice acquired in the past from other people ("go to a high point").

Given the great scope and relevance of the role of the collective in spatial cognition, we think that it is a bit surprising how rarely spatial cognition research has taken the co-presence and potential influence of other people into account (Fernández Velasco, 2022). A few recent studies have examined social interaction during collective navigation (Bae & Montello, 2019; Forlizzi et al., 2010; Haddington, 2012; He et al., 2015; Reilly et al., 2009). Some important work on spatial cognition that explicitly incorporates the social is that of Hutchins on "cognition in the wild" (Hutchins, 1995). His work goes far in conceptualizing the social in spatial cognition, but the spatiality of his work is a rather unintentional outcome of his larger aim to socially contextualize human cognitive research more broadly. There is a substantial body of research on wayfinding communication ("route directions") (e.g., Allen, 1997; Denis et al., 1999; Klein, 1983; Wunderlich & Reinelt, 1982). But most of this work focuses on the content of the instructions from one person to another, only implicitly involving social aspects of spatial cognition. Finally, a rather extensive body of research has specifically examined the spatial behavior of groups or crowds, using both empirical behavioral observation and computational modeling (Faria et al., 2009; Helbing & Johansson, 2011; Kinateder et al., 2014; Moussaid et al., 2009). Most of these studies concern themselves with unintentional collectives—what Dalton et al. (2019), discussed later, labeled "weak synchronous social wayfinding." Much of this work has addressed various forms of herding behavior by nonhuman animals, such as the tendency of herd members to stick together, which by itself can lead groups to navigate more accurately than constituent members (the so-called "many-wrongs principle"). The work in this tradition involving humans has mostly focused on the spatial patterns of emergency egress. All of this work offers a promising start, but we believe that the present book shows that the existing work only scratches the surface of potential research on CSC, an oversight that distorts our understanding of spatial cognition.

It may be of value to consider why this relative neglect has occurred. Dalton et al. (2019), in their discussion of social wayfinding, speculated on some reasons for it. The disciplinary tradition of scientific psychology, historically the main science of human mind and behavior, has been to focus on the individual person as the level of analysis. Disciplinary culture aside, it is clear that adding people beyond the dyad to the analysis of decision-making factorially increases the complexity of potential interaction combinations—a group of five people, for instance, might have ten different pairs of interacting dyads, in addition to subgroups of three or four individuals—and potentially creates emergent properties that go well beyond a simple sum of the cognitions of individuals or even pairs of individuals. A third reason for the relative neglect of CSC may be the increasing popularity of virtual reality (VR) as a research tool and setting for spatial cognition research since the 1990s. Additional actors and interactants can be implemented in VR (e.g., Kim et al., 2017; Lamarche & Donikian, 2004), but this adds technical and computational challenges on top of the intellectual and methodological ones. Most research in spatial cognition that uses VR includes only individual research participants, notably including the recently celebrated and important work utilizing the video game *Sea Hero Quest* (e.g., Coutrot et al., 2022). When other people are implemented in a VR, they are typically depicted as avatars with little cognitive or communicative sophistication. Finally, when cognition is conceived of as rooted in the functional activity of the brain (at least much of its cerebrum), as it so often is, it seems obvious that one should focus on analyzing the activity of individual brains in the pursuit of understanding cognition. Indeed, this continues to characterize the approach of cognitive neuroscience research (e.g., Ekstrom et al., 2018;

Burgess, 2014; Maguire et al., 2000). While cognitive neuroscience research has likely provided the most significant advances in our scientific understanding of spatial cognition in decades, it has rarely incorporated issues of collectivity or social influence, if at all.

Aside from spatial tasks, of course, there has been a very large body of work on judgment and decision-making in general (Camerer et al., 2004; Gigerenzer & Gaissmaier, 2011; Kahneman, 2011; Plous, 1993). Bruch and Feinberg (2017) claimed that this literature has mostly minimized the role of social context in decision processes. They ascribe this to researchers' desires to control the research context by "desocializing" it, allowing a simplified and unconfounded empirical situation. At the same time, we recognize that the long and substantial tradition of research that introduced cognition into the study of social behavior—the extensive literature on collective cognition we reviewed earlier—has rarely focused on spatial tasks, even though they are among the most universal and frequent of cognitive tasks facing humans. The example scenarios with which we start this chapter suggest the commonness of spatial tasks, and such examples could easily be multiplied; see the discussion of spatial tasks by Montello and Raubal (2012) below.

Most of the distinctions relevant to understanding collective problem-solving in general are likely to be relevant to understanding collective spatial problem-solving. We need to scrutinize the degree, nature, and temporal frame of the influence exerted by groups on individuals and by individuals on groups. There are a host of task contexts to scrutinize, in different types of environments and at different scales. Group composition is probably critical, in terms of group size, member familiarity, gender, age, intellectual and personality traits, and so on. It is informative to look at social processes of influence and interaction during the performance of spatial tasks, including different modes of interaction, whether visual, verbal, motoric, or others. As in work on collective decision-making generally, we should investigate how and why group performance is better or worse than the performance by an individual or sum of individuals.

Likewise, nearly, all of the distinctions relevant to understanding spatial problem-solving in general are likely to be relevant to understanding collective spatial problem-solving. We need to analyze the similarities and differences of spatial cognitive activities as carried out by groups versus individuals, the latter of which has garnered the great majority of research attention, as we have noted. Clearly, when solving a spatial problem with others, people will typically look to others for ideas, to get feedback on one's own ideas or acts, and to clear plans of actions that have implications for the group. A fascinating contrast of individual to group spatial cognition concerns the monitoring of orientation that goes on during navigation. When navigating alone or while ignoring others, travelers monitor their sense of orientation, an example of metacognition during problem-solving; indeed, geographic disorientation or getting lost is the subjective belief (metacognition) that you are uncertain as to your location or which route to take (Montello, 2020).

In a collective, group members monitor not only their own sense of orientation but also, to some extent, that of other group members—a fascinating concept Fernández Velasco (2022) identifies as "collective metacognition." The great importance to CSC of metacognition about what others know is suggested by the fact that the authors of at least three chapters of our book discuss it as a central component of what is known as "transactive memory systems" (TMSs; see chapters by Baumann et al., Maupin et al., and Park and Hinsz).

Issues Defining the Problem Space of Collective Spatial Cognition

All five of the scenarios that introduced this chapter exemplify CSC, but they also suggest some of the myriad ways that instances of CSC differ from each other. This effectively points to the issues that define the problem space for conducting research on CSC. In Table 1.1, we group the issues into five areas: (1) the nature of collectives and their members, (2) the nature of collective interaction/influence, (3) the nature of spatial cognitive tasks/problems, (4) the nature of the space in which the collective cognition occurs, and (5) the diversity of conceptual/empirical issues in studying CSC. With respect to each of these five areas, there are relatively "pure" instances of CSC that involve only a single variant within each issue area, but many instances undoubtedly include multiple variants occurring in a great variety of different combinations. For example, collectives might consist of teams of different sizes, each with different types of members and different subtasks.

Issues Defining the Problem Space of Research on Collective Spatial Cognition

Table 1.1 The Problem Space

Issue	*Elements*
Nature of collectives and their numbers	Number of people in collective Composition of collective Characteristics of individuals making up collective
Nature of collective interaction/influence	Intentionality of collectivity Temporal synchrony of collective interaction Nature of mediation of collective interaction
Nature of spatial cognitive tasks/problems	e.g., wayfinding, search, evacuation, and many more
Nature of the space/place in which the collective cognition occurs	Scale of the space/place Dynamic nature of the space/place Place as a humanistic perspective on space
Diversity of conceptual and empirical approaches to the study of CSC	Diverse conceptual/theoretical approaches Diverse empirical/methodological approaches Qualitative, quantitative, mixed methods

1. Nature of Collectives and Their Members

Starting with the nature of collectives and their members, it is relevant to consider the size of the collective—how many individuals are involved in it. Of course, collectives vary in many other characteristics, such as their function and history, whether different members have different roles, and whether the collective has an explicit leadership hierarchy. Individual members also have different characteristics, such as personality, intellectual skills, motor skills, training, life experiences, interests, gender, age, and language. A humanistic perspective further points to the ambivalent binary concept of crowding/privacy as relevant terms reliant on the human experience and emotion in specific settings (Tuan, 1976). Two can be in a crowd or one can feel alone in a sea of people. In this regard, we can ask how the characteristics of individuals influence their own experience and performance as members of a collective, as well as how the characteristics of individuals influence the experience and performance of the collective. Most of the research discussed in this book focuses on small collectives, many being dyads or triads. This likely reflects the historical preponderance of research on individual cognition, including spatial cognition. The most obvious immediate extension of work on individuals is to move to pairs and then triplets, although this may mislead us in understanding the cognition and behavior of larger groups if they emerge as more than simply sums of the cognition and behavior of smaller groups.

2. Nature of Collective Interaction/Influence

As mentioned earlier, the discussion of what constitutes a collective was one of the more contentious intellectual issues that arose in the process of defining this nascent research area. This issue revolves mainly around the nature of collective interaction/influence, that is, whether members contribute to the cognition of the collective intentionally or only unintentionally. Dalton et al. (2019) presented a typology of social wayfinding (Figure 1.1) that could potentially be applied to all CSC, even to all collective cognition that is nonspatial. Of relevance to our discussion of intentionality, Dalton et al. referred to intentional interaction within collectives as "strong" social wayfinding and unintentional interaction, such as inadvertently sending cues to others, as "weak" social wayfinding. A second aspect of collective interaction or influence, also recognized by Dalton et al., concerns its temporality. The cognitive interaction of collectives can occur synchronously or asynchronously—either in real time or with substantial time lags between the actions of different collective members. For example, two people can discuss their route while they travel together through the environment (synchronous), or one person can draw a sketch map that the other person uses the next day to follow a route (asynchronous). Finally, Dalton et al. recognized that interaction between or among collective members can be direct—in-person—or technologically mediated in different ways (i.e., with communication technologies such as phones or computers). Direct interaction depends greatly on physical closeness, which strongly influences visibility, audibility, tangibility, and the like.

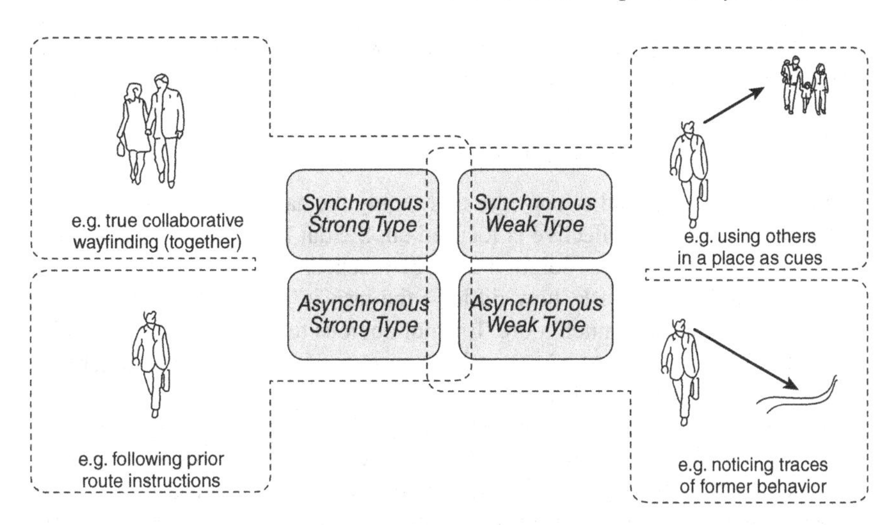

Figure 1.1 Collective perspectives including asynchronous and weak types as classed by Dalton et al. (2019) in the context of social wayfinding: synchronous and strong (upper left); asynchronous and strong (lower left); synchronous and weak (upper right); and asynchronous and weak (lower right). Figure reprinted from Dalton et al. (2019) Copyright © 2019 Dalton, Hölscher and Montello.

A contrasting view among our contributors is that a group of individuals operating with no intentional interaction with each other does not constitute a human collective in any meaningful sense of the term, particularly for studies of cognition. This view maintains that examples of weak synchronous collectives, such as a person navigating through a shopping mall or an airport who picks up clues or prompts to wayfinding decisions from observing other people, are not cases of learning about the space together with others, that is, learning *from* other people is not the same as learning *with* other people. According to this view, a necessary condition for a human collective is a shared goal that can be cognitively addressed (i.e., not reflexive or implicitly guided movement). The people who make up all of those moving through an airport (even coming off of a specific flight) don't have a collectively *shared* goal, although the possibility space of what their individual goals are—to board another flight, pick up bags, locate a restroom, or acquire food—is more limited because they are located together in the controlled space of an airport. A similar argument holds for crowds, where many individuals are gathered together but often with only a focus on their individual goal ("find my seat") and little to no direct interaction with others in the crowd other than purposeful avoidance. For the purpose of this volume, however, we recognize that there are important social aspects to CSC even when the collective is not intentionally collaborating. Because we think that early research on CSC will benefit from a broad perspective that includes as many aspects of the problem space as possible, we include unintentional as well as intentional collectives in our purview. We do respect, however, that some members of the research community will insist the term "collective" be reserved for intentional groups only.

3. Nature of Spatial Cognitive Tasks/Problems

Certainly, researchers need to consider thoroughly the nature of the spatial cognitive tasks/problems the collective is focused on. Spatial tasks vary in many ways, such as what specific goal or solution must be reached, what information is made available to address the task, how difficult the task is, what prior knowledge it requires or assumes, and much more. It is not obvious to us what is the most useful way to classify these task characteristics or even if it is definitely valuable to attempt such an a priori classification.

That said, Table 1.2 presents a general categorization of spatial cognitive tasks proposed by Montello and Raubal (2012). These categories of tasks are characterized primarily by what they accomplish for people (i.e., their function) but also by the modality of the spatial information they involve, that is, they include tasks that involve direct sensorimotor experience with the physical spatiality of objects and environments, such as navigation and acquisition of knowledge directly from the environment, and those that involve indirect symbolically mediated experience of spatiality, such as reasoning with mental models and learning from maps or language. The first category of tasks, wayfinding, is the cognitive component of navigation, in which navigators coordinate to the distal environment, the environment not immediately accessible to sensorimotor systems. This contrasts with locomotion, which is coordination with the proximal or surrounding environment while navigating. Wayfinding includes tasks, such as creating and choosing routes, establishing and maintaining orientation while traveling, and locating landmarks in the environment. The second category of tasks is acquiring and using spatial knowledge learned directly from sensorimotor experience in the world. The particular sensory and motor systems involved in direct learning clearly depend on the scale of the spatial entity being learned. Haptic manipulation allows us to learn the spatial layout of handheld objects, whole-body locomotion gives us access to the layout of a building, and vision alone is used to apprehend the location of distant landmarks on the horizon (audition sometimes plays a role). The third category of tasks is acquiring and using spatial knowledge learned indirectly from spatially iconic symbolic representations. These are graphical (maps, graphs, and pictures)

Table 1.2 Categories of Collective Spatial Cognitive Tasks*

1. Wayfinding/Navigation
2. Acquiring/Using Spatial Knowledge From Direct Experience
3. Acquiring/Using Spatial Knowledge From Spatially Iconic Symbolic Representations
4. Acquiring/Using Spatial Knowledge From Natural Language
5. Imagining Places/Reasoning With Mental Models
6. Location Allocation

* from Montello and Raubal (2012)

and volumetric (physical models and globes) symbolic representations that represent spatial information at least partially via their own spatial properties. For example, places further away from each other in the world are typically shown as further apart on a map (with some caveats). Very large and very small spatial entities are only learned indirectly via symbolic representations (we do not learn the relative locations of atoms in molecules or planets within solar systems from direct experience); of course, all scales of spaces are sometimes learned indirectly. Our fourth category of tasks also involves indirect acquisition and use of spatial knowledge, but via natural language such as Mandarin, English, or Quechua. Languages represent spatiality noniconically, for the most part (sign language is a clear exception, a hybrid of task categories 3 and 4). Instead, natural language mostly represents spatiality abstractly, by employing semantics—verbal meaning—to express spatiality. Given this noniconicity, language operates equally easily at any spatial scale to acquire and use spatial knowledge. Task category 5 is imagining places and reasoning with the mental representations (mental models) generated thereof in working memory. Such mental models are stimulated by both linguistic and nonlinguistic input (i.e., from language and direct environmental experience). But memory and imagination also stimulate mental models so that one can even reason about spatial entities and situations that do not exist and never have; truly innovative spatial design requires such creative reasoning (e.g., Gero, 2015). Our sixth and final task category is location allocation, finding good locations to put facilities such as fire stations, hospitals, and schools. Although this task is now mostly accomplished in a professional context in non-cognitive ways (by computer routines), for most of human history it was accomplished informally, involving extensive spatial thinking. Even today, individuals make locational and siting decisions (such as choosing where to live) that involve a great deal of informal spatial thinking, thinking that can be quite challenging.

Another approach to classifying tasks is suggested by the typology of spatial cognitive tasks proposed by Newcombe and Shipley (2015) (see also Uttal et al., 2013). In the style of cognitive psychology, it is based on a theory of the underlying mental skills associated with spatial cognition. Newcombe and Shipley's typology is more abstract than Montello and Raubal's (2012) list, distinguishing two fundamental dimensions of spatial cognitive tasks: intrinsic versus extrinsic and static versus dynamic (creating a 2 × 2 classification). The first dimension distinguishes between cognition involving intrinsic spatial information and that involving extrinsic spatial information. Intrinsic information specifies the spatial properties of parts of an object and the relations between the parts. Extrinsic information is about the spatial relations among members of a set of objects, relative to each other or to an external reference system such as cardinal directions. For example, shape is an intrinsic spatial property that distinguishes different garden tools from each other, while location is an extrinsic spatial property that describes how the tools are spatially related to each other in the shed. The second dimension distinguishes among skills that are static and those that are dynamic. Static tasks involve identifying and reasoning with nonmoving spatial relations, such as deciding where to locate an object so it would be visible from particular viewpoints. Dynamic tasks involve identifying and reasoning with changing spatial relations, such as deciding where to move an object so it would be continuously visible from the changing viewpoints of a traveler over time. Newcombe (2018) modified her framework a little by highlighting intrinsic and extrinsic spatial cognition as fundamental, adding symbolically mediated (indirect) spatial cognition as a third fundamental category. Newcombe identifies tool use as the prototypical case of intrinsic cognition and navigation as the prototypical case of extrinsic. Given the abstracting power of symbolically mediated cognition, such as its power to transcend scale, indirect spatial cognition operates across task types.

In the end, neither the list by Montello and Raubal nor the typology by Newcombe and her colleagues presents distinct task categories that are mutually exclusive. We doubt that any attempt at a comprehensive list could succeed at this. For example, wayfinding might involve direct experience, iconic spatial symbols, or spatial language—or all three. Similarly, wayfinding involves the intrinsic problem of recognizing landmarks and the extrinsic problem of knowing how landmarks are located relative to each other. But maybe more telling for our purpose here, neither Montello and Raubal's list nor Newcombe and Shipley's typology includes any explicit reference to social reasoning about space or to reasoning about social space. In this edited collection, we hope to shed light on the under-addressed issues of social aspects of spatial cognitive tasks.

4. Nature of the Space/Place in Which the Collective Cognition Occurs

Issues that define the problem space of CSC include the objective characteristics of the spaces or places themselves. Relevant are both properties of where collective members are located (size, layout, appearance, etc.) and the nature of the spatiality involved in the cognitive task the collective focuses on. This is an exceptionally large and heterogeneous set of issues to consider. It includes the scale and dimensionality of the space or place, the shape and other geometric aspects of spatial layouts, the precision or vagueness of spatial properties, the geometric sophistication of spatiality involved (topological, metric, etc.), whether environments are urban or rural or wilderness, indoor or outdoor, and much more. Moreover, as some of our motivating examples suggest, there can be dynamism in the space that is independent of the actors within that space, but that fundamentally changes the environment and the ability of collectives to complete their task. The example of a high-rise fire that changes the space where operations can safely occur and even changes the shape of the building itself throughout the course of the fire demonstrates this dynamism inherent in the space rather than in the task. In many studies of spatial cognition, the variety of spaces that are considered is extremely limited. Numerous studies employ even a single "neighborhood" space with perhaps no more than two routes through that neighborhood. When conducting human subjects experiments in this domain, the complications of modeling and testing many different spaces are clear. This is a set of issues where simulated environments can potentially contribute significantly to the research literature.

Taking a humanistic perspective on space as place, one can focus on the idea of place as a way of seeing, knowing, and understanding the world. For instance, instead of focusing on the objective qualities or attributes of a space, we can focus on the way one chooses to think about it—what is emphasized and what is deemed unimportant. Seeing space through the lens of place can bring about reactionary or exclusionary tendencies such as xenophobia, racism, and bigotry—for example, my place is threatened, others must be excluded here (Cresswell, 2015). This humanistic thread points to how place may be considered as territory (Sack, 1983; Tuan, 1976). Human attitudes of places as territories vary in conceptualization based on their spatial attributes and how they are used. For instance, territory may be a network of paths and places permeable by others or there may be a strong sense of property as a bounded space (Storey, 2012; Tuan, 1976). The role of emotion and thought in attachment to place influences the nature of experience and perception. Implications of this premise within the research area of CSC suggest that a humanist geography viewpoint could help to address the complexities of emotion and identity in space and place, for groups as well as individuals.

5. *Diversity of Conceptual and Empirical Approaches to the Study of Collective Spatial Cognition*

Finally, research on CSC takes diverse approaches to various conceptual and empirical issues. Both conceptual/theoretical and empirical/methodological approaches are addressed elsewhere in this chapter and throughout the book. This is obviously a substantial and very heterogeneous set of issues defining our problem space. Especially given the thorough multidisciplinarity we embrace in this book, the research area of CSC potentially includes ideas about causality ranging from cultural practices to individual motivations, preferences to knowledge, beliefs to physiology, and neural activity; it also includes ideas about spatial properties of galaxies to continents to urban neighborhoods, from architectural structures to room layouts to tabletops, and from cartographic maps to electrical circuits and molecular structures. Methods include physical and physiological measurements, surveys and tests, behavioral observation, data archives and social media records, computational modeling, and so on. Both real spaces and simulated spaces are involved. A question of particular interest in this respect is what, if any, concepts and methods are specifically relevant to CSC.

Potential Contributions to the Basic and Applied Study of Collective Spatial Cognition: Research Questions in Ten Fundamental Domains

The study of CSC involves a host of basic and applied scientific research issues, many of which are raised in the ensuing chapters of this book. We hope that this book will help stimulate and further develop research on these issues, helping to build a foundation for a broad multidisciplinary research program. In this section, we present our views of many of these basic and applied issues by offering several research questions in the area of CSC, organized into ten fundamental domains of research.

1. Conceptual and Empirical Issues
    a. How are spatial cognitive tasks the same as nonspatial cognitive tasks, and how are they different, in ways that are relevant to collective cognition?
    b. What are the most important ways that work on solitary (unaccompanied) spatial cognition does and does not apply to CSC?
    c. How can collaboration be facilitated among researchers from the many disciplines with an interest in CSC?
2. The Nature of Spatial Entities Involved
    a. How do aspects of CSC operate similarly or differently across different types of environments, such as urban, rural, and wilderness environments?
    b. How do aspects of CSC operate on physical spatial entities such as objects and places versus on symbolic representations of spatial entities, such as maps and natural language?
    c. How does spatial scale matter to collective spatial tasks?
3. Group Size
    a. How does the nature of spatial cognitive tasks change qualitatively and quantitatively as group size increases?
    b. Under what conditions do multiple agents improve or facilitate spatial cognitive performance, and under what conditions do they hurt or impede it?
4. Individual Differences Among Collective Members
    a. What are the differences among individual people that contribute to variation in group spatial cognition and behavior, and how do they do so?
    b. Are there individual personality or intellectual traits that are particularly conducive to group success or failure at collective spatial cognitive tasks?
5. Leadership and Role Differentiation
    a. What task and situational factors influence the informal emergence of leadership within collectives performing spatial tasks?
    b. What individual difference variables influence not only who emerges as a leader but how members of a collective embrace emergent leadership or not?
    c. How do we best characterize the structure of roles within collectives beyond the simplistic binary of "leader/follower"?
    d. What are factors that induce collective metacognition about spatial knowledge within a collective and influence whose spatial opinions are attended to or ignored?

6. The Role of Familiarity
   a. What role does the familiarity of group members with each other play in CSC?
   b. What role does the familiarity that collective members have with the spatial context play in spatial cognitive tasks, such as familiarity with the place where the tasks occur or with the material available for accomplishing them?
7. Intra-Group Communication
   a. What are the most effective ways for collective members to communicate spatial information among themselves?
   b. In specific instances when relevant spatial information is divided among group members, what is the best way for members to integrate that information?
   c. Are there identifiable factors that influence when individuals do or do not notice and follow spatial cues when in an unintentional collective?
8. Development/Change
   a. How does interaction among members solving a collective spatial task change over time, as a function of factors such as member experience and group modifications?
   b. How does CSC vary over the course of the lifespan development of its members?
9. The Role of Technological Aids
   a. How do spatial information technologies play a facilitative or impeding role in collective spatial tasks?
   b. How can geospatial techniques and technologies offer methods or metrics to help us understand how the attributes of space and spatial behavior influence collective cognition?
10. Application Areas
    a. What are important application areas for collective spatial cognitive research, and how is this research similar or different across areas? Major areas include military; public safety/emergency response; medical epidemiology; urban, architectural, and transportation planning; and international aid.
    b. How can we train groups to perform collective spatial tasks more effectively?
    c. How can the design of spatial information technologies be improved specifically for collective spatial tasks?

Synopsis of This Book

As we have noted, we attempt to take a strongly multidisciplinary approach in this book. Our major hope is to integrate at least two major bodies of research:

the science of spatial cognition, as developed by cognitive psychologists, behavioral geographers, neuroscientists, linguists, computer scientists, biologists, architects, and others; and the science of team cognition, as developed by industrial/organizational psychologists, social psychologists, sociologists, anthropologists, communication researchers, management researchers, and others. We recognize that research problems transcend disciplines, in the tradition of Campbell's (1969) "fish-scale model of omniscience." Scholars have also written about the development of a "multiscience" or, more recently, the concept of "transdisciplinarity" (Hadorn et al., 2008). We agree that transdisciplinary efforts like the one we pursue in this book offer great potential to push outward the known boundaries of the sciences. But such an effort comes with several well-known challenges that follow when researchers and scholars with different academic and scientific cultures attempt to interact. Vocabularies, research goals, ontological and epistemological assumptions, and more need to be made explicit and shared. Common and uncommon ground needs to be identified. To do this, we think that efforts such as our own should be continued and expanded, such as meetings, workshops, and publishing in interdisciplinary outlets. Balanced teams of scholars and researchers should be instigated and nurtured, including domain scientists, cognitive scientists, spatial researchers, and various species of practitioners. Our hope is that this edited book contributes to this effort.

This book consists of 13 chapters, which we have organized into five sections. The first section includes the present chapter, which introduces and provides the background for—and an overview of—the topic of CSC. The second section, titled *Navigation,* consists of four chapters. Chapter 2, by Toru Ishikawa, is titled "Navigation in Collaboration." It reviews the small number of studies that have examined groups of interacting people navigating together to destinations, particularly a study he and his colleagues completed on dyadic navigation. The chapter concludes by considering questions for future research on collective navigation, including the implications of large differences among individual navigators for the design of effective navigation assistance systems. Chapter 3, by Tad T. Brunyé, Dalit Hendel, Aaron L. Gardony, Erika K. Hussey, and Holly A. Taylor, is titled "Personality Traits and Spatial Skills Are Related to Group Dynamics and Success During Collective Wayfinding." The chapter considers the complex roles of interpersonal dynamics and group composition in the planning, initiating, updating, and completion of navigation. The authors identify gaps in the research on collective navigation and present the results of a preliminary study involving individuals, dyads, and triads wayfinding in a virtual-reality environment. Chapter 4 in this section, by Ioannis Giannopoulos and Daniel R. Montello, is titled "Facilitating Collective Cognition During Group Wayfinding Through the Human Eye." It points out that wayfinding during navigation typically relies on integrating novel spatial knowledge with existing knowledge of the environment. They discuss the intriguing prospect of creating navigation assistance systems that monitor eye movements in order to help groups of navigators acquire, organize, and use spatial knowledge to more effectively reach destinations and support knowledge acquisition. Chapter 5, by Peter Khooshabehadeh, Kimberly A. Pollard, Ashley H. Oiknine,

Benjamin T. Files, Bianca Dalangin, Anne M. Sinatra, Steven D. Fleming, and Tiffany R. Raber, is titled "Virtual Humans and Their Influence on Navigation." It examines the intriguing possible effect of virtual humans (VHs) on the performance of real humans carrying out a spatial learning and navigation task in a virtual environment (VE). Human interaction with virtual humans depends, at least in some contexts, on the social fidelity of those virtual humans—the degree to which they sound and look like real human beings (e.g., in their speech and gestures). The authors propose a study in a virtual environment to examine the effect of the social fidelity of a virtual human on spatial learning and wayfinding.

The third section, *Knowledge Acquisition and Reasoning*, consists of three chapters. Chapter 6, by Jessica Andrews-Todd and David N. Rapp, is titled "Adverse Consequences of Collaboration on Spatial Problem Solving." It examines groups collaborating to solve spatial problems when one member provides accurate or inaccurate information to the other members; existing research has examined this scenario with nonspatial tasks only. In their study, the authors have dyads attempt to recall the names and locations of U.S. states and capital cities by placing them on an outline map. The verbal expressions of partners were analyzed to detect instances when inaccurate information would or would not be incorporated into solutions to the spatial problem. Chapter 7, by Elizabeth R. Chrastil and You (Lily) Cheng, is titled "Central Coordination and Integration of Diverse Information to Form a Single Map." Their chapter examines the hypothetical scenario of a platoon of soldiers exploring a destroyed city and retrieving observations for the platoon leader to integrate into a representation—essentially a map—of the city's features and layout. The authors consider the factors that would influence this group spatial learning task, including characteristics of platoon members, the nature of spatial communication, the nature of information integration, the presence of risk factors and associated stress, and the complexity of the city's layout. They consider alternative empirical approaches to address these issues in research. Chapter 8 in this section, by Cynthia K. Maupin, Neil G. MacLaren, Gerald F. Goodwin, and Dorothy R. Carter, is titled "Collective Spatial Cognition: Improving Wayfinding Through Transactive Memory Systems." It addresses spatial cognition in teams, intentional collectives of people working together to solve spatial problems. The chapter is one of the three in the book that discusses the important concept of "transactive memory systems" we introduced earlier. The term refers to metacognition about what other group members know—in the case of this chapter, what others know about spatial properties. The authors propose it to be crucial to the success of teams performing spatial tasks.

The fourth section, *Teams*, consists of two chapters. Chapter 9, by Ernest S. Park and Verlin B. Hinsz, is titled "The Dynamics and Performance of Groups as Spatial Information Processors." It presents a conceptualization of groups as spatial information processors, apt as a description of intentional collectives collaboratively solving spatial problems. As their paradigmatic case, the authors consider teams of unmanned aerial vehicle (UAV) operators working together to process spatial information and produce group actions. They suggest ways that differences may emerge in ways tasks are performed when done socially instead of solitarily.

Chapter 10, by Michael R. Baumann, Donald R. Kretz, and Qiliang He, is titled "A Review of Multiteam Systems With an Eye Toward Applications for Collective Spatial Reasoning." This chapter reviews the literature on how members of a team share information with each other to form shared representations of a task and how this affects the team performance. It extends this to collections of teams known as "multiteam systems," specifically addressing a gap in the literature on such systems addressing spatial reasoning tasks.

The fifth and final section, *Applications and Techniques*, consists of three chapters. Chapter 11, by Thomas J. Cova and Frank A. Drews, is titled "Wildfire Protective Actions and Collective Spatial Cognition." It discusses the danger of wildfires near populated areas and the protective actions of evacuation and shelter-in-place. Decisions concerning these protective actions fundamentally incorporate spatial information and, despite the critical role of the chief decision-maker designated by the incident command system (ICS), these decisions must be made collectively, particularly including the critical involvement of the public. Furthermore, such scenarios are typically both highly dynamic and highly uncertain. Chapter 12, by Dario Esposito, Davide Schaumann, Megan Rondinelli, Domenico Camarda, Yehuda E. Kalay, Kevin M. Curtin, and Penelope Mitchell, is titled "Modeling and Simulating the Impact of Human Spatial and Social Behavior on Infection Spread in Hospitals." This chapter deals with the timely problem of infectious disease transmission, long recognized to have an essential spatial component. The authors discuss their research on the impact of human collective social and spatial behavior on infection spread in healthcare environments like hospitals. Their research includes the development of a multi-agent system that represents how spatial cognition of the built environment and agents' perception of others sharing a common space influences human decision-making and spatial behavior in the context of hospital wards. This lays the groundwork for the development of a ready-to-use simulator to serve as a spatial decision support system (SDSS) for real-time infection risk management and long-term hospital planning and layout. The final chapter, Chapter 13, by Kevin M. Curtin, Penelope Mitchell, and Megan Rondinelli is titled "Spatial Analytic Tools and Techniques to Inform Research of Team Spatial Cognition." This chapter speculates on the many ways in which the evolving set of spatial analytic tools and methods could be brought to bear on the topic of CSC. They review both long-standing and recently developed methods in the areas of cartographic technique, network analysis, spatial statistics, raster analysis, 3D analysis, and more.

Conclusions

This chapter introduced and provided an overview of the topic of this edited book: CSC. Our chapter began by presenting five scenarios in which groups or collectives of people had to reason about and make decisions centrally involving space and spatiality. We believe that this topic is theoretically interesting and practically important, but distinctly under-researched. Furthermore, many research disciplines have an interest in aspects of CSC and have potential insights to contribute to its study, and yet the actual and potential efforts of these many disciplines have operated largely in isolation. Thus, the motivation for this volume is to raise awareness of the topic of CSC; help outline its basic concepts, research questions, and methodologies; and promote a conversation among the multiple relevant disciplines.

Our analysis of the topic of CSC proceeded by considering the three conceptual pillars of collective, spatial, and cognition. The first pillar, "cognition," refers to structures and processes of knowledge and knowing, whether by human beings, nonhuman animals, or intelligent machines. Because the mind emerges from the brain and nervous system operating within the physical body, acting within a physical and sociocultural world, valuable research on cognition occurs not only at the physiological and behavioral levels, but at functional, semantic, experiential, computational, environmental, and social levels. Our second conceptual pillar, "spatial," is a concept everyone intuitively understands even while formally defining it is very challenging. We can more readily list spatial properties such as location, distance, size, and connectivity, and we identify spatial problems or situations as those that involve spatial properties (spatial information) in a substantial way. Our third conceptual pillar, "collective," simply refers to a situation involving more than one person. Collectives can be as small as two people or as large as hundreds of people or more. They form accidentally or are planned and are made up of members that interact/influence one another intentionally or unintentionally. Different researchers are of differing opinion as to whether it only makes sense to refer to collectives when they are intentional. In this book, we approach the concept of collectives broadly by including both intentional and unintentional groups (such as crowds).

We then continued our analysis of CSC by considering the two binary concepts of spatial cognition and collective cognition. These binary topics have been the subject of research for decades. "Spatial cognition" is cognition that is primarily about spatiality—that serves primarily to solve problems involving spatial properties as a central component, whether properties of places or environments, objects or events, or symbolic representations such as maps, language, or VRs. Spatial cognition is central to many human activities, useful in solving many human problems, and a fundamental component of human experience. A key concept in the study of spatial cognition is the "cognitive map"; although this term is an appealing metaphor, it has frequently been considered in misleading ways, both because properties of cartographic maps are overextended to cognitive maps and because the possible range of cartographic-map properties is understood too narrowly. "Collective cognition" is the performance of cognitive tasks by two or more individuals. There are varying degrees and types of collective influence or interaction, but when considered broadly, one can argue that all cognition is collective. That recognized, this book focuses on more overt and explicit examples of collective cognition. Research on group behavior and collective cognition has historically been very important in social psychology and sociology, with several thousand published books and articles on the topic, going back to the 19th century. A host of group characteristics and cognitive tasks has been addressed in this vast literature. A notable issue has been whether and how group performance is better or worse than individual performance. A subset of this literature has concerned itself with the important topic of cognition by teams, including multiteam systems (MTSs).

Our conceptual analysis then considered the three-part concept—collective spatial cognition—that is the focus of our book. We believe that social influences on spatial cognition are very common and virtually ubiquitous. Besides the direct influence of people on other people engaged in spatial cognition, social influence is reflected in navigation artifacts and technologies, built spaces and signage, and direct instruction and behavioral observation in the past. The prominence of the social in spatial cognition is belied by the relative paucity of research on it, a claim we discuss and provide evidence for; we also propose reasons for this paucity. The study of CSC will involve many of the issues common to both spatial cognition and collective cognition, but we assume that it may invoke some unique issues. We went on to list and discuss in some detail issues we think define the problem space of CSC, grouped into the five areas of (1) the nature of collectives and their members, (2) the nature of collective interaction/influence, (3) the nature of spatial cognitive tasks/problems, (4) the nature of the space in which the collective cognition occurs, and (5) the diversity of conceptual/empirical issues in studying CSC.

Finally, our chapter offered a list of 26 research questions we hope will contribute to the basic and applied study of CSC. We organized these questions into ten domains: (1) conceptual and empirical issues, (2) the nature of spatial entities involved, (3) group size, (4) individual differences among collective members, (5) leadership and role differentiation, (6) the role of familiarity, (7) intra-group communication, (8) development/change, (9) the role of technological aids, and (10) application areas. The chapters in the book address some of these research questions but, of course, only begin to suggest answers to those questions and others. We briefly introduced those chapters in a final section that provides a synopsis of the book.

References

Allen, G. L. (1997). From knowledge to words to wayfinding: Issues in the production and comprehension of route directions. In S. C. Hirtle & A. U. Frank (Eds.), *Spatial information theory: A theoretical basis for GIS* (pp. 363–372). Berlin: Springer.

Allen, G. L. (Ed.). (2004). *Human spatial memory: Remembering where*. Mahwah, NJ: Lawrence Erlbaum Associates.

Allen, G. L. (Ed.). (2007). *Applied spatial cognition: From research to cognitive technology*. Mahwah, NJ: Lawrence Erlbaum Associates.

Asch, S. E. (1951). Effects of group pressure on the modification and distortion of judgments. In H. Guetzkow (Ed.), *Groups, leadership and men* (pp. 177–190). Pittsburgh, PA: Carnegie Press.

Bae, C. J., & Montello, D. R. (2019). Dyadic route planning and navigation in collaborative wayfinding. In S. Timpf, C. Schlieder, M. Kattenbeck, B. Ludwig, & K. Stewart (Eds.), *14th International conference on spatial information theory* (pp. 24:1–24:20). Article No. 24, Proceedings of COSIT '19. LIPIcs 142. Leibniz International Proceedings in Informatics. Leipzig, Germany: Dagstuhl Publishing.

Beck, S. J., & Keyton, J. (2012). Team cognition, communication, and message interdependence. In E. Salas, S. M. Fiore, & M. P. Letsky (Eds.), *Theories of team cognition: Cross-disciplinary perspectives. Series in applied psychology* (Vol. xxv, pp. 471–494). New York, NY: Routledge and Taylor & Francis Group.

Bell, B. S., & Kozlowski, S. W. J. (2002). A typology of virtual teams: Implications for effective leadership. *Group & Organization Management*, *27*(1), 14–49.

Bloom, P., Peterson, M. A., Nadel, L., & Garrett, M. F. (Eds.). (1996). *Language and space*. Cambridge, MA: The MIT Press.

Brandstätter, H., Davis, J. H., & Stocker-Kreichgauer, G. (Eds.). (1982). *Group decision making*. London: Academic.

Bruch, E., & Feinberg, F. (2017). Decision making processes in social contexts. *Annual Review of Sociology*, *43*.

Burgess, N. (2014). The 2014 Nobel prize in physiology or medicine: A spatial model for cognitive neuroscience. *Neuron*, *84*(6), 1120–1125.

Byrne, P., Becker, S., & Burgess, N. (2007). Remembering the past and imagining the future: A neural model of spatial memory and imagery. *Psychological Review*, *114*, 340–375.

Camerer, C., Loewenstein, G., & Rabin, M. (Eds.). (2004). *Advances in behavioral economics*. Princeton, NJ: Princeton University Press.

Campbell, D. T. (1969). Ethnocentrism of disciplines and the fish-scale model of omniscience. In M. Sherif & C. W. Sherif (Eds.), *Interdisciplinary relationships in the social sciences* (pp. 328–348). Chicago: Aldine.

Coleman, J. S. (1959). The mathematical study of small groups. In H. Solomon (Ed.), *Mathematical thinking in the measurement of behavior: Small groups, utility, factor analysis, part I* (pp. 1–149). Glencoe, IL: Free Press.

Coutrot, A., Manley, E., Goodroe, S., Gahnstrom, C., Filomena, G., Yesiltepe, D., Dalton, R. C., Wiener, J. M., Hölscher, C., Hornberger, M., & Spiers, H. J. (2022). Entropy of city street networks linked to future spatial navigation ability. *Nature*, *604*(7904), 104–110.

Cresswell, T. (2015). *Place: An introduction* (2nd ed.). Chichester and Oxford: Wiley Blackwell.

Curseu, P. L., Schalk, R., & Wessel, I. (2008). How do virtual teams process information? A literature review and implications for management. *Journal of Managerial Psychology*, *23*(6), 628–652.

Dalton, R. C., Hölscher, C., & Montello, D. R. (2019). Wayfinding as a social activity. *Frontiers in Psychology*, *10*(142).

Davidson, R. A., & Hollenbeck, J. R. (2012). Boundary spanning in the domain of multiteam systems. In S. J. Zaccaro, M. A. Marks, & L. A. DeChurch (Eds.), *Multiteam systems: An organizational form for dynamic and complex environments* (pp. 323–363). New York, NY: Routledge and Taylor & Francis Group.

DeChurch, L. A., & Marks, M. A. (2006). Leadership in multiteam systems. *Journal of Applied Psychology*, *91*(2), 311–329.

Denis, M. (2018). *Space and spatial cognition: A multidisciplinary perspective*. London and New York: Routledge.

Denis, M., Pazzaglia, F., Cornoldi, C., & Bertolo, L. (1999). Spatial discourse and navigation: An analysis of route directions in the city of Venice. *Applied Cognitive Psychology*, *13*(2), 145–174.

Downs, R. M., & Stea, D. (1973). Cognitive maps and spatial behavior: Process and products. In R. M. Downs & D. Stea (Eds.), *Image and environment* (pp. 8–26). Chicago: Aldine.

Ekstrom, A. D., Spiers, H. J., Bohbot, V. D., & Rosenbaum, R. S. (2018). *Human spatial navigation*. Princeton, NJ: Princeton University Press.

Ellis, D. G., & Fisher, B. A. (1994). *Small group decision making: Communication and the group process* (4th ed.). New York: McGraw-Hill.

Faria, J. J., Codling, E. A., Dyer, J. R. G., Trillmich, F., & Krause, J. (2009). Navigation in human crowds: Testing the many-wrongs principle. *Animal Behaviour*, *78*(3), 587–591.

Fernández Velasco, P. (2022). Group navigation and procedural metacognition. *Philosophical Psychology*, 1–19. doi:10.1080/09515089.2022.2062316

Fiore, S. M., Smith-Jentsch, K. A., Salas, E., Warner, N., & Letsky, M. (2010). Towards an understanding of macrocognition in teams: Developing and defining complex collaborative processes and products. *Theoretical Issues in Ergonomics Science*, *11*(4), 250–271.

Fisher, S. G., Hunter, T. A., & Macrosson, W. D. K. (1998). The structure of Belbin's team roles. *Journal of Occupational and Organizational Psychology*, *71*(3), 283–288.

Forlizzi, J., Barley, W. C., & Seder, T. (2010). Where should I turn: Moving from individual to collaborative navigation strategies to inform the interaction design of future navigation systems. In *Proceedings of the SIGCHI conference on human factors in computing systems* (pp. 1261–1270). New York: ACM.

Friedman, A., & Brown, N. R. (2000). Reasoning about geography. *Journal of Experimental Psychology: General*, *129*(2), 193–219.

Galton, F. (1907). Vox populi (the wisdom of crowds). *Nature*, *75*(1949), 450–451.

Gazzaniga, M. S. (2000). *Cognitive neuroscience: A reader*. Malden, MA: Blackwell.

Gerard, H. B., & Miller, N. (1967). Group dynamics. *Annual Review of Psychology*, *18*(1), 287–332.

Gero, J. S. (Ed.). (2015). *Studying visual and spatial reasoning for design creativity*. Dordrecht, The Netherlands: Springer.

Gigerenzer, G., & Gaissmaier, W. (2011). Heuristic decision making. *Annual Review of Psychology*, *62*, 451–482.

Glass, A. L., & Holyoak, K. J. (1986). *Cognition* (2nd ed.). New York: Random House.

Golledge, R. G. (1987). Environmental cognition. In D. Stokols & I. Altman (Eds.), *Handbook of environmental psychology* (pp. 131–174). New York: Wiley.

Golledge, R. G. (Ed.). (1999). *Wayfinding behavior: Cognitive mapping and other spatial processes*. Baltimore, MD: Johns Hopkins University Press.

Golledge, R. G. (2004). Spatial cognition. In C. Spielberger (Ed.), *Encyclopedia of applied psychology* (Vol. 3, pp. 443–452). Oxford and Boston: Elsevier Academic Press.

Grünbaum, A. (1973). *Philosophical problems of space and time* (2nd ed.). Dordrecht, The Netherlands: D. Reidel.

Haddington, P. (2012). Movement in action: Initiating social navigation in cars. *Semiotica*, *191*, 137–167.

Hadorn, G. H., Biber-Klemm, S., Grossenbacher-Mansuy, W., Hoffmann-Riem, H., Joye, D., Pohl, C., Wiesmann, U., & Zemp, E. (2008). The emergence of transdisciplinarity as a form of research. In G. H. Hadorn, H. Hoffmann-Riem, S. Biber-Klemm, W. Grossenbacher-Mansuy, D. Joye, C. Pohl, U. Wiesmann, & E. Zemp (Eds.), *Handbook of transdisciplinary research* (pp. 19–39). Dordrecht, The Netherlands: Springer.

Hall, E. J., Mouton, J. S., & Blake, R. R. (1963). Group problem solving effectiveness under conditions of pooling vs. interaction. *Journal of Social Psychology*, *59*(1), 147–157.

He, G., Ishikawa, T., & Takemiya, M. (2015). Collaborative navigation in an unfamiliar environment with people having different spatial aptitudes. *Spatial Cognition and Computation*, *15*(4), 285–307.

Helbing, D., & Johansson, A. (2011). Pedestrian, crowd and evacuation dynamics. In R. A. Meyers (Ed.), *Extreme environmental events* (pp. 697–716). New York: Springer.

Hillier, B., & Hanson, J. (1984). *The social logic of space*. Cambridge: Cambridge University Press.

Hinds, P. J., & Bailey, D. E. (2000). Virtual teams: Anticipating the impact of virtuality on team process and performance. *Academy of Management Proceedings*, *2000*(1), C1–C6.

Hinds, P. J., & Bailey, D. E. (2003). Out of sight, out of sync: Understanding conflict in distributed teams. *Organization Science*, *14*(6), 615–632.

Huggett, N. & Hoefer, C. (2018, Spring). Absolute and relational theories of space and motion. In E. N. Zalta (Ed.), *The Stanford encyclopedia of philosophy*. http://plato.stanford.edu/archives/sum2016/entries/mental-imagery/

Hutchins, E. (1995). *Cognition in the wild*. Cambridge, MA: The MIT Press.

Kahneman, D. (2011). *Thinking, fast and slow*. New York: Farrar, Straus and Giroux.

Kameda, T., Tindale, R. S., & Davis, J. H. (2002). Cognitions, preferences, and social sharedness: Past, present, and future directions in group decision making. In S. L. Shneider & J. Shanteau (Eds.), *Emerging perspectives on judgment and decision research* (pp. 458–485). Cambridge, UK: Cambridge University Press.

Kerr, N. L., & Tindale, R. S. (2004). Group performance and decision making. *Annual Review of Psychology*, *55*, 623–655.

Keyton, J., Ford, D. J., & Smith, F. L. (2012). Communication, collaboration, and identification as facilitators and constraints of multiteam systems. In S. J. Zaccaro, M. A. Marks, & L. A. DeChurch (Eds.), *Multiteam systems: An organization form for dynamic and complex environments* (Vol. xxi, pp. 173–190). New York, NY: Routledge and Taylor & Francis Group.

Kim, K., Maloney, D., Bruder, G., Bailenson, J. N., & Welch, G. F. (2017). The effects of virtual human's spatial and behavioral coherence with physical objects on social presence in AR. *Computer Animation and Virtual Worlds*, *28*, e1771.

Kinateder, M., Müller, M., Jost, M., Mühlberger, A., & Pauli, P. (2014). Social influence in a virtual tunnel fire—Influence of conflicting information on evacuation behavior. *Applied Ergonomics*, *45*(6), 1649–1659.

Kitchin, R., & Blades, M. (2002). *The cognition of geographic space*. New York and London: IB Tauris Publishers.

Klein, W. (1983). Deixis and spatial orientation in route directions. In H. L. Pick & L. P. Acredolo (Eds.), *Spatial orientation: Theory, research, and application* (pp. 283–311). New York: Plenum Press.

Kozlowski, S. W. J., & Bell, B. S. (2003). Work groups and teams in organizations. In *Handbook of psychology, Volume 12: Industrial and organizational psychology*. Hoboken, NJ: John Wiley & Sons, Inc.

Lamarche, F., & Donikian, S. (2004). Crowd of virtual humans: A new approach for real time navigation in complex and structured environments. *Computer Graphics Forum*, *23*(3), 509–518.

Levine, J. M., & Moreland, R. (1990). Progress in small group research. *Annual Review of Psychology*, *41*(1), 585–634.

Maguire, E. A., Gadian, D. G., Johnsrude, I. S., Good, C. D., Ashburner, J., Frackowiak, R. S. J., & Frith, C. D. (2000). Navigation-related structural change in the hippocampi of taxi drivers. *Proceedings of the National Academy of Sciences of the United States of America*, *97*, 4398–4403.

Mathieu, J. E., Marks, M. A., & Zaccaro, S. J. (2002). Multiteam systems. In N. Anderson, D. S. Ones, H. K. Sinangil, & C. Viswesvaran (Eds.), *Handbook of industrial, work and organizational psychology 2* (pp. 289–313). London: Sage.

McGrath, J. E., & Kravitz, D. A. (1982). Group research. *Annual Review of Psychology, 33*(1), 195–230.

McNamara, T. P., Sluzenski, J., & Rump, B. (2008). Human spatial memory and navigation. In H. L. I. Roediger (Ed.), *Cognitive psychology of memory* (Vol. 2, pp. 157–178). Oxford: Elsevier.

Montello, D. R. (2008). Cognitive science. In K. K. Kemp (Ed.), *Encyclopedia of geographic information science* (pp. 40–43). Thousand Oaks, CA and London: SAGE Publications.

Montello, D. R. (Ed.). (2018). *Handbook of behavioral and cognitive geography*. Cheltenham, UK and Northampton, MA: Edward Elgar Publishing.

Montello, D. R. (2020). Geographic orientation, disorientation, and misorientation: A commentary on Fernandez Velasco and Casati. *Spatial Cognition and Computation, 20*(4), 306–313.

Montello, D. R., & Raubal, M. (2012). Functions and applications of spatial cognition. In D. Waller & L. Nadel (Eds.), *Handbook of spatial cognition* (pp. 249–264). Washington, DC: American Psychological Association.

Moussaid, M., Garnier, S., Theraulaz, G., & Helbing, D. (2009). Collective information processing and pattern formation in swarms, flocks, and crowds. *Topics in Cognitive Science, 1*(3), 469–497.

Newcombe, N. S. (2018). Three kinds of spatial cognition. In J. Wixted (Ed.), *Stevens' handbook of experimental psychology and cognitive neuroscience* (4th ed., Vol. 3, pp. 521–552). Hoboken, NJ: Wiley.

Newcombe, N. S., & Huttenlocher, J. (2000). *Making space: The development of spatial representation and reasoning*. Cambridge, MA: The MIT Press.

Newcombe, N. S., & Shipley, T. F. (2015). Thinking about spatial thinking: New typology, new assessments. In J. S. Gero (Ed.), *Studying visual and spatial reasoning for design creativity* (pp. 179–192). Dordrecht, The Netherlands: Springer.

Ochsner, K. N., & Kosslyn, S. M. (2014). *The Oxford handbook of cognitive neuroscience*. Oxford and New York: Oxford University Press.

Ostrom, T. M. (1984). The sovereignty of social cognition. In R. S. Wyer, Jr. & T. K. Srull (Eds.), *Handbook of social cognition* (Vol. 1, pp. 1–38). Hillsdale, NJ: Lawrence Erlbaum Associates Publishers.

Plous, S. (1993). *The psychology of judgment and decision making*. Philadelphia: Temple University Press.

Reilly, D., Mackay, B., Watters, C., & Inkpen, K. (2009). Planners, navigators, and pragmatists: Collaborative wayfinding using a single mobile phone. *Personal and Ubiquitous Computing, 13*(4), 321–329.

Rico, R., & Cohen, S. G. (2005). Effects of task interdependence and type of communication on performance in virtual teams. *Journal of Managerial Psychology, 20*(3/4), 261–274.

Rogoff, B. (1990). *Apprenticeship in thinking: Cognitive development in social context*. New York: Oxford University Press.

Sack, R. D. (1983). Human territoriality: A theory. *Annals of the Association of American Geographers, 73*, 55–74.

Salas, E., Cooke, N. J., & Rosen, M. A. (2008). On teams, teamwork, and team performance: Discoveries and developments. *Human Factors: The Journal of the Human Factors and Ergonomics Society, 50*(3), 540–547.

Salas, E., Fiore, S. M., & Letsky, M. P. (2013). *Theories of team cognition: Cross-disciplinary perspectives*. New York: Routledge.
Senior, B. (1997). Team roles and team performance: Is there 'really' a link? *Journal of Occupational and Organizational Psychology, 70*(3), 241–258.
Shah, P., & Miyake, A. (Eds.). (2005). *The Cambridge handbook of visuospatial thinking*. Cambridge: Cambridge University Press.
Steiner, I. D. (1964). Group dynamics. *Annual Review of Psychology, 15*(1), 421–446.
Storey, D. (2012). *Territories: The claiming of space* (2nd ed.). London: Routledge.
Swap, W. C. (1984). *Group decision making*. Beverly Hills, CA: Sage Publications.
Tindale, R. S., Kameda, T., & Hinsz, V. B. (2003). Group decision making: Review and integration. In M. A. Hogg & J. Cooper (Eds.), *Sage handbook of social psychology* (pp. 381–403). London: Sage.
Tolman, E. C. (1948). Cognitive maps in rats and men. *Psychological Review*, 55, 189–208.
Trowbridge, C. C. (1913). On fundamental methods of orientation and "imaginary maps". *Science, 38*, 888–897.
Tuan, Y.-F. (1976). Humanistic geography. *Annals of the Association of American Geographers, 66*(2), 266–276.
Tversky, B. (2000). Remembering spaces. In E. Tulving & F. I. M. Craik (Eds.), *The Oxford handbook of memory* (pp. 363–378). Oxford: Oxford University Press.
Uttal, D. H., Meadow, N. G., Tipton, E., Hand, L. L., Alden, A. R., Warren, C., & Newcombe, N. S. (2013). The malleability of spatial skills: A meta-analysis of training studies. *Psychological Bulletin, 139*(2), 352–402.
Waller, D., & Nadel, L. (Eds.). (2012). *Handbook of spatial cognition*. Washington, DC: American Psychological Association.
Weisman, J. (1981). Evaluating architectural legibility: Way-finding in the built environment. *Environment and Behavior, 13*(2), 189–204.
Wertsch, J. V. (1985). *Vygotsky and the social formation of mind*. Cambridge, MA: Harvard University Press.
Wilson, R. A., & Keil, F. C. (Eds.). (1999). *The MIT encyclopedia of the cognitive sciences*. Cambridge, MA: MIT Press.
Wunderlich, D., & Reinelt, R. (1982). How to get there from here. In F. J. Jarvella & W. Klein (Eds.), *Speech, place, and action* (1st ed., Vol. 1, pp. 183–201). New York: John Wiley & Sons Ltd.
Xiao, D., & Liu, Y. (2007). Study of cultural impacts on location judgments in Eastern China. In S. Winter, M. Duckham, L. Kulik, & B. Kuipers (Eds.), *Spatial information theory* (pp. 20–31). Berlin: Springer.
Yoo, Y., & Alavi, M. (2004). Emergent leadership in virtual teams: What do emergent leaders do? *Information and Organization, 14*(1), 27–58.
Zaccaro, S. J., & DeChurch, L. A. (2012). Leadership forms and functions in multiteam systems. In S. J. Zaccaro, M. A. Marks, & L. A. DeChurch (Eds.), *Multiteam systems: An organizational form for dynamic and complex environments* (pp. 253–288). New York: Routledge and Taylor & Francis Group.
Zaccaro, S. J., DeChurch, L. A., & Marks, M. A. (2012). Multiteam systems: An introduction. In S. J. Zaccaro, M. A. Marks, & L. A. DeChurch (Eds.), *Multiteam systems: An organizational form for dynamic and complex environments* (pp. 3–32). New York: Routledge and Taylor & Francis Group.
Zaccaro, S. J., Rittman, A. L., & Marks, M. A. (2001). Team leadership. *The Leadership Quarterly, 12*(4), 451–483.

Navigation

2 Navigation in Collaboration

Toru Ishikawa

Introduction

Human spatial cognition has been the focus of systematic research since at least the 1960s (Golledge & Timmermans, 1990), and researchers across disciplines have addressed various issues, including cognitive maps, spatial orientation, route navigation, spatial language, qualitative reasoning, and individual differences (Montello & Raubal, 2013). The issues of cognitive maps and spatial orientation, in particular, concern how people learn the surrounding environment and understand where they are in it, at a spatial scale larger than the human body. It is one of the most fundamental human everyday activities but poses difficulties for some people, especially people with low spatial aptitudes (Newcombe, 2019). People poor at mental rotation and perspective taking have trouble with map use and spatial orientation in the environment (Liben & Downs, 1993; Nazareth et al., 2018). People with a poor sense of direction have difficulty comprehending the environment with sufficient metric and configurational accuracy and acquire only impoverished spatial knowledge; that is, their "cognitive maps" are inaccurate and poorly structured (Hegarty et al., 2006; Kozlowski & Bryant, 1977). And, importantly, recent research into human spatial cognition builds on the finding that the accuracy and extent of cognitive maps show large individual differences (Ishikawa & Montello, 2006; Weisberg et al., 2014) and seeks to understand why and how people differ in their cognitive mapping skills (Ishikawa, 2022).

Interestingly enough, past research into spatial cognition viewed spatial orientation and navigation as *individual*, rather than group or *social*, activities. Study participants typically walked a route in an unfamiliar environment and conducted spatial tasks about the route individually, such as estimating distances and directions, drawing a sketch map, or finding a new path between pairs of places. In reality, people conduct these spatial orientation and navigation tasks *in collaboration with* other people, for example, walking in a foreign country in a group of friends or helping tourists find their way to a series of sightseeing spots. Successful spatial orientation and navigation, in general, require accurate knowledge of the environment and adaptive use of the knowledge (internal representations) in coordination with the physical environment and navigation assistance (external representations), such as maps, verbal directions, or mobile navigation tools.

DOI: 10.4324/9781003202738-4

Successful *collaborative* spatial orientation and navigation, furthermore, require coordination with others, that is, efficient communication of one's knowledge and flexible adjustment of how the knowledge is communicated to the partner (e.g., their spatial ability, sense of direction, navigation preferences, spatial anxiety) and context (e.g., environmental structure, availability of landmarks, navigation situation). Thus, the focus in the existing literature on spatial orientation and navigation as individual activities leaves the dynamic and interactive aspect of collectively conducted navigation understudied.

What is, then, the interactive, or *social*, aspect of collaborative navigation? Imagine that you and your partner visit a new place. You two are now at different locations in the unfamiliar environment and want to meet at a certain third location; you do not know how to get there, but your partner does, or vice versa. In such a situation, you can help each other to find the way to the destination, telling in real time on a mobile phone what is seen and which way to turn. In another situation, you two navigate in a novel place toward a series of sightseeing spots, guiding each other in direct contact together. Then, you can talk to each other while walking, as well as consult a guidebook or ask another person. In these situations, how do you conduct the task of navigation in collaboration with the partner? For example, do you give different navigational directions when the partner is a child, an older person, or a visitor? Do you change the directions according to the partner's psychological attributes, such as personality traits (e.g., outgoing, cautious, nervous), familiarity with the environment (e.g., a newcomer or long-term resident), navigational preferences (e.g., attending to landmarks, left–right turns, or cardinal directions), and spatial aptitudes (e.g., spatial ability, sense of direction, map skills)? Do you use different landmarks when you are in urban and rural environments or in built and natural environments? Most likely, you adjust your directions to your and your partner's knowledge of the environment, your perception of the partner's navigation skills and performance, and the situation in which a conversation takes place; this points to the dynamic and interactive (or one can use the term *social*; see Dalton et al., 2019) nature of navigation in collaboration with others. Contemporary navigation assistance (notably, satellite navigation systems), by contrast, takes a fixed, generalized approach, primarily providing turn-by-turn instructions, with little consideration of the dynamic interaction between the user, tool, and environment. As a consequence, empirical research has shown that many advanced mobile navigation tools do not necessarily assist the user to effectively navigate or accurately learn space (Gardony et al., 2013; Ishikawa et al., 2008; Willis et al., 2009).

Types of Collaborative Navigation

One can think of different types of navigation that are conducted collectively. For example, a group of people walk toward a destination together; they all know where they are going but not how to get there. In this case, a person who is poor at wayfinding could simply follow other people (hopefully they are better wayfinders), leaving the effortful task of route planning to them. Such behavior is not of particular interest to the study of collaborative navigation because it lacks elaborate cognitive processing (see Buck et al., 2020). One can point out that this sort of passive following (or lack of decision-making) is identified as one of the reasons for deteriorated spatial memory after the use of mobile navigation tools (Ishikawa, 2019; Ruginski et al., 2019; Sugimoto et al., 2022).

There is another situation in which where to go (a destination) and how to get there (a route) are known to some, but not all, members of a group. In this case, the members who are informed will direct other members to the destination. In the existing literature, this type of collaborative navigation has been studied in the context of verbal navigational directions, namely, when a knowledgeable person gives verbal directions to another person who is unfamiliar with the route. This line of research contributes to the understanding of effective navigational instructions, particularly the salience of different types of landmarks (e.g., Denis et al., 1999; Forlizzi et al., 2010; Klippel & Winter, 2005; Raubal & Winter, 2002), but does not directly address the dynamics of collaborative navigation.

A third form of collaborative navigation, which is the main focus of this chapter, concerns a group of people traveling in an unfamiliar environment assisting each other, each member playing an active role in the collectively conducted navigation. Questions of interest concern the difficulties people may have in the communication of navigational directions and the quality of spatial knowledge that is acquired by individuals or shared by the group. Past studies addressed these questions in an empirical setting of paired navigation: for example, with a pair of people sharing the same navigation information (Reilly et al., 2009), one informed person directing the other person who has no prior knowledge (He et al., 2015), or two people exploring in a (partially) new environment (Buck et al., 2020). Below I look at how the results from these studies provide insights about the communication and processing of spatial information in collaborative navigation. Before doing that, I briefly overview existing research in the literature on collective learning and behavior in general to clarify the major concepts and frameworks of group decision-making that are suggestive for the study of collaborative navigation as a group activity.

Collective Learning and Decision-Making

Research into collective learning and behavior focuses on people who are making decisions in a group and, hence, considers a group as a basic information processing unit (Hinsz & Tindale, 1997). People work together as a collective to accomplish an objective that is shared by the team in a social, as opposed to solitary, context. In doing so, they process two kinds of information: task-specific and social information (see the chapter by Park & Hinsz in this volume). The former includes the missions of the team, the roles of team members, and the structure of environments that they work in; the latter includes an understanding of preferences, identities, or abilities of team members. Through the processing of these kinds of information, the team constructs a shared mental model, which, in turn, enables the team to engage in the joint effort efficiently with an appropriate situational understanding of the task at hand. For successful group decision-making, therefore, the mental models possessed by individual team members need to converge into a shared and accurate team mental model, an important property of collective learning identified as *social sharedness* (Burtscher & Manser, 2012; Tindale & Kameda, 2000).

Existing studies of collective learning have shown that people tend to do better on various tasks as a group than individually. For example, in a task of reasoning about the mapping of 10 letters (A–J) to 10 numbers (0–9) given in a series of equations, Laughlin et al. (2002) found that groups of participants working together did better than high-performance individuals working alone. In the animal behavior literature, which group navigation research often refers to (see below), Liker and Bókony (2009) similarly observed that sparrows did better on a foraging task in large groups than in small groups.

Researchers have discussed that the superiority of group performance is related to the diversity of skills, personalities, and experiences among group members. For example, Aggarwal et al. (2019) found an inverted U-shaped relationship between performance in group learning and the degree of diversity in cognitive styles: Group learning performance was highest when the degree of diversity was moderate. Ein-Dor et al. (2011) likewise found that group performance in threat detection was positively affected by greater diversity of personality. At the same time, researchers have pointed to some possible caveats concerning group thinking. Minson (2012) showed that people working as a group were less willing to use peer input with a higher degree of confidence in the accuracy of their performance. Aggarwal and Woolley (2013) found that the formation of a strategic consensus was negatively correlated with the degree of heterogeneity in cognitive style, observing increased errors in learning in a heterogeneous group.

Navigation and Spatial Orientation in a Group

As a major example of collective decision-making in the everyday spatial context, one can point out the tasks of group navigation and spatial orientation, particularly when they are conducted as a social activity with intentional and direct communication among group members (Dalton et al., 2019). Similar to many studies in the literature of collective learning and behavior, characteristics of group navigation and spatial orientation are often examined based on computer simulations and animal models (e.g., Helbing et al., 2000; Saloma et al., 2003). For example, Codling et al. (2007) showed that animals performed spatial orientation and movement more accurately when they flocked than was predicted from individual error rates, an effect known as the "many-wrongs principle." In the task of locational choice in human crowds, Dyer et al. (2008) showed that a group of people chose a destination and headed to it faster and more directly when they were guided by informed people than when all group members had no prior information. When the information given to a subset of group members was contradictory, the group tended to make a collective decision in favor of the information held by the majority of informed individuals. These findings point to the social aspects of information processing in group navigation. In group decision-making about which path to follow,

Flack et al. (2015) found that a group's joint movement was mostly influenced by the social connectedness and structure of group members, such as the roles of leaders and followers or the degree of interaction.

Another social aspect of group navigation that is examined in empirical studies is the choice of spatial frames of reference in verbal communication. Galati and Avraamides (2013) examined a collaborative spatial task in which one person (the director) studied an array of objects and described it from memory to the other person (the matcher), who then reconstructed the array on the basis of the director's descriptions. The director and the matcher were seated at two different positions on a round table, being separated by 90°, 135°, or 180°. The researchers found that people conducted the task flexibly, adjusting spatial descriptions by taking situational variables into consideration. With no prior information about the matcher's viewpoint, the director described the array egocentrically from his or her own perspective. When the matcher's viewpoint was known in advance, however, the director encoded and described the array from the matcher's viewpoint by mentally rotating it to an imagined heading aligned with the matcher. People adjusted their descriptions to the degree of cognitive demand of mental perspective taking as well. When the paired members were separated by 90°, mentally rotating the array of objects was relatively easy, and the director chose to use descriptions that were centered on the matcher. When they were separated by an oblique angle of 135°, mentally rotating the array was cognitively more effortful, and the director chose to describe it from his or her own perspective. Thus, knowledge of the partner's viewpoint enabled the paired member to adjust spatial memory and descriptions to the social cues available and to situational contexts. Galantucci (2005) similarly found an effect of communication on collective decision-making. The researcher observed, in a series of collaborative video game tasks, that pairs of people developed their own communication systems and that pairs who achieved successful communication performed better than pairs who failed at communication.

Collaborative Navigation as a Social Activity

Group decision-making in people's everyday activities is seen when they navigate in the environment toward a common destination as part of a group. Navigation in this sense is identified as a social activity: Dalton et al. (2019) discussed four variations of social wayfinding based on (a) the degree of intentionality in communication among group members and (b) the copresence of the sender and the receiver of information during navigation. With respect to intentionality, *strong* social navigation involves direct and intentional communication of navigation information within a group, while *weak* social navigation uses the existence or traces of people indirectly and unintentionally as wayfinding clues. With respect to copresence, *synchronous* social navigation involves group members giving and receiving navigation information in contact in real time, while *asynchronous* social navigation refers to cases where the sender and the receiver of navigation information are not copresent during the exchange of the information. In this framework, the type of navigation that is considered specifically *collaborative* and of particular interest to the present chapter is the category of *strong, synchronous* social navigation.

Some studies have examined this type of collaborative navigation empirically. Reilly et al. (2009) examined how people in pairs process and communicate information while navigating collaboratively in indoor environments with a shared mobile navigation tool. The researchers found that the pairs used different navigation strategies, such as sync-and-go and plan-and-go strategies, and that they dynamically changed strategies depending on their roles (e.g., a leader or a follower) and the stage of navigation (e.g., planning, walking, or validation). These results indicate that to communicate navigation information efficiently, one needs to take the different possible strategies into account and tailor the method of instructions to user preference and navigational context.

Another question of concern to research into collaborative navigation is the information that people in a group attend to during navigation, namely, which part of the environment individual members look at and how they communicate and share the information with other members. Schwarzkopf et al. (2017) discussed the fact that people who navigate collaboratively in pairs need to establish a common ground for their cognitive cooperation so that they can construct a shared mental model. In doing so, paired navigators can either divide or share their attention to the surrounding environment. In the former approach, paired navigators would divide their roles in navigation and search different, nonoverlapping parts of the environment, whereas in the latter approach, they would share common, meaningfully redundant parts to search. To examine these two possibilities, Schwarzkopf et al. examined people's gaze behavior during paired collaborative navigation in a complex indoor environment. They found that people in pairs visually tracked the partner's perspectives, instead of dividing their search space. This indicates that achieving social sharedness (Tindale & Kameda, 2000) is significant in real-world collaborative navigation tasks.

Buck et al. (2020) examined how people communicate spatial information during paired navigation in a virtual environment and how accurately they learn a traversed environment both individually and collectively. First, the researchers asked participants to navigate in a virtual environment individually and then assessed their knowledge of the environment in terms of relative direction and straight-line distance judgments. Then, they asked the participants to navigate in pairs in another virtual environment and conduct the direction and distance judgment tasks collaboratively. Results showed the participants estimated relative directions more accurately when they did it in pairs than individually, revealing that better spatial orientation performance is achieved as a result of collaborative learning and thinking. The researchers also reported that the majority of pairs performed in the direction and distance judgment tasks better than individual members did.

These findings in the existing literature foster further empirical studies of collaborative navigation to compare individual and group performance in various tasks of spatial navigation and orientation. In the next section, I describe a study of paired collaborative navigation that I conducted previously in a real-world setting

(He et al., 2015) and, then, based on the results from the study, discuss questions for future research into strong and synchronous social navigation.

Empirical Study of Paired Navigation

The study reported in this section examined people's navigation performance and interactive communication when they traveled routes in an unfamiliar environment guiding each other. In the study, each member of a pair learned a separate route individually first and then directed the partner along the learned route over the cellphone while, at the same time, traveling the other novel route being directed by the partner (details below). This experimental design was employed to examine the process of information communication and decision-making, with participants engaging in the collaborative navigation task actively and interactively.

Method

Forty-four adults (27 male and 17 female) participated in an experiment that was designed to examine people's wayfinding performance and information communication when navigating in pairs in an unfamiliar residential neighborhood in Tokyo. Prior to the experiment, participants took the Santa Barbara Sense-of-Direction (SBSOD) Scale (Hegarty et al., 2002). Based on the scores on the scale, the participants were grouped into 22 pairs, varied by the combination of good and poor levels of sense of direction: Six were pairs of people who both had a good sense of direction (high–high pairs), 10 were pairs of people who had a good and a poor sense of direction (high–low pairs), and six were pairs of people who both had a poor sense of direction (low–low pairs). Of the 22 pairs, seven were male–male pairs, two were female–female pairs, and 13 were male–female pairs.

In the study area, two routes were selected that shared a common starting point (Figure 2.1). In the experiment, each pair was taken to the study area, one pair at a time, and each member of the pair first walked a route (either Route 1 or Route 2, which was randomly assigned) individually to the end point, guided by the experimenter (an initial guided walk). At the start of the initial walk, the experimenter pointed out the direction of north to participants, so that they could use the information if they wished to guide their partner along the route in the task of collaborative navigation later. When they reached the end point of the route in the initial walk, they retraced it back to the starting point. Then they switched routes and walked the other route, this time being directed by the partner through conversations over a cellphone. That is, a person who had first traveled Route 1 (or Route 2) now traveled Route 2 (or Route 1) being guided by the partner and, at the same time, guided the partner along Route 1 (or Route 2) by talking on the cellphone. The photographs in Figure 2.2 show examples of views and major landmarks that were available to participants along the two routes.

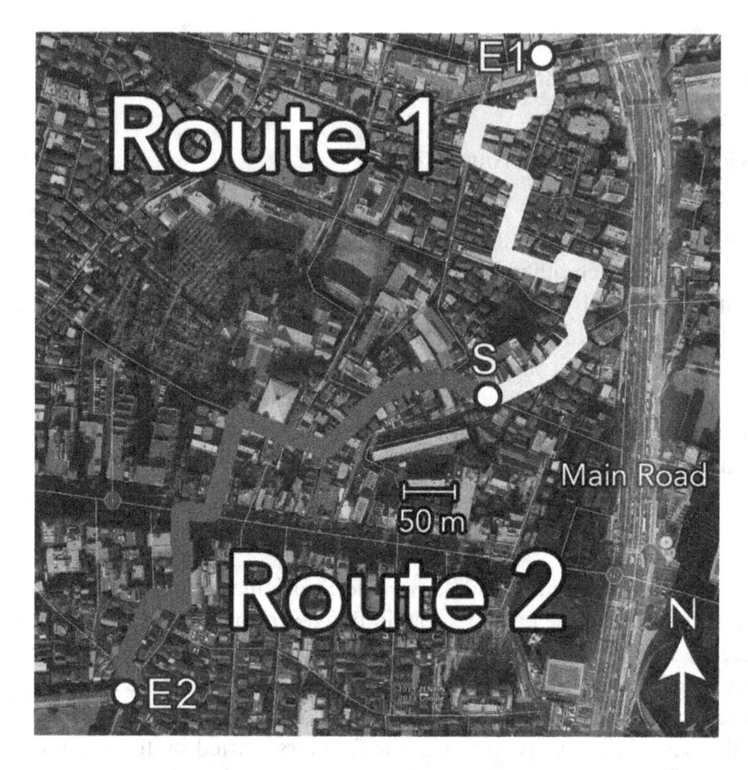

Figure 2.1 A map of the study area. The two routes, Routes 1 and 2, share the same starting point (denoted by S in the figure) and have their respective end points (E1 and E2). Map data ©2013 Google, Zenrin (Reprinted by permission of Taylor & Francis from He et al., 2015, p. 290, Figure 1)

Figure 2.2 Major landmarks along the two routes: a hospital building on Route 1 (*left*) and a temple on Route 2 (*right*). Photographs by G. He (Reprinted by permission of Taylor & Francis from He et al., 2015, p. 290, Figure 2)

Verbal Protocol Analysis

Participants' verbal reports (records of conversations on the cellphone) from the initial guided walk and the subsequent collaborative navigation were analyzed to reveal the information that they used and attended to, in terms of the categories of landmarks (roads, permanent landmarks, ephemeral landmarks, and signs), additional descriptions (colors, text information, and other), and the type of navigational directions (go straight, make a turn, cardinal directions, distance, and time).

Statements that participants made while giving directions in collaborative navigation were classified into four categories (Table 2.1): (a) giving determinate directions, (b) giving less determinate and exploratory directions, (c) asking for cooperation, and (d) being lost. Likewise, statements that participants made while receiving directions were classified into four categories: (a) following determinately the directions given by the partner, (b) following directions in an exploratory way while interacting with the partner, (c) trying to find a place in response to the request for cooperation, and (d) being lost. The four categories (a)–(d) for direction giving and receiving are in the decreasing order of confidence and were given four, three, two, and one points, respectively, in the assessment of the level of confidence below.

Navigational Instructions by Efficient and Inefficient Groups

As performance measures of collaborative navigation, the distance traveled and time taken for each pair to reach the goals were observed. Based on a composite

Table 2.1 Categories of statements of direction giving and receiving during collaborative navigation

Direction giving:

(a) Giving determinate directions
"Go straight for 100 feet and turn left at the hospital."
(b) Giving less determinate and exploratory directions
"I think you should turn at the sign and there should be a white building afterwards."
(c) Asking for cooperation
"Where are you right now? I think there is a staircase around there somewhere, so tell me when you find it."
(d) Being lost
"Where are you right now? I don't know where you are."

Direction receiving:

(a) Following determinately the directions given by the partner
"Yeah I see that. I'm going there now."
(b) Following directions in an exploratory way interacting with the partner
"I'm at the street corner. I see a stop sign to my left. What do I do now?"
(c) Trying to find a place in response to the request of cooperation
"I will go look for a parking lot. Where should I turn after that?"
(d) Being lost
"I don't know where I am. I don't understand what you want me to do."

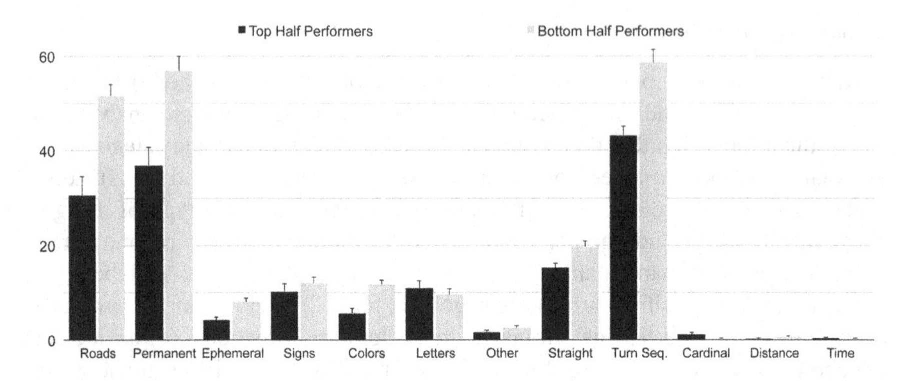

Figure 2.3 Frequency distributions of the types of navigation information used by the efficient and inefficient groups. Bars indicate the mean numbers of times that each information appeared in vernal directions by the two groups. Vertical lines depict standard errors of the means. (Reprinted by permission of Taylor & Francis from He et al., 2015, p. 295, Figure 4)

score of the two measures, the 22 pairs were dichotomized into an *efficient* (or top-half) group and an *inefficient* (or bottom-half) group. With respect to the types of navigation information used by the two groups of pairs, both of their verbal directions frequently mentioned roads, permanent landmarks, and turn sequences (Figure 2.3).

On average, pairs in the inefficient group gave a larger number of instructions, $F(1, 20) = 19.18, p < .001$, and mentioned roads, permanent landmarks, ephemeral landmarks, colors, and turn sequences more frequently than pairs in the efficient group, $t(20) = 4.48, 4.18, 3.72, 4.28$, and 4.60, respectively, $p < .005$ (Bonferroni). Together with the inefficient pairs' poor navigation performance in terms of travel distance and time, these findings indicate that although they communicated a large amount of information in collaborative navigation, their verbal directions were not effective in guiding the partner to a destination.

Sense of direction (SBSOD) scores did not correlate with travel distance or travel time when analyzed at the individual level, but the difference in scores between paired members correlated significantly with travel time, $r = .42, p = .049$ (Figure 2.4). That is, pairs of people whose sense of direction differed greatly (i.e., the high–low pairs) found it difficult to communicate navigation information and took a longer time to find their way to the destination. A longer travel time, in turn, was associated with lower confidence levels in direction giving and receiving, $r = -.77, p < .001$.

Conversation Dynamics in Paired Navigation

The verbal reports from the 22 pairs were further analyzed to understand the characteristics of interactive information communication and group decision-making during collaborative navigation. For the analysis, the pairs were classified into six

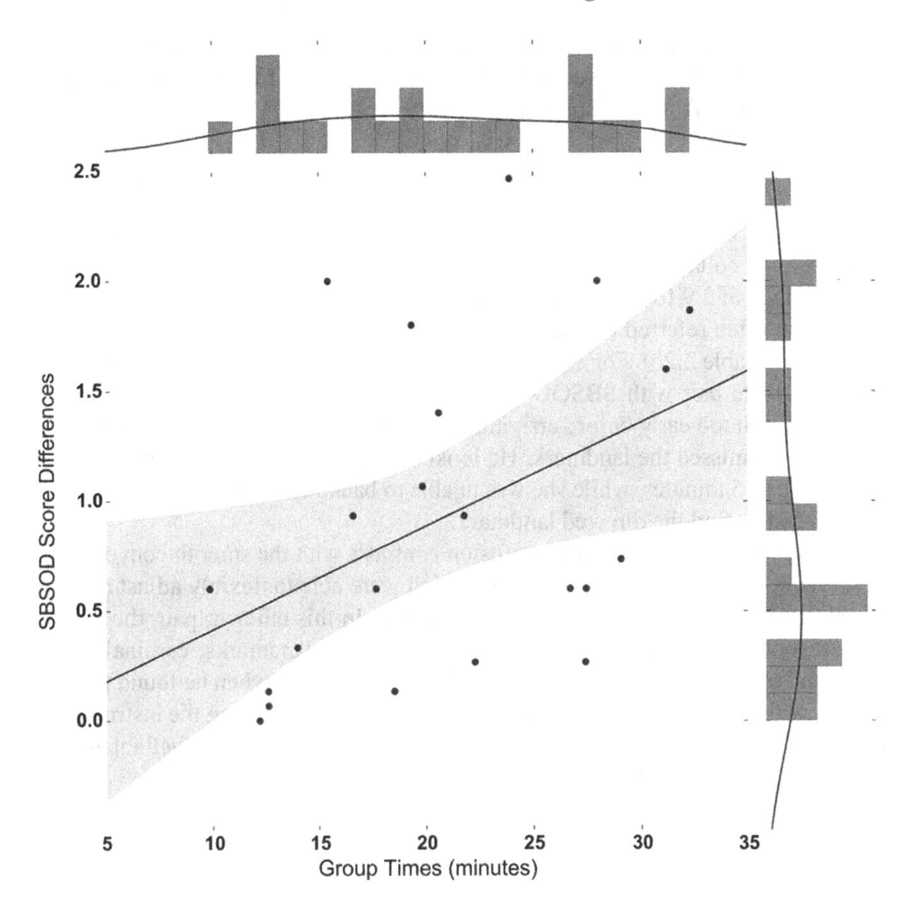

Figure 2.4 Relationship between travel time and the difference in SBSOD scores between paired members. Shading denotes a 95% confidence interval for the linear regression line. (Reprinted by permission of Taylor & Francis from He et al., 2015, p. 296, Figure 5)

groups based on the levels of sense of direction (the high–high, high–low, and low–low pairs) and navigation performance (the efficient and inefficient pairs). Patterns of conversations and navigation performance for some representative pairs are described below.

High–High Pairs

Five of the six high–high pairs (people who both had a good sense of direction) conducted collaborative navigation with a high level of confidence (mean confidence ratings of 3.6 for direction giving and 3.3 for direction receiving on a four-point scale). Their conversations consisted predominantly of statements like "yeah I see that," with only minimal feedback requests and interruptions (Table 2.2a).

They identified landmarks that were salient and meaningful, and communicated navigational directions in a succinct, efficient, and evenly distributed manner over the course of collaborative navigation.

High–Low Pairs

Six of the 10 high–low pairs (people with a good and a poor sense of direction) did poorly on collaborative navigation with low levels of confidence (mean confidence ratings of 2.9 for direction giving and 2.7 for direction receiving). People in these pairs often referred to different landmarks in the environment and confused the partner (Table 2.2b). For example, in one inefficient high–low pair (which was a male–female pair with SBSOD scores of 5.2 and 3.3, respectively), a woman directed a turn too early before arriving at a certain landmark and, consequently, a male partner missed the landmark. He looked for the landmark in the nearby area for as long as 5 minutes, while she was unable to backtrack her directions and simply insisted he find the directed landmark.

This kind of inefficiency and confusion contrasts with the smooth conversation observed for an efficient high–low pair, which were able to flexibly adjust navigational directions during the course of interaction. In this efficient pair, the person with a good sense of direction mentioned ephemeral landmarks, cardinal directions, and text information in verbal directions initially. But when he found that the partner with a poor sense of direction had difficulty understanding the instructions, he shifted to using permanent landmarks and turn sequences, which facilitated their collaborative navigation thereafter (Table 2.2c).

Low–Low Pairs

Four of the six low–low pairs (people who both had a poor sense of direction) had trouble communicating with each other in collaborative navigation. Their mean confidence level was low, with a rating of 2.9 for direction giving and 3.0 for direction receiving. Their verbal directions were unclear, with landmarks mentioned in the wrong order or without consideration of navigational context. When a directed landmark was not found, backtracking was uncommon in these pairs, and they typically instructed the partner to skip ahead to the next landmark, saying, "Look for it, and tell me when you see it" (Table 2.2d). This resulted in reduced conversation and fast walking, without properly taking the time to reorient themselves in the environment.

Some low–low pairs did well. In particular, the efficient performance by two low–low pairs contrasts with the inefficient performance by the above four low–low pairs. These efficient low–low pairs showed a high confidence level with a mean rating of 3.5 for direction giving and 3.4 for direction receiving (these values were comparable to the confidence level observed for the efficient high–high pairs). The performance by one of the pairs was in the top 5% of all 22 pairs in terms of travel distance and time, with no navigational errors throughout the collaboration process (Table 2.2e). These high performance and confidence scores suggest that people

Table 2.2 Examples of verbal navigational directions for major pairs, broken down by SOD combinations and navigation performance (Reprinted by permission of Taylor & Francis from He et al., 2015, pp. 300–301, Table 2)

(a) *Efficient high–high pair* (clear instructions with cardinal directions and landmarks):

"You want to go left there, so your next thing after that is another left, so it should be basically to a point where you have to turn. You went down the stairs, take a right, heading north, then you hit a T, then you went left. It should not be toward the stairs."

"OK."

"Have you crossed the street? Take the first left, on the first right, take the pedestrian pathway to the right around the parking lot. When you get to the other side of the parking lot, make the first left, you are going in the exact same direction you were going, you zigzagged the same way around the parking lot, you are just one street over."

"I have crossed the street, I am standing at the corner of the large black building 'Leo-palace.' So I'm looking for the parking."

(b) *Inefficient high–low pair* (describing different landmarks):

"Look for a nice looking house with flowers hanging outside. You will see a gravel road right next to the house."

"I don't see the house with a nice look and flowers hanging outside. The houses all look the same, but I did walk past a four-way crossing after turning left from where you last told me."

"OK, go back to the crossing and describe to me again what you see. I remember there was a left turn there, and another left should bring you to the house. Just go back."

"Here is a poster with some kanji on it, I think it is something political, is that the one?"

"I can't read kanji. I'm not too sure, but look for a cone close to there."

(c) *Efficient high–low pair* (able to adjust and tell the partner to go backward):

"Alright, you should be on the right path, there should be a small road coming up ahead, go straight then take a right."

"I think I'm lost, I probably took a turn too early. I took two rights and a left as you described, after the house with the staircases on the outside and the Mercedes parked in the front."

"Did you see a stop sign? No? Where are you now, near a parking lot? I think you might have walked past it, go back a few blocks and tell me when you see a red sign with kanjis written on it. . . . What do you see now? Are there any lamps in front of you? Go there, is a small road, follow that way. Is like an S thing."

(d) *Inefficient low–low pair* (simply telling the partner to search for the next visible landmark):

"I think you should be coming up on a turn and you should be able to see a building from there, pinkish. Then there is like a turn after that, I can't remember. Tell me what you see right now."

"What do I see now? I'm coming to another junction now, is like a pinkish building. There is like a notice board, I turn right at the notice board. I think you lost me. A park? There is a what?"

"You should just try to find the main road again, just walk until you see the main road, the main road. Once you get there, you want to turn left toward the Chinese restaurant, toward the main road. I don't remember the turns before that, but if you find the main road and the Chinese restaurant, then you are good."

"OK, I'll just try to find the main road. I'll tell you when I get there."

(e) *Efficient low–low pair* (a rare case using cardinal directions and time-based instructions):
"If you are at the clinic, go straight, there is a red upside-down triangle. The clinic is where you turn left, heading west, and the other side of the street should be a sign with a bicycle. Head north from there."
"Got it."
"You are going to look for, . . . the easiest thing is a building faintly painted pink or purple, also a green car, but you are going to turn right in between the house and the car."
"I see Hossen Hall. I'm going left, so I'll stay on the main road for a minute or two."

with a poor sense of direction, who as individuals tend to be poor at configurational spatial learning and landmark choice (Ishikawa & Montello, 2006; Ishikawa & Nakamura, 2012), can assist each other effectively in navigation in collaboration. Having the experience of spatial disorientation due to a poor sense of direction presumably enables these people to understand the problem that poor navigators may have and provide assistance in a way that is particularly helpful to them. Such ability is identified as a strength of people with a poor sense of direction, which encourages the discussion of cognitive mapping from the perspective of a personality trait (Ishikawa, 2022).

Summary of the Empirical Study

Results from the empirical study show that performance on collaborative navigation is related to the spatial aptitude of sense of direction, when considered at the group level in terms of the difference in SBSOD scores for paired members. When members in a pair differ in sense of direction, performance in navigation as a group becomes poor, being less efficient in the communication of navigation information and taking more time to direct the partner to a destination.

Conversations during collaborative navigation show that people with a poor sense of direction tend to misunderstand correctly given directions or confuse the partner with incorrect directions. However, when people with a poor sense of direction navigate in pairs, they can communicate with each other more easily and feel more confident than when they are paired with a person with a good sense of direction.

It should be pointed out, at the same time, that people with a good sense of direction can flexibly adjust navigational directions to the partner when their communication turns out not to be effective. Successful navigational directions include salient landmarks and necessary information, and people with a good sense of direction can give extra descriptions or new landmarks when asked by the partner. Also, they effectively tell the partner to return to a previous known location when lost, in contrast to people with a poor sense of direction, who tell the partner to find a directed landmark or proceed to the next one.

A study with a different experimental design would contribute to a further understanding of collaborative navigation. For example, pairs of participants could alternate the roles of a traveler and a director (in the study described above, paired members acted as a traveler and a director simultaneously); then, the effect of sense of direction could be examined separately for the traveler and the director (e.g., the high–low pair can be further broken down into a "high-traveler and low-director" pair and a "low-traveler and high-director" pair). One should note that in such a study, the paired members' roles are fixed, and the setting is basically the same as the one used in the studies of verbal navigational directions (see the section of "Types of Collaborative Navigation"). Another design could ask pairs of participants to navigate in an unfamiliar environment together (as in Buck et al., 2020). In such a design, the experimenter needs to make sure that both members participate in the navigation task actively without simply following the partner.

In sum, people with a good sense of direction attend to various information in the traveled environment (some of which may be redundant) and adjust their instructions to the needs of the partner. By contrast, people with a poor sense of direction tend to give instructions that are incorrect (e.g., landmarks mentioned in the wrong order), less understandable (e.g., obscure or unstable landmarks, such as pedestrians, parked vehicles, and small posters), or inflexible (e.g., directions not adjusted to the partner or situation). In some cases, however, people with a poor sense of direction do well in pairs, guiding each other efficiently with a high level of confidence, with an understanding of the difficulties that the partner may have.

Questions for Future Research

Profiles of Group Navigation

To gain a further understanding of collaborative navigation, future research needs to collect and analyze basic data on the frequency and situation of collectively conducted navigation in everyday life, just as a large amount of data has been accumulated about the use of online digital maps in navigation (e.g., Zenrin, 2018). How often do people engage in the task of navigation in collaboration and coordination with other people, for example, when planning a route, finding their way in a local neighborhood, or exploring an unfamiliar environment? As long as navigation is considered a social activity (Dalton et al., 2019), how common is socially situated navigation in daily life, in comparison to individually conducted navigation? How much varied are the forms and situations of collaborative navigation? It may range from simply following the crowd to asking a stranger for directions, helping others to find their way, and sightseeing with a group of friends. Also, an analysis of professional activities that require group decision-making in navigation (e.g., firefighting, emergency rescue, or military reconnaissance) would contribute to the understanding of collaborative navigation from the perspective of expert reasoning.

Individual Versus Group Performance

A major question of concern to researchers of collective spatial cognition is how and whether group performance in navigation differs from individually conducted navigation. To address this question, different measures of navigation performance and spatial learning that tap into different types of spatial knowledge need to be distinguished. One type of navigation performance concerns whether a person can accurately navigate toward a destination, which is enabled by an understanding of the sequence of decision points along a route (where and which direction to turn). This type of spatial knowledge is called *route knowledge* (Golledge, 1991), and it lacks configurational accuracy and metric precision that more elaborated spatial knowledge possesses (as seen below). Navigation performance at the level of route knowledge may be facilitated by collaboration in a group, as implied by the accuracy of movement in animal and human crowds (Codling et al., 2007; Dyer et al., 2008).

A detailed examination of wayfinding based on active group decision-making, beyond passive following of an informed person, would shed light on the effects of psychological attributes and navigational contexts on collaborative navigation. Better group performance in wayfinding is reasonable when some members of a group are informed (Dyer et al., 2008), but what if the prior information is not correct? Is the group open to peer input, flexibly altering their decisions? Such openness and flexibility are important for accurate group navigation when one considers that location information tends to involve uncertainty.

Now, suppose that you are faced with a more spatially demanding task, such as taking a shortcut, making a detour, or pointing to an unseen landmark. You need to know how far two places are apart, where a place is located in relation to another place, or how a set of places are spatially arranged in a neighborhood area. Namely, you need to understand distance, direction, and spatial layout. Thus, beyond the sequential (non-metric) understanding required at the level of route knowledge, you need to comprehend the environment with sufficient metric and configurational accuracy (the acquisition of *survey knowledge*; Montello, 1998). At the individual level, performance on these "survey" tasks varies to a great degree: Some people estimate distances and directions with accuracy at chance level, while others draw a sketch map of traveled routes in very close agreement with a correct cartographic map (Ishikawa & Montello, 2006). When conducted collectively, then, how good is the performance on such survey tasks? Does a group do better than individuals? To what degree is the accuracy or the extent of configurational spatial knowledge acquired by group members interrelated; namely, are the group members' cognitive maps shared or independent? For example, does the receiver of navigational directions acquire a cognitive map that is similar to the one possessed by a person who gives the directions? Do they acquire equally accurate and detailed, or similarly inaccurate and distorted, cognitive maps? Is the knowledge acquired and shared by a group more accurate than the knowledge acquired by individual members?

Also, does a group perform better on spatial navigation tasks with higher metric and configurational accuracy than is predicted from individual members' accuracy levels, as a result of each member's knowledge and skills contributed to the group? Put differently, is the cognitive map shared by a group the sum (or union) of individual members' cognitive maps, or more than that? Buck et al. (2020) found that paired travelers estimated relative directions between landmarks more accurately than individuals, and past research on collective learning in general similarly observed the superiority of group performance because of diversities in abilities and personalities (Aggarwal et al., 2019; Ein-Dor et al., 2011). A question for future research, then, is how group and individual performances are similar or different in the context of collaborative navigation. In this regard, it is also of interest to examine how and whether the size of a group affects navigation performance at the levels of route knowledge and survey knowledge. As observed in group learning (Aggarwal et al., 2019), is there an inverted U-shaped relationship between group size and group performance as mediated by the degree of diversity? In other words, is there an optimal group size that gives the highest accuracy, flexibility, and efficiency of collaborative navigation?

Effects of Psychological Attributes

A major focus of interest in the research into human spatial cognition is the existence of large individual differences (Ishikawa & Montello, 2006; Weisberg et al., 2014). Researchers discussed various psychological attributes that affect performance in cognitive mapping (Ishikawa, 2022), including spatial aptitudes (e.g., spatial ability and sense of direction; Hegarty et al., 2006), navigational preferences (e.g., use of local and global landmarks, egocentric and allocentric frames of reference, and route and orientation strategies; Lawton & Kallai, 2002), the level of stress (e.g., trait and state anxiety, wayfinding anxiety, fear of getting lost, or a growth mindset; He & Hegarty, 2020), and personalities (e.g., conscientiousness, extraversion, emotional stability, openness to change, and tendencies to be a leader and a follower; Condon et al., 2015). Experiential variables may also have influence, for example, familiarity with traveled environments, residential experiences in different types of environments (rural versus urban environments, or pedestrian, public transport, and automobile based environments), and the experience of using various navigation tools (maps, mobile navigation tools, or in-car navigation systems). These variables have been shown to affect spatial learning and navigation performance at the individual level (see Ishikawa, 2022); it is of interest to examine how variations in these attributes among group members influence the accuracy, efficiency, and flexibility of group spatial learning and navigation.

Physical and Social Variables

A major premise of research into spatial cognition from the cognitive-behavioral perspective posits that people's spatial behavior is based on psychological environments as well as physical environments (Gärling & Golledge, 1989; Moore & Golledge, 1976). Psychological environments are environments that are perceived by people and represented in their minds (internally represented environments or "cognitive maps"). In fact, people choose among a set of alternative mailboxes the one that *they think* is closest (not necessarily the one that is objectively closest), or people feel like returning to a place that they are *emotionally attached* to (pointing to the significance of experience and affect). The foregoing discussion, therefore, looked at the potential influence of psychological and experiential variables on group navigation.

In addition to these psychological variables, physical and social characteristics of environments also have effects. That is, environments and situations in which the task of individual or group navigation is conducted affect performance. First, the structure of the environment correlates with the manner in which people mentally represent it and, accordingly, how they describe spatial relations and conduct spatial reasoning. For example, people find it easier to orient themselves on an orthogonal street network than on oblique streets (Montello, 1991), although the cultural universality of this is still open to debate (Davies & Pederson, 2001). Likewise, within local street networks that are orthogonal but slightly tilted from the north–south direction, people tend to have a sense of disorientation, especially when they daily use the phrases "up north" and "down south" unconsciously (see Brunyé et al., 2010).

Second, mental representations of environments have a preferred orientation, as revealed by accurate and fast pointing performance associated with a specific direction, and, importantly, the preferred orientation is aligned with a reference system that is defined by salient environmental features or intrinsic spatial layouts (McNamara et al., 2003; Mou et al., 2008; Werner & Schmidt, 1999). The use of relative and absolute frames of reference in the description of spatial relations is also correlated with the existence of environmental landmarks or salient topographic features (Li & Gleitman, 2002; Palmer et al., 2019).

Finally, one can consider the influence of social attributes of groups that people belong to or of environments that they live in. For example, what is the dominant spatial frame of reference that people use to communicate navigation information in their language? Is preference given to a relative frame of reference, such as front–back–left–right, or to an absolute frame of reference, such as north–south–east–west (Levinson, 1996)? Also, what is the degree of relationship and connectedness among the members of a group? Do they know each other well, including their spatial aptitudes and navigational tendencies? Can they judge and adjust to other members' navigation abilities and skills on the basis of their responses and behavior? Is there a conflict of learning style among group members? If some members prefer a visual-spatial style and others prefer a verbal style, do they come to a consensus on how to communicate navigation information in a group? What are the roles of group members (e.g., a leader or a follower)? Are the roles taken spontaneously or imposed by others? How much are they inclined to trust group members' decisions? These questions are important and require careful consideration when group navigation is a social activity.

Concluding Remarks

This chapter has looked at collaborative navigation as a major mode of social and collective spatial cognition. The literature on group learning clarifies the basic idea that a group of people is an information processing unit in collective cognition. It forms the basis for research into collaborative navigation, which aims to understand spatial learning, decision-making, and communication in a group. Also, the literature on spatial cognition offers research findings that can stimulate researchers of collective learning to apply and extend their insights to the spatial domain, particularly the effects of psychological, physical, and social variables on spatial cognition and variations in types of knowledge about spatial environments.

The effects of the various variables discussed above—psychological, physical, and social—are interrelated, which makes it challenging to conduct a controlled experiment on collaborative navigation. But the possible next step to further our understanding of collective spatial cognition is a continued empirical examination of paired navigation in more detail, including extending the above-referenced studies to the case of a trio rather than a pair. Findings from such empirical studies would shed more light on the existing research into individual differences (Ishikawa & Montello, 2006), sense of direction (Hegarty et al., 2006), verbal route directions (Denis et al., 1999), landmark selection (Klippel & Winter, 2005), and spatial anxiety (He & Hegarty, 2020) in the context of collaborative and interactive spatial thinking.

Research along this line would also provide insights into the development of navigation assistance for people with poor spatial aptitudes, who comprise a significant portion of the population, and the design of personalized navigation tools that are adaptive to the attributes of the user, the characteristics of the environment, and the social context of navigation. A question of ultimate interest concerns how navigation tools influence collaborative navigation and how and whether collaboration with other people in group navigation differs from interaction with a navigation tool. In navigation assistance that is truly user friendly and context aware, interaction with a human and interaction with a machine would not make a difference. In this regard, theoretical understanding of collaborative navigation will enable the society to help and encourage people with various aptitudes and preferences in the difficult task of spatial orientation.

References

Aggarwal, I., & Woolley, A. W. (2013). Do you see what I see? The effect of members' cognitive styles on team processes and errors in task execution. *Organizational Behavior and Human Decision Processes*, *122*, 92–99.

Aggarwal, I., Woolley, A. W., Chabris, C. F., & Malone, T. W. (2019). The impact of cognitive style diversity on implicit learning in teams. *Frontiers in Psychology*, *10*, 112.

Brunyé, T. T., Mahoney, C. R., Gardony, A. L., & Taylor, H. A. (2010). North is up(hill): Route planning heuristics in real-world environments. *Memory & Cognition*, *38*, 700–712.

Buck, L. E., McNamara, T. P., & Bodenheimer, B. (2020). Dyadic acquisition of survey knowledge in a shared virtual environment. In *VR 2020: Proceedings of the 2020 IEEE conference on virtual reality and 3D user interfaces* (pp. 579–587). Washington, DC: IEEE Computer Society.

Burtscher, M. J., & Manser, T. (2012). Team mental models and their potential to improve teamwork and safety: A review and implications for future research in healthcare. *Safety Science, 50*, 1344–1354.

Codling, E. A., Pitchford, J. W., & Simpson, S. D. (2007). Group navigation and the "many-wrongs principle" in models of animal movement. *Ecology, 88*(7), 1864–1870.

Condon, D. M., Wilt, J., Cohen, C. A., Revelle, W., Hegarty, M., & Uttal, D. H. (2015). Sense of direction: General factor saturation and associations with the big-five traits. *Personality and Individual Differences, 86*, 38–43.

Dalton, R. C., Hölscher, C., & Montello, D. R. (2019). Wayfinding as a social activity. *Frontiers in Psychology, 10*, 142.

Davies, C., & Pederson, E. (2001). Grid patterns and cultural expectations in urban wayfinding. In D. R. Montello (Ed.), *Spatial information theory* (pp. 400–414). Berlin: Springer.

Denis, M., Pazzaglia, F., Cornoldi, C., & Bertolo, L. (1999). Spatial discourse and navigation: An analysis of route directions in the city of Venice. *Applied Cognitive Psychology, 13*, 145–174.

Dyer, J. R. G., Ioannou, C. C., Morrell, L. J., Croft, D. P., Couzin, I. D., Waters, D. A., & Krause, J. (2008). Consensus decision making in human crowds. *Animal Behaviour, 75*(2), 461–470.

Ein-Dor, T., Mikulincer, M., & Shaver, P. R. (2011). Effective reaction to danger: Attachment insecurities predict behavioral reactions to an experimentally induced threat above and beyond general personality traits. *Social Psychological and Personality Science, 2*(5), 467–473.

Flack, A., Biro, D., Guilford, T., & Freeman, R. (2015). Modelling group navigation: Transitive social structures improve navigational performance. *Journal of the Royal Society Interface, 12*, 20150213.

Forlizzi, J., Barley, W. C., & Seder, T. (2010). Where should I turn? Moving from individual to collaborative navigation strategies to inform the interaction design of future navigation systems. In *CHI 2010: Proceedings of the SIGCHI conference on human factors in computing systems* (pp. 1261–1270). New York, NY: ACM.

Galantucci, B. (2005). An experimental study of the emergence of human communication systems. *Cognitive Science, 29*, 737–767.

Galati, A., & Avraamides, M. N. (2013). Flexible spatial perspective-taking: Conversational partners weigh multiple cues in collaborative tasks. *Frontiers in Human Neuroscience, 7*, 618.

Gardony, A. L., Brunyé, T. T., Mahoney, C. R., & Taylor, H. A. (2013). How navigational aids impair spatial memory: Evidence for divided attention. *Spatial Cognition and Computation, 13*, 319–350.

Gärling, T., & Golledge, R. G. (1989). Environmental perception and cognition. In E. H. Zube & G. T. Moore (Eds.), *Advances in environment, behavior, and design* (Vol. 2, pp. 203–236). New York, NY: Plenum Press.

Golledge, R. G. (1991). Cognition of physical and built environments. In T. Gärling & G. W. Evans (Eds.), *Environment, cognition, and action: An integrated approach* (pp. 35–62). New York, NY: Oxford University Press.

Golledge, R. G., & Timmermans, H. (1990). Applications of behavioural research on spatial problems I: Cognition. *Progress in Human Geography*, *14*, 57–99.

He, C., & Hegarty, M. (2020). How anxiety and growth mindset are linked to navigation ability: Impacts of exploration and GPS use. *Journal of Environmental Psychology*, *71*, 101475.

He, G., Ishikawa, T., & Takemiya, M. (2015). Collaborative navigation in an unfamiliar environment with people having different spatial aptitudes. *Spatial Cognition and Computation*, *15*, 285–307.

Hegarty, M., Montello, D. R., Richardson, A. E., Ishikawa, T., & Lovelace, K. (2006). Spatial abilities at different scales: Individual differences in aptitude-test performance and spatial-layout learning. *Intelligence*, *34*, 151–176.

Hegarty, M., Richardson, A. E., Montello, D. R., Lovelace, K., & Subbiah, I. (2002). Development of a self-report measure of environmental spatial ability. *Intelligence*, *30*, 425–447.

Helbing, D., Farkas, I., & Vicsek, T. (2000). Simulating dynamical features of escape panic. *Nature*, *407*, 487–490.

Hinsz, V. B., & Tindale, R. S. (1997). The emerging conceptualization of groups as information processors. *Psychological Bulletin*, *121*(1), 43–64.

Ishikawa, T. (2019). Satellite navigation and geospatial awareness: Long-term effects of using navigation tools on wayfinding and spatial orientation. *The Professional Geographer*, *71*, 197–209.

Ishikawa, T. (2022). Individual differences and skill training in cognitive mapping: How and why people differ. *Topics in Cognitive Science*, *15*, 163–186. https://doi.org/10.1111/tops.12605

Ishikawa, T., Fujiwara, H., Imai, O., & Okabe, A. (2008). Wayfinding with a GPS-based mobile navigation system: A comparison with maps and direct experience. *Journal of Environmental Psychology*, *28*, 74–82.

Ishikawa, T., & Montello, D. R. (2006). Spatial knowledge acquisition from direct experience in the environment: Individual differences in the development of metric knowledge and the integration of separately learned places. *Cognitive Psychology*, *52*, 93–129.

Ishikawa, T., & Nakamura, U. (2012). Landmark selection in the environment: Relationships with object characteristics and sense of direction. *Spatial Cognition and Computation*, *12*, 1–22.

Klippel, A., & Winter, S. (2005). Structural salience of landmarks for route directions. In A. G. Cohn & D. M. Mark (Eds.), *Spatial information theory* (pp. 347–362). Berlin: Springer.

Kozlowski, L. T., & Bryant, K. J. (1977). Sense-of-direction, spatial orientation, and cognitive maps. *Journal of Experimental Psychology: Human Perception and Performance*, *3*, 590–598.

Laughlin, P. R., Bonner, B. L., & Miner, A. G. (2002). Groups perform better than the best individuals on letters-to-numbers problems. *Organizational Behavior and Human Decision Processes*, *88*, 605–620.

Lawton, C. A., & Kallai, J. (2002). Gender differences in wayfinding strategies and anxiety about wayfinding: A cross-cultural comparison. *Sex Roles*, *47*, 389–401.

Levinson, S. C. (1996). Frames of reference and Molyneux's question: Cross-linguistic evidence. In P. Bloom, M. A. Peterson, L. Nadel, & M. F. Garrett (Eds.), *Language and space* (pp. 109–169). Cambridge, MA: MIT Press.

Li, P., & Gleitman, L. (2002). Turning the tables: Language and spatial reasoning. *Cognition*, *83*, 265–294.

Liben, L., & Downs, R. (1993). Understanding person-space-map relations: Cartographic and developmental perspectives. *Developmental Psychology*, *29*, 739–752.

Liker, A., & Bókony, V. (2009). Larger groups are more successful in innovative problem solving in house sparrows. *Proceedings of the National Academy of Sciences of the United States of America, 106*(19), 7893–7898.

McNamara, T. P., Rump, B., & Werner, S. (2003). Egocentric and geocentric frames of reference in memory of large-scale space. *Psychonomic Bulletin & Review, 10*, 589–595.

Minson, J. A. (2012). The cost of collaboration: Why joint decision making exacerbates rejection of outside information. *Psychological Science, 23*(3), 219–224.

Montello, D. R. (1991). Spatial orientation and the angularity of urban routes: A field study. *Environment and Behavior, 23*(1), 47–69.

Montello, D. R. (1998). A new framework for understanding the acquisition of spatial knowledge in large-scale environments. In M. J. Egenhofer & R. G. Golledge (Eds.), *Spatial and temporal reasoning in geographic information systems* (pp. 143–154). New York, NY: Oxford University Press.

Montello, D. R., & Raubal, M. (2013). Functions and applications of spatial cognition. In D. Waller & L. Nadel (Eds.), *Handbook of spatial cognition* (pp. 249–264). Washington, DC: American Psychological Association.

Moore, G. T., & Golledge, R. G. (1976). Environmental knowing: Concepts and theories. In G. T. Moore & R. G. Golledge (Eds.), *Environmental knowing: Theories, research, and methods* (pp. 3–24). Stroudsburg, PA: Dowden, Hutchinson, & Ross.

Mou, W., Fan, Y., McNamara, T. P., & Owen, C. B. (2008). Intrinsic frames of reference and egocentric viewpoints in scene recognition. *Cognition, 106*, 750–769.

Nazareth, A., Weisberg, S. M., Margulis, K., & Newcombe, N. S. (2018). Charting the development of cognitive mapping. *Journal of Experimental Child Psychology, 170*, 86–106.

Newcombe, N. S. (2019). Navigation and the developing brain. *Journal of Experimental Biology, 222*(suppl 1), jeb186460.

Palmer, B., Blythe, J., Gaby, A., Hoffmann, D., & Ponsonnet, M. (2019). Geospatial natural language in indigenous Australia: Research priorities. In K. Stock, C. B. Jones, & T. Tenbrink (Eds.), *Proceedings of the workshop on speaking of location 2019* (pp. 17–27). Regensburg, Germany: CEUR-WS.org.

Raubal, M., & Winter, S. (2002). Enriching wayfinding instructions with local landmarks. In M. J. Egenhofer & D. M. Marks (Eds.), *Geographic information science* (pp. 243–259). Berlin: Springer.

Reilly, D., Mackay, B., Watters, C., & Inkpen, K. (2009). Planners, navigators, and pragmatists: Collaborative wayfinding using a single mobile phone. *Personal and Ubiquitous Computing, 13*, 321–329.

Ruginski, I. T., Creem-Regehr, S. H., Stefanucci, J. K., & Cashdan, E. (2019). GPS use negatively affects environmental learning through spatial transformation abilities. *Journal of Environmental Psychology, 64*, 12–20.

Saloma, C., Perez, G. J., Tapang, G., Lim, M., & Saloma, P. (2003). Self-organized queuing and scale-free behavior in real escape panic. *Proceedings of the National Academy of Sciences of the United States of America, 100*, 11947–11952.

Schwarzkopf, S., Büchner, S. J., Hölscher, C., & Konieczny, L. (2017). Perspective tracking in the real world: Gaze angle analysis in a collaborative wayfinding task. *Spatial Cognition and Computation, 17*, 143–162.

Sugimoto, M., Kusumi, T., Nagata, N., & Ishikawa, T. (2022). Online mobile map effect: How smartphone map use impairs spatial memory. *Spatial Cognition and Computation, 22*, 161–183.

Tindale, R. S., & Kameda, T. (2000). 'Social sharedness' as a unifying theme for information processing in groups. *Group Processes and Intergroup Relations, 3*(2), 123–140.

Weisberg, S. M., Schinazi, V. R., Newcombe, N. S., Shipley, T. F., & Epstein, R. A. (2014). Variations in cognitive maps: Understanding individual differences in navigation. *Journal of Experimental Psychology: Learning, Memory, and Cognition, 40*, 669–682.

Werner, S., & Schmidt, K. (1999). Environmental reference systems for large-scale spaces. *Spatial Cognition and Computation, 1*, 447–473.

Willis, K. S., Hölscher, C., Wilbertz, G., & Li, C. (2009). A comparison of spatial knowledge acquisition with maps and mobile maps. *Computers, Environment and Urban Systems, 33*, 100–110.

Zenrin. (2018). *Survey on the use of maps 2018*. Tokyo: Zenrin Co. www.zenrin.co.jp/product/article/map-18/index.html

3 Personality Traits and Spatial Skills Are Related to Group Dynamics and Success During Collective Wayfinding

Tad T. Brunyé, Dalit Hendel, Aaron L. Gardony, Erika K. Hussey, and Holly A. Taylor

Introduction

Wayfinding involves a complex interplay between individuals and environmental features, demanding myriad perceptual, cognitive, and emotional processes to initiate and sustain effective performance. This interplay is responsible for effectively planning routes, starting a journey, updating plans, communicating about space, and successfully finding the way to one or more destinations. While sometimes wayfinding is done independently, in many cases, it is a social activity (Dalton et al., 2019; Haghani & Sarvi, 2017; Hutchins, 1995a). Examples of social wayfinding include touring a city with friends in search of historical landmarks, finding a sequence of destinations as a small team of military personnel or first responders, or finding the way back to a conference hotel with colleagues. In each of these cases, the complexity of wayfinding and its component mental processes is compounded by social interactions that emerge surrounding the planning and execution of a collaborative wayfinding task. Even with shared goals, group members may disagree regarding how to accomplish the goals (Alper et al., 1998), leading to social interactions that shape the decision-making process and alter wayfinding behavior. Because most theories and computational models of wayfinding describe and predict wayfinding behavior among individuals, there is an absence of empirical research and theory regarding groups of collaborating navigators. This relative absence of research contrasts with how common and important collaborative wayfinding is in real-world activities (Helbing et al., 2000; Xia et al., 2009).

This chapter is motivated by a triad collectively navigating a novel environment, asking how individual differences and group interactions shape wayfinding performance and memory. A small pilot study was conducted to shed light on these questions, focusing specifically on how personality traits and spatial skills influence virtual wayfinding as individuals, dyads, or triads. To motivate our study and develop testable hypotheses, we review literature from three research domains. First, we begin by discussing current theoretical positions regarding individual wayfinding, and how personality traits and spatial skills may influence the performance. Second, we expand this discussion to include a few studies examining collaborative wayfinding, and some recently proposed theories examining the social

DOI: 10.4324/9781003202738-5

aspects of wayfinding. Finally, we review research from the group dynamics literature that can help frame examinations of collaborative wayfinding. We conclude with a description of our study and its results, a discussion of its relevance to extant research, and by proposing some directions for continued research in this domain.

Wayfinding: Individual Behavior

When discussing navigation, in line with earlier definitions (Montello, 2005), it is important to distinguish the terms locomotion and wayfinding. Locomotion is the process of physically moving within a local environment, including processes such as visual perception, motor planning, postural control, and obstacle avoidance (Drew & Marigold, 2015). Wayfinding occurs when moving through an environment becomes goal-oriented, involving not only locomotion but also tracking location, orientation, landmarks, and distances in the pursuit of one or more objectives (Allen, 1999; Golledge, 1999; Raubal & Worboys, 1999). In many cases, navigation occurs in large-scale spaces wherein proximal landmarks will occlude distal landmarks and goal locations; this characteristic can make real-world environments challenging for localizing yourself, orienting yourself toward an unseen objective, and tracking progress along the way (Hurlebaus et al., 2008; Iaria et al., 2003). Several models exist to account for these and other challenges inherent to wayfinding.

Among many models of wayfinding, most detail a sequence of processes that includes planning a route and executing it. Distinctions among models are typically made at the levels of feature availability and reliance, mental processes or brain structures emphasized, or the types of environments and situations considered. For instance, Passini's early models used verbal reports and emphasized the phases of route planning and execution, with formal relationships between top-level goals and sub-goals (e.g., to get to the *goal*, turn right at a specific landmark) (Passini, 1981, 1984, 1992). Other early models by Gärling focused on action plans and travel plans, again emphasizing distinctions between route planning and execution; these models also tended to focus on how cognitive maps were developed and how routes were planned between multiple destinations in familiar (e.g., pedestrian shopping) versus unfamiliar environments (Gärling et al., 1984; Gärling & Gärling, 1988). The TOUR model (Kuipers, 1978) emphasized the importance of partial knowledge states (e.g., a partial mental network of roads and landmarks), how new percepts are incorporated into emergent mental representations, and how interactions between partial knowledge and the environment are used to solve wayfinding problems. Extensions of the TOUR model include a focus on the "spatial semantic hierarchy" to express states of partial knowledge (Kuipers, 2000) and the concept of a skeleton, a mental representation that emphasizes major paths and boundaries in familiar environments (Kuipers et al., 2003; Pailhous, 1984).

Wolbers and Hegarty defined an emerging framework that captures three broad classes of mechanisms involved in spatial knowledge acquisition and navigation: the spatial environmental and self-motion cues, spatial and executive computational mechanisms, and online and memory-based representations (Wolbers & Hegarty,

2010). This framework capitalizes on the myriad internal and external factors identified as contributing to localization, orientation, and wayfinding. With regard to spatial cues, the model considers the importance of several environmental and self-motion cues in guiding navigation, including visual landmark salience and distal orientation cues (Klippel & Winter, 2005; Sandstrom et al., 1998), the inherent geometric structure of environments (K. Cheng et al., 2013), symbolic representations and guidance technologies (Hirtle, 2017), and self-motion cues (Britten, 2008). Spatial and executive processes include how transient states of uncertainty guide wayfinding decisions (Brunyé et al., 2017; Giannopoulos et al., 2014; Joietz & Kiefer, 2017), the importance of perspective transformations and distance and direction computations (Kozhevnikov et al., 2006; Weisberg et al., 2014), and the complexity of planning and goal maintenance (Bailenson et al., 2000; Brunyé et al., 2010, 2015a; Dalton, 2003; Hochmair, 1995; Wiener et al., 2004; Wiener & Mallot, 2003). With regard to spatial representations, included in this framework are understandings of the hierarchical stages and nature of emergent mental representations in children and adults (Montello, 1998; Siegel & White, 1975; Thorndyke & Hayes-Roth, 1982), the influence of spatial knowledge and experience interacting with environmental constraints to guide navigation (Wiener et al., 2009), and the iterative and recursive online representations of self-position and orientation, action planning, and goal updating (Raubal & Worboys, 1999). Finally, the model emphasizes the complex and interactive roles of brain regions and networks in supporting and constraining wayfinding performance (Maguire et al., 1998, 2015; Spiers & Maguire, 2006).

While the spatial cognition literature has a rich history of theories accounting for environmental, strategy-based, and neuroscientific origins of wayfinding behavior, these theories are largely descriptive rather than predictive, but more relevantly, they limit their focus to the mental representations and behavior of individuals rather than groups.

Wayfinding Behavior: Individual Differences

Compounding the diversity of wayfinding theory is the diversity of wayfinding behavior across individuals and the variability seen in spatial memory resulting from matched environmental experiences (Hölscher et al., 2006; Ishikawa & Montello, 2006; Tenbrink et al., 2011). To explain this variability, researchers have turned to measuring individual differences across wayfinders, including differences in sex, personality traits, spatial ability and strategy preferences, and culture. Each of these factors has been linked, independently and interactively, with a range of strategies and behaviors during navigation.

Decades of research demonstrate sex differences across a range of visuospatial tasks, including object visualization, mental rotation, spatial working memory, and wayfinding (Halpern, 2004; Halpern & Collaer, 2009). Meta-analytic examinations of sex differences in spatial cognition suggest that men generally tend to outperform women on a range of spatial tasks (Nazareth et al., 2019). One difference between men and women during wayfinding is relative reliance on varied environmental features and reference systems: Some studies show that females tend to rely more on landmarks and turn directions, and males more on environmental configuration and cardinal directions (Lawton, 1994, 2001). For example, in the virtual dual-solution paradigm (DSP), men are more likely to take shortcuts that rely on allocentric representations of space, whereas females are more likely to follow known routes (Boone et al., 2018). Varied feature and reference system reliance may alter preferences for varied navigation guidance, with females preferring verbal turn-by-turn instructions and males preferring maps (Anacta & Schwering, 2010; Dabbs et al., 1998). Additional evidence for this dichotomy comes from studies demonstrating that females tend to recall more landmarks and make fewer landmark-related errors than males and males tend to make smaller directional errors than females (Galea & Kimura, 1993; Montello et al., 1999). In contrast to hypotheses based on these findings, a recent meta-analysis examining sex differences in human navigation suggested that males show advantages in landmark positioning tasks performed in the route perspective; however, most of the above-cited papers were not included in the meta-analysis, making it challenging to evaluate how they might contribute to the reported pattern (Nazareth et al., 2019). Thus, the sexes appear to differ in the strategies they use to process and represent environments in memory, though this pattern remains up for debate. Interestingly, the extent to which sex differences emerge in spatial task performance may be related to gender inequality of the culture being sampled (Coutrot et al., 2018). When an international sample of over 2.5 million people completed the Sea Hero Quest navigation game, sex differences were most pronounced in countries with higher levels of gender inequality.

Personality traits are complex and multifaceted, and several theoretical models and accompanying questionnaires exist for quantifying and categorizing individuals into personality-based typologies (Goldberg, 1993; Matthews et al., 2009). One early study while there are many ways to quantify personality traits, two specific constructs and measures have been used to relate traits to spatial behavior: the California Personality Inventory (CPI) and the Big-Five Inventory (BFI). Using the CPI, one early study (Bryant, 1982) demonstrated several personality traits positively associated with pointing accuracy: sociability, self-acceptance, capacity for status, and social presence. In general, these traits tend to relate to the single factor assessing effectiveness in interpersonal relationships, suggesting associations between pointing accuracy and a tendency to be "outgoing, participative, at ease in social situations, ambitious, self-assured, and independent" (Bryant, 1982, p. 1320). Of course, it is unknown whether any causal relationship exists between these factors and, if so, in which direction. Another study examined how the Big-Five traits, including neuroticism, extraversion, openness, agreeableness, and conscientiousness, correlate with a sense of direction (Condon et al., 2015). The authors found that conscientiousness and extraversion were associated with a higher subjective sense of direction, though no measures of wayfinding were taken. In another study, higher extraversion was linked to altered strategy use during route selection, namely, a stronger reliance on routes with initially straight path segments and lower reliance on south-going (vs. north-going) routes (Brunyé et al., 2015a).

No formal mechanistic explanation has been proposed for these findings, but it could be the case that those with higher extraversion are more concerned with the global direction of a destination when selecting routes [leading to an initial segment strategy; (Bailenson et al., 2000)], rather than using other spatial heuristics such as southern route preferences (Brunyé et al., 2010).

Complementing these Big-Five trait findings, research has also demonstrated the importance of spatial anxiety when accounting for wayfinding behavior. As its name suggests, spatial anxiety describes an individual's anxiety toward performing spatial tasks, such as understanding or following directions and navigating through novel environments (Hund & Minarik, 2009; Lawton & Kallai, 2002). In her seminal work on spatial anxiety and navigation (also see Bryant, 1982), Lawton (1994) demonstrated that sex differences in navigation strategies (females using a route strategy and males using an orientation strategy) were correlated with spatial anxiety. Specifically, women showed higher levels of spatial anxiety. Furthermore, higher spatial anxiety was negatively correlated with adopting an orientation-based strategy during navigation. Similar research supported this pattern (Lawton, 1996), but later studies suggested that this pattern was either non-existent or relatively weak (Prestopnik & Roskos-Ewoldsen, 2000; Saucier et al., 2002). More recent work in this domain shows that females may not have significantly higher spatial anxiety than men, but higher spatial anxiety is correlated with longer navigation times and an increased number of errors during navigation (Hund & Minarik, 2009; Walkowiak et al., 2015). In a related study, Hund and colleagues found that higher spatial anxiety was associated with the number of times a navigator mentioned landmarks and left–right turns, supporting the tendency for higher spatial anxiety to be related to a route-based navigation strategy (Hund & Padgitt, 2010).

Recent theory suggests that the common symptoms and social and communication impairments accompanying autism spectrum disorders lie on a continuum that extends into the normal population (Baron-Cohen et al., 2001). This conceptualization provides novel opportunities to assess and understand how autism-related personality traits may be correlated with behavior in neurotypical populations. Autism-related personality traits may prove important for understanding group wayfinding behavior as they can modulate the effectiveness of interpersonal communication (Bishop et al., 2006) and are negatively associated with spatial perspective-taking (Brunyé et al., 2012) and egocentric transformations (Pearson et al., 2014). Instruments, such as the autism-spectrum quotient (AQ; Baron-Cohen et al., 2001), provide a method for non-diagnostically assessing sub-clinical autistic traits, leveraging a less categorical and more quantitative approach to understanding how these traits extend into the healthy adult population (Frith, 2001). Given how group settings introduce complex communication, coordination, perspective taking, theory of mind, and flexibility demands, we believe that these may be especially challenging for groups with high heterogeneity in autism-related traits. Recent research examining groups has highlighted the importance of group adaptability for accommodating differences in communication style that emerges when group members vary in autism-related traits (Crompton et al., 2019; Morrison et al., 2019). While higher autism-related traits tend to be related to superior spatial abilities (Caron et al., 2004), group settings may introduce challenges among those with higher autism-related traits when it comes to communicating and cooperatively implementing spatial strategies.

Individual wayfinding behavior is thus under the influence of not only environmental percepts and knowledge but also the unique contributions of individual differences in at least sex, personality, and spatial anxiety. However, the exact nature of these contributions and how they interact with environmental constraints, knowledge, and strategies is still being discovered. Further complicating this relationship is that personality traits influence not only individual behavior but also interpersonal interactions and performance outcomes when wayfinding within a group.

Wayfinding: Collaborative Behavior

Many wayfinding activities occur within groups of two or more people, organizing their collective thinking and behavior to accomplish one or more spatial goals. Theory to explain and predict group wayfinding behavior is relatively lacking, though one recent proposal takes a step in this direction (Dalton et al., 2019). Dalton and colleagues broadly defined social wayfinding as "any situation where the presence and/or activities of others, now or in the past, has an observable impact on wayfinding behavior and cognition" (p. 2). Realizing the extraordinary breadth of such a definition, the authors narrowed the scope to focus on situations where others influence, in a direct or indirect manner, the routes selected by a navigator. A multitude of scenarios fall into consideration here, with influences occurring both with and without another being present or engaged during wayfinding. For instance, others could provide advice on how to get somewhere or areas to avoid before you begin traveling (e.g., *the traffic is terrible on Main St., so take Elm St., and then merge back onto Main St. north of the town center*), an acquaintance or stranger could offer you discrete guidance during travel (e.g., *keep going that way and turn right at the coffee shop*), you could monitor the behavior of others to help you select routes (e.g., going with the flow or avoiding busy areas), or you could be jointly planning and executing a path with one or more partners.

On the basis of these and other social wayfinding scenarios, Dalton and colleagues defined a 2 × 2 typology of social wayfinding that parses activities into synchronous versus asynchronous types and strong versus weak types (Dalton et al., 2019). Synchronicity considers social interactions relied upon during (synchronous) or prior to (asynchronous) the act of wayfinding, and the strength of the social wayfinding considers the directness and collaborative nature of the social interaction in guiding decisions. For instance, synchronous behavior would include wayfinding with another person (strong) or monitoring the behavior of others to guide your own choices (weak). Asynchronous behavior would include the following instructions previously provided by someone who is not present (strong) or leveraging traces of others' previous behavior (weak). In this study, we consider synchronous social wayfinding: participants navigated collaboratively in a shared space with shared goals, while they directly communicated (verbally and non-verbally) with one another. Of course, there is no reason to believe that synchronous wayfinding will always proceed smoothly, as the dynamics of groups will inevitably introduce complexity; indeed, some group experiences may be relatively weak such as when merely following a leader [i.e., a leader/follower style; (Reilly et al., 2009)], whereas other experiences may be characterized by strong collaborative behavior.

As detailed by Dalton and colleagues, the characteristics of each individual within a group, including their environmental knowledge, gender, spatial ability, social status, and personality, will continuously guide individuals and their interactions with others (Dalton et al., 2019). A recent study provides some evidence of this: when dyads described routes to one another over the phone, difficulties tended to arise most when the two people differed widely in spatial skills (sense of direction) (He et al., 2015). Social roles, such as driver versus passenger, can also influence how collaborative navigation unfolds (Forlizzi et al., 2010), and for groups unfamiliar with one another, one might assume that personality traits and leadership styles would prove influential.

In situations when one individual becomes a leader due to explicit assignment or high familiarity with an environment, a leader/follower dynamic will likely emerge. In these cases, the success of the group is highly dependent on the navigation skills of the leader, and the ability for others to follow. But, when all group members are similarly unfamiliar with the environment, several possible group dynamics could emerge (Toader & Kessler, 2018). First, an implicit leader may emerge on the basis of perceived social status, stereotypical, perceived, or self-declared spatial skills, perspective-taking skills, and/or leadership traits. In this case, communication is more likely to be unidirectional, with a leader dictating route plans to the rest of the group, demonstrating a case of weak synchronous wayfinding. Second, the group could show mutual planning and decision-making, with a high level of reciprocal communication, demonstrating a case of strong synchronous wayfinding. We suggest that the latter is only possible when group members are homogeneous in terms of leadership style and spatial skills and have similar social skills such as empathic perspective taking. High heterogeneity in leadership traits and spatial and social skills within a group will likely lead to sub-optimal navigation behavior (He et al., 2015).

While research specifically examining group wayfinding is limited, motivations and predictions can be derived from theory and empirical work in distributed cognition (Hutchins, 1995a). Distributed cognition posits, among other things, that cognitive processes are not solely the purview of an individual but also can be distributed across group members and environmental tools (Saloman, 1993). These distributed cognitive processes include planning, problem-solving, decision-making, and creative discovery. Because the cognitive properties of groups can be very different from those of individuals, groups may show either superior or inferior, or simply different, cognitive performance during complex tasks. In some cases, cognitive processes initiated by individuals based on memory alone are insufficient for solving real-world problems without also considering manipulation of environmental tools (e.g., a compass, sextant, and map) and communication with others. When a dyad or triad is navigating together toward common goals, the cognitive processes necessary for accomplishing the task are distributed across team members and technology (Hutchins, 2006). For example, one individual may plan a route using a map, point in a direction, acknowledge an environmental feature, receive feedback from another group member, revise the plan, and reach a shared decision; in this manner, the team is using not only the cognitive processes of an individual but also the interactions among individuals and between individuals and elements of the environment. These distributed aspects of real-world cognitive processing have been demonstrated in several applied domains, including the coordinated work activities of teams navigating large ships (Hutchins, 2014) and aircraft pilots understanding and making decisions about flight information (Hutchins, 1995b). In other words, cognition is distributed across individuals, environmental artifacts, and members of a group; examining cognition solely from the perspective of one individual performing a cognitive task (such as wayfinding) in a vacuum is likely not representative of real-world thought and behavior.

Group Dynamics

Because so few studies have been conducted in the domain of synchronous social wayfinding, it is valuable to consider relatively domain-general research and theories of group dynamics and leadership emergence. Coordinated action is necessary to succeed in both physical and cognitive group tasks, but a large body of literature suggests that groups generally do not function well in reaching mutually acceptable decisions (Wilson et al., 2004). Prominent theories of cooperative behavior posit that groups are not better than individuals at producing more or better ideas during brainstorming; They tend to seek concurrence, lack vigilance, expend less effort, and prefer risks; and they are not necessarily better than individuals at making appropriate decisions (Janis, 1982; Karau & Williams, 1993; Wilson et al., 2004). In other words, the whole is not always greater than the sum of its parts; in fact, the whole may often be less than the sum of its parts (Stroebe & Diehl, 1994).

This general pattern has prompted research examining what aspects of groups and the individuals that comprise them alter group dynamics and lead to performance deficits (V. R. Brown & Paulus, 2002; Mullen et al., 1991). In many groups, a leader organically emerges during task planning or execution, and the leader is thought to play a pivotal role in determining group success. Leader emergence and selection is a complex and dynamic group phenomenon that evolves among group members from the time they meet to when they part ways (Bormann, 1990). According to emergence theory, leaders are generally selected because they conform to the expectations of followers, including a set of beliefs about which attributes are most important to solve group goals; in other words, the way an individual expresses themselves and the extent to which this expression conforms to leadership conceptions will dictate whether they become and remain a leader (Brown & Lord, 2001; Foti et al., 1982). The process by which leaders emerge and persist is thought to involve competition and elimination, with inflexible and/or unknowledgeable leaders being eliminated in favor of other individuals (Johnson & Bechler, 1998). The attributes and behaviors perceived as essential for solving group goals may vary by context and the tasks at hand, resulting in leadership styles being more appropriate or less appropriate for different situations.

The attributes that predict leader emergence and leadership success may be only partially overlapping. Meta-analyses indicate that personality traits, including extraversion, emotional stability, authoritarianism, and social perspective-taking skills, predict emergence as a leader in a previously leaderless group (Ensari et al., 2011). It could be the case that these traits lead certain people to appear to others as more like a leader, but to sustain a leadership role, this image of leadership ability would need to be maintained throughout a task (Chemers, 2001). The traits that make a stable and effective leader are slightly different. Successful leaders tend to be more cooperative and optimistic (Gächter et al., 2012), persuasive and communicative, and are good listeners and lead by example. Some believe that these traits converge on effective leaders showing higher *emotional intelligence*, with higher self-awareness, self-regulation, motivation, empathy, and social skills (Goleman, 1998).

The Present Study

We conducted a small, preliminary study to examine how individuals and small groups of two or three solve collective wayfinding challenges. Inspired by research and theory from wayfinding and group dynamics, participants completed a battery of individual differences measures to characterize personality traits, social perspective-taking skills, and spatial skills. Participants then completed an individual or group virtual wayfinding task that involved searching for a sequence of landmarks in an unfamiliar virtual city (Figure 3.1). Critically, participants were provided with a digital map that *only one person could view at a time* as an on-screen "mini-map" overlay. In keeping with Wiener's taxonomy, participants completed an aided and goal-directed search through an unfamiliar environment (Wiener et al., 2009), with a map as an aid, and landmark objectives as goals. To succeed at this wayfinding task, all group members needed to arrive at an objective to receive the next goal; in this manner, the task facilitated synchronous wayfinding. However, because only one group member could view the map at a time, the task also encouraged coordinated behavior within the group and possible leader emergence and competition.

We measured several aspects of individual and group behavior, including which individual(s) viewed the map over the course of trials, the efficiency of each individual's navigated route between objectives, the distances and heading differences between individuals during group wayfinding, and path efficiency differences between individuals in groups. At the end of the experiment, we asked participants to independently place landmarks onto a map of the navigated environment from memory. Finally, for dyads and triads, we asked each participant to rate the

Figure 3.1 An example view from within the virtual environment, facing one of the goal landmarks (hotel; note flag positioned in front). Indicated at upper left are the current goal, current trial number, and how many team members are currently at the goal location (if any). Upper right is compass, and lower right shows minimap overlay.

other participant(s) on leadership attributes. Analyses focused on answering eight questions:

Do group size and individual differences influence wayfinding dynamics? Larger group sizes may show different wayfinding dynamics, perhaps different patterns of map sharing (Reilly et al., 2009), varied distances between group members, and varied differences in angular heading between group members. Individual differences in leadership-oriented personality traits, or differences in spatial skills, may also be associated with map possession. Though this possibility has not been explored previously, it could be the case that group members will cede the map to an individual with higher perceived spatial skills or leadership ability, allowing him or her to take control of the wayfinding task. In contrast, an individual with poorer spatial skills may desire to possess the map more frequently, perhaps to reduce their anxiety related to navigating unfamiliar spaces (Hund & Minarik, 2009).

Does group heterogeneity influence wayfinding dynamics? We predicted that more heterogeneous groups, in terms of individual differences in gender, personality traits, and/or spatial skills, would show relatively uncoordinated spatial behavior (He et al., 2015). This may manifest as high spacing or angular heading differences among group members during navigation.

Do group size and individual differences influence wayfinding efficiency? We predicted that increasing group size would influence wayfinding efficiency, and this effect may interact with individual differences in personality or spatial skills (Janis, 1982; Karau & Williams, 1993; Stroebe & Diehl, 1994).

Do leadership traits influence group wayfinding efficiency? Perceived leadership ability, whether self-assessed or other-assessed, may influence path efficiency in dyads and triads, with groups performing better when members have higher leadership traits (Zaccaro et al., 2018). In contrast, it could be the case that groups with many leaders could develop more competitive disagreements or power struggles among members (Greer & Chu, 2020).

Does group heterogeneity influence wayfinding efficiency? More heterogeneous groups, in terms of individual differences in gender, personality traits, and/or spatial skills, may show relatively low path efficiency during wayfinding (He et al., 2015). In contrast, a single leader with high spatial skills could potentially emerge and effectively guide the group.

Does group behavior influence wayfinding efficiency? We predicted that groups with strong synchronous wayfinding (Dalton et al., 2019), characterized by more map sharing, higher group cohesion as measured in distance, and heading differences among group members, would show higher group path efficiency outcomes.

Do group size and individual differences influence final map memory? We predicted that larger groups would show relatively poor final map memory based on research showing suboptimal performance of larger teams (Stroebe & Diehl, 1994).

Do wayfinding dynamics influence final map memory? We predicted that more exposure to the map during wayfinding would be related to lower distance error during map reconstruction (Brunyé et al., 2020; Zhang et al., 2014).

Participants and Design

A total of 89 participants were recruited to participate in the study in exchange for monetary (US$20/hour) compensation, of which 74 completed all tasks and were included for analysis (32 male and 42 female; $M_{age} = 21.3$), primarily comprised of undergraduate and graduate students at Tufts University. The 15 participants who did not complete the task either: (1) took longer than 2 hours to complete, causing the experimenter to terminate the session given the upper limit of our approved data collection window; (2) did not cooperate appropriately with each other, for example refusing to speak to one another or being inappropriately aggressive with one another, and were therefore dismissed, or (3) there was a technical error with network connectivity between the computers and therefore incomplete data. Remaining participants ($N = 74$) were randomly assigned to individual ($N = 29$), dyad ($N = 12$), or triad ($N = 33$) sessions. An experimental session consisted of pre-wayfinding individual differences measures, the wayfinding task, and a post-wayfinding leadership questionnaire.

Materials

Individual Differences Measures: We included measures to assess each participant's self-perception of their personality, social skills, leadership potential, and social status. These included the Big Five Inventory (BFI) (John & Srivastava, 1999), the autism-spectrum quotient (AQ) (Baron-Cohen et al., 2001), the leadership traits questionnaire (LTQ) (Northouse, 2019), and the dominance-prestige scale (DPS) (Cheng et al., 2010).

The BFI is a 44-item inventory that measures an individual on the Big Five personality dimensions (extraversion vs. introversion, agreeableness vs. antagonism, conscientiousness vs. lack of direction, neuroticism vs. emotional stability, and openness vs. closedness to experience). For each item on the inventory, individuals indicated to what degree they believed the statement applied to them, on a scale from 1 (disagree strongly) to 5 (agree strongly). The reliability coefficient previously reported for the BFI is relatively good, $\alpha = 0.83$ (extraversion = 0.88, agreeableness = 0.79, conscientiousness = 0.82, neuroticism = 0.84, and openness = 0.81) (John & Srivastava, 1999).

The AQ is designed to determine to what degree an individual may exhibit traits resembling autism-spectrum disorder by examining five areas of functioning: social skills, attention switching, attention to detail, communication, and imagination (Baron-Cohen et al., 2001). The AQ is a 50-item forced choice questionnaire in which participants must mark to what degree each statement describes themselves, using a scale anchored at definitely agree, slightly agree, slightly disagree, and definitely disagree. Cronbach's alpha coefficients previously reported are moderate: social skills = 0.77, attention switching = 0.67, attention to detail = 0.63, communication = 0.65, and imagination = 0.65 (Baron-Cohen et al., 2001).

The LTQ measures the personal leadership characteristics and involves participants rating how well each of the 14 leadership characteristics describes themselves (or another), on a scale from 1 (strongly disagree) to 5 (strongly agree). Overall Cronbach's alpha data for the LTQ were not found in the extant literature, though one report suggests that it is typically above 0.8 (Ginzburg et al., 2018).

The DPS is a 17-item questionnaire intended to gauge each participant's perceived social status. Participants indicate how well each of the statements describes themselves (or another) on a scale from 1 (not at all) to 7 (very much). Reliability coefficients for the DPS are generally good, with trait dominance at 0.83 and trait prestige at 0.80 (Cheng et al., 2010).

To measure self-reported and objective spatial ability, we administered a card rotation task, a perspective-taking task, a sense of direction scale, and a video game experience assessment. The first was a card rotation task that involved 80 mental rotation trials, each trial presented a pair of two-dimensional geometric figures and the participant determined whether they were the same or different, and accuracy and response times were recorded (Ekstrom et al., 1976; Hegarty & Waller, 2004). The second was a road map perspective-taking test for which participants had to use mental rotation to make egocentric turn decisions (i.e. right or left at intersections) to describe a meandering path overlaid onto a map (Money et al., 1965; Zacks et al., 2000). The third was the self-report Santa Barbara Sense of Direction (SBSOD) scale, which asked participants to answer 15 questions about their spatial and navigational abilities, preferences, and experiences from 1 (strongly agree) to 7 (strongly disagree) (Hegarty et al., 2002). Finally, we asked participants whether they consider themselves video gamers, which is previously shown to correlate with spatial ability (Spence & Feng, 2010).

Virtual Environment: As an adaptation of environments and tasks used in our previous research (Brunyé et al., 2020; Brunyé, Gardony et al., 2018), we developed a navigable large-scale urban desktop virtual environment and a corresponding map of the environment using the Unity 3D Gaming Engine (Unity Technologies; San Francisco, CA). The square environment covered an approximately 1.3 km^2 area with 87 intersections and approximately 400 buildings. Eleven of the buildings served as landmarks, one as a starting landmark for wayfinding (*The Ford Center*) and the others as goal landmarks for wayfinding. Each of the 11 landmarks was clearly labeled on the façade and had a small white flag positioned in front of the building; when a participant's avatar touched the correct goal's flag, it served as the trigger for a subsequent trial (i.e., *You have found DynoTech Systems, now find Sky Bank*).

During navigation, a mini-map could be toggled on and off using the F1 keyboard key, with the map appearing at the bottom right-hand side of the screen; on the map, the 11 landmarks (including the starting landmark) were clearly labeled, but there was no real-time locational indicator for any participants (Figure 3.1).

The software was programmed such that only one participant could view the mini-map at a time; for example, if a participant in a group toggled their mini-map on, the mini-map would correspondingly disappear from the screen of another participant. The virtual environment was run on a local area network with one computer as the server and the others (up to 3) as clients using the native Unity 3D networking capability. This allowed us to include avatars for each participant labeled with the participant's real name floating above the avatar's head; in this manner, participants could see the other group members' locations and identities if they were within their virtual field of view (Figure 3.2). As a participant moved and rotated their avatar, it displayed a walking animation and body rotation, respectively. The horizontal field of view of the player was approximately 105°.

Landmark Placement Task: To test memory for the environment, we created a landmark placement test using the Unity 3D gaming engine. This test consisted of a road map of the environment devoid of landmarks and landmark names, with a list of 21 landmark labels that participants were asked to position on the map. Eleven of these landmark labels were originally learned, and ten were possibly incidentally learned during navigation. Participants would drag and drop each label and marker onto the blank map; once all 21 markers were placed, the landmark placement task was completed.

Procedure

Participants were randomly assigned to one of three group conditions: individual, dyad, or triad. Participants visited the laboratory for one session that ranged from 30 to 120 minutes, depending on wayfinding performance.

Following consent, the participants were seated at a table, each with a laptop and a computer mouse (Figure 3.3). Participants were not allowed to look at each other's screens but were encouraged to communicate with each other verbally.

Figure 3.2 An example view from within the virtual environment, showing an individual group member's perspective on the other two members of a triad, including labels depicting participant names.

Figure 3.3 A triad navigating the virtual environment, with faces obscured for privacy.

Participants completed the individual differences measures, and then a practice virtual wayfinding task. The practice task involved wayfinding among five landmarks (e.g., *pool*, *dog*, and *car*) in a small-scale virtual environment. They were instructed on how to move their avatars, identify and reach landmark goals and toggle the mini-map on and off the screen. In dyads and triads, practice also gave the group members the opportunity to navigate together and learn about the task rules (e.g., all members must arrive at a goal location before the next goal is triggered). Once all five landmarks were reached successfully and any remaining questions were answered, participants began the primary virtual wayfinding task.

Participants were then placed into the large-scale urban virtual environment and began the series of wayfinding tasks. In dyads or triads, participants needed to work together as a team to find each landmark, as the trials would only progress once all team members reached a goal. At the white flag in front of a goal landmark, all group members' avatars needed to be simultaneously within a 1-meter radius of the flag to trigger the next trial. If one participant found the goal independently, the next trial would not be triggered and the participant was still free to navigate the environment (i.e., they were free to leave and locate/guide the other participants to the goal).

All participants in dyads and triads were given the same series of instructions, they were told:

> *You will now be navigating through a virtual environment on the computer. For this, you will have to navigate as a group through a city to find the landmarks listed on the upper left-hand side of the screen. You will work together as a team to find a sequence of ten landmarks in a city environment. In the upper left-hand side of the screen, you will be instructed on the landmark that you must walk to. You will be provided a map that only one of you can view on your screen at a time. Press F1 for the map to appear on your screen. By doing this the map will disappear from the screen it was previously on. You can also press F1 to remove a map from your screen. Press the 'W' key to move your avatar forward and move your mouse to change the direction. Your avatar can only move along the roads and sidewalks. Each landmark goal in the environment has a white flag in front of it. All three of you must reach the white flag in front of each landmark to move on to the next trial.*

Participants in the individual group condition were given a similar series of instructions, without the collaborative aspects. Note that individual participants could keep the mini-map displayed for the entire wayfinding experience without risk of another participant requesting it. At the beginning of the first trial in a session, no mini-map was displayed until a participant requested it by pressing the F1 key; at the beginning of each wayfinding trial, the mini-map would remain on the screen of the participant who was displaying it when the last trial was completed.

After participants completed the wayfinding task, they each individually completed a final memory test. This computerized test consisted of participants dragging and dropping 21 listed landmark names onto a road map of the navigated environment. Participants were not permitted to look at each other's screens or communicate for this final test. All 21 landmarks had to be placed on the blank map in order to complete the test, and if they were uncertain of a landmark's location, they were instructed to take their best guess. The task could not be finished until all 21 landmarks were placed on the map.

Finally, participants in dyads or triads completed the LTQ and DP post-questionnaires, answering questions about the other participants' leadership and dominance. Participants who were in the individual group condition did not participate in the post-questionnaire.

Data Scoring

Questionnaires were scored using standard published procedures for each instrument. For the LTQ and DP, recall that pre-wayfinding was a self-assessment and post-wayfinding was an other-assessment for those in groups; as such, for each group member, we averaged others' ratings of that member to evaluate others' perceptions of their leadership and dominance/prestige. Accuracy and response times (for correctly answered trials) were recorded for the card rotation and perspective-taking tasks.

For wayfinding, we calculated four measures of navigation behavior. First, for each participant and trial, we calculated the proportion of trial duration (in ms) that the map was on an individual's screen. Second, for each trial, we calculated the mean Euclidean distance to other team member(s); this allowed us to assess whether team members tended to be clustered together or spread apart ("group distance"). Third, for each participant and trial, we calculated the mean absolute heading difference between the participant and other group member(s) as they navigated; this allowed us to assess whether team members tended to face the same direction while navigating or if they tended to head in different directions than one another ("group heading"). Fourth, for each participant and trial, we calculated path efficiency, which relates optimal path length (calculated using the A* path finding algorithm; Hart et al., 1968) to actual navigated path length; this measure has a maximum of 1 and a minimum approaching 0 ("path efficiency"). Note that the second and third measures could only be calculated for group wayfinding, not individual wayfinding experiences.

For landmark placement, recall that the landmarks to be positioned on the map were either goal landmarks during wayfinding (11 landmarks) or landmarks that could only be learned incidentally by passing them during wayfinding (ten landmarks). This allowed us to assess three measures: intentional learning of the 11 landmarks, incidental learning of landmarks passed during wayfinding, and guessing behavior for positioning landmarks that were never passed during wayfinding. To assess each of these, we calculated three measures. First, to measure intentional learning, for each of the 11 target landmarks, we calculated the mean Euclidean distance between the correct location of a landmark and the position where the participant placed the landmark. Second, to measure incidental learning, we analyzed each participant's wayfinding path to check whether one or more of the ten non-goal landmarks were passed; if so, we coded them as candidate landmarks for incidental learning. Third, to compare these two measures with guessing behavior, we coded all landmarks that never served as a goal and were never passed during navigation as "non-learned" landmarks.

Results

Data Pre-Processing

Four measures were log(10) corrected for negative skew: card rotation response time, perspective-taking response time, group distance, and group heading. Note that pre-transform data are detailed in figures and tables, whereas all statistical analyses used post-transform data.

To reduce the number of dimensions inherent to our individual differences measures, we conducted a principal components analysis (with varimax rotation) with our 19 measures. The analysis revealed seven components with eigenvalue > 1, accounting for 73.5% of variance. These components converged on the following seven constructs, listed in descending order of variance accounted for.

Component 1 (Others' perception of leadership qualities): This component accounted for 20.71% of variance and weighted primarily on how others rate an individual's leadership qualities, dominance, and prestige, the three outcomes of the LTQ and DP scales.

Component 2 (Self-perception of leadership qualities): This component accounted for 14.2% of variance and weighted primarily on how individuals rated their own agreeableness and conscientiousness (Big-5 measures), leadership qualities (LTQ), and prestige (DP). It also negatively weighted on neuroticism (Big-5).

Component 3 (Autism-related traits): This component accounted for 10.3% of variance and weighted primarily on the autism quotient (AQ) but also negatively weighted on extraversion from the Big-5 personality inventory.

Component 4 (Spatial processing speed): This component accounted for 9.3% of variance and weighted primarily on response times to the card rotation and perspective-taking tasks.

Component 5 (Dominance and openness): This component accounted for 7.1% of variance and weighted primarily on self-rated dominance (DP) and openness to new experiences (Big-5).

Component 6 (Spatial anxiety): This component accounted for 6.6% of variance and weighted primarily on spatial anxiety, with a strong negative weighting of sense of direction (SBSOD).

Component 7 (Mental rotation ability): This component accounted for 5.3% of variance and weighted primarily on accuracy during the card rotation task, and whether someone considered themselves a video gamer. Given known associations between mental rotation and video gaming, we refer to this component as mental rotation ability (De Lisi & Wolford, 2002; Quaiser-Pohl et al., 2006; Terlecki & Newcombe, 2005).

Factor scores were output using the regression method, and these scores were used to enter each component as a covariate in most analyses. For all analyses, effect sizes are provided using eta-squared (η^2) or Cohen's *d*. For the eight questions posed below, the results are summarized in Table 3.1.

Question 1: Do Group Size and Individual Differences Influence Wayfinding Dynamics?

To examine this question, we focused on dyads and triads only (not individuals) and conducted three analyses. All three used a mixed analysis of covariance (ANCOVA) with gender (two levels: male and female) and group size (two levels: dyad and triad) as between-participant factors, trial set as a within-participants factor (two levels: first five trials and second five trials), and seven covariates corresponding to Components 1–7. In the first analysis, the outcome of interest was the proportion of time that individuals possessed the map during navigation. The second analysis examined the distance between individuals during navigation, and the third examined angular differences between individuals.

The first ANCOVA examining map possession patterns revealed no main or interactive effect of trial set, gender, or group size (p's > .05). However, Component 6 (spatial anxiety) was associated with map possession means, $F(1, 34) = 4.0$, $p < .05$, and $\eta^2 = .09$; positive parameter estimates (β) demonstrated that group members with higher anxiety tended to possess the map for a higher proportion of trial time.

The second ANCOVA examining the distance between individuals during navigation revealed no main effect of gender or group size (p's > .05). A main effect of trial set demonstrated higher distances between group members during Trial Set 1 versus 2, $F(1, 34) = 7.15$, $p < .05$, and $\eta^2 = .12$. This effect was qualified by two interactions. First, there was an interaction between trial set and group size, $F(1, 34) = 5.0$, $p < .05$, and $\eta^2 = .09$. Group members became generally closer together from Trial Set 1 to Trial Set 2; however, this effect was stronger in dyads than triads. Second, there was an interaction between trial set and Component 3 (autism-related traits), $F(1, 34) = 7.74$, $p < .01$, and $\eta^2 = .13$. Comparing parameter estimates across trial sets, Component 3 was positively related to the distance between group members in Trial Set 1 ($\beta = .08$ and $p = .08$), but this pattern was relatively flat in Trial Set 2 ($\beta = -.02$ and $p = .27$). Finally, the ANCOVA revealed the effects of Component 1 (other perception of leadership qualities), $F(1, 34) = 7.01$, $p < .05$, and $\eta^2 = .13$ and Component 5 (dominance and openness), $F(1, 34) = 4.5$, $p < .05$, and $\eta^2 = .08$. Parameter estimates demonstrated that people who remained closer to their other group members were rated as having higher leadership qualities, and people with higher dominance and openness tended to be farther from other group members.

The third ANCOVA examining the angular heading difference between individuals during navigation revealed no main effect of trial set ($p > .05$). However, there were main effects of group size, $F(1, 34) = 9.47$, $p < .01$, and $\eta^2 = .15$, and gender, $F(1, 34) = 4.43$, $p < .05$, and $\eta^2 = .07$. Triads tended to show a higher angular

Table 3.1 Summary of Research Findings Across the Eight Questions Posed

Question Posed	*Level of Analysis*	*Overall Answer*	*Summary of Findings*
Do group size and individual differences influence wayfinding dynamics?	Dyads and Triads	Yes	Higher anxiety levels were associated with more map possession time. Higher leadership qualities associated with closer distances between group members. Higher dominance and openness associated with farther distances between group members. Triads showed higher angular heading differences than dyads, and females higher than males. In trial set 2 (trials 6–10), lower spatial processing speeds were associated with higher angular heading differences between group members.
Does group heterogeneity influence wayfinding dynamics?	Dyads and Triads	Yes	Heterogeneity of spatial anxiety in a group was associated with higher distance between group members. Heterogeneity of spatial processing speed and mental rotation ability was associated with higher angular heading differences between group members.
Do group size and individual differences influence wayfinding efficiency?	Individuals, Dyads, and Triads	Yes	In Trial Set 1 (trials 1–5), dyads showed lower path efficiency relative to individuals and triads. This effect dissipated in Trial set 2 (trials 6–10). In Trial Set 1 (trials 1–5), higher autism-related traits were associated with lower path efficiency. This effect dissipated in Trial Set 2 (trials 6–10). Slower spatial processing speed was associated with lower path efficiency.

Do leadership traits influence group wayfinding efficiency?	Dyads and Triads	No	There was no significant association between self-perceived leadership-related traits and wayfinding efficiency. There was no significant association between other-perceived leadership-related traits and wayfinding efficiency.
Does group heterogeneity influence wayfinding efficiency?	Dyads and Triads	Yes	Heterogeneity of gender in a group was associated with lower path efficiency. Heterogeneity of spatial processing speed was associated with lower path efficiency. Heterogeneity of mental rotation ability was associated with lower path efficiency. Heterogeneity of autism-related traits was marginally associated with lower path efficiency.
Does group behavior influence wayfinding efficiency?	Dyads and Triads	Yes	Lower angular heading differences among group members were associated with higher path efficiency. Lower map sharing among group members was associated with higher path efficiency.
Do group size and individual differences influence final map memory?	Dyads and Triads	Yes	Triads showed lower intentional learning than individuals and dyads. Dyads showed lower incidental learning than individuals and triads.
Does wayfinding behavior influence final map memory?	Dyads and Triads	Yes	Higher map possession during navigation was associated with higher intentional memory performance.

difference in heading direction relative to dyads, as did females relative to males. The ANCOVA also revealed an interaction between the trial set and Component 4 (spatial processing speed), $F(1, 34) = 6.14$, $p < .05$, and $\eta^2 = .11$. Comparing parameter estimates across the trial sets revealed that slower spatial processing speeds were associated with higher angular heading differences between group members, but this effect was specific to Trial Set 2 ($\beta = .11$ and $p < .001$).

Question 2: Does Group Heterogeneity Influence Wayfinding Dynamics?

To examine this question, for each component, we calculated the standard deviation of component scores within each group, arriving at seven group heterogeneity scores (one for each component). The same was done for gender. These heterogeneity scores were used as covariates in an ANCOVA, along with group size as a between-participants factor (two levels: dyad and triad) and trial set as a within-participants factor (two levels: Trial Sets 1 and 2). Using this design, we reproduced two of the three ANCOVAs described in Question 1: first for distance between navigators and a second for angular heading differences. In examining the results, we only considered main and interactive effects of gender and the seven heterogeneity scores (main and interactive effects trial set and group size are detailed in Question 1).

Map possession proportion could not be considered as an outcome because, when collapsed across group members, it averages to $1/N$ (i.e., the map was usually displayed at any given point during the experiment). Instead, for this analysis, we used the standard deviation of map possession within a group, which indexes map sharing behavior within a group; lower or higher scores indicate less or more map possession ("hogging the map").

The first ANCOVA, examining map sharing patterns, revealed no significant main or interactive effects of group heterogeneity (all p's $> .05$).

The second ANCOVA, examining distance between navigators, revealed one significant effect of Component 6 (spatial anxiety), $F(1, 7) = 8.97$, $p < .05$, and $\eta^2 = .41$; positive parameter estimates showed that more group heterogeneity in spatial anxiety was associated with higher distances between group members.

The third ANCOVA, examining angular heading differences between navigators, revealed two significant effects, the first with Component 4 (spatial processing speed), $F(1, 7) = 6.21$, $p < .05$, and $\eta^2 = .23$, and second with Component 7 (mental rotation ability), $F(1, 7) = 8.17$, $p < .05$, and $\eta^2 = .31$. Positive parameter estimates showed that more group heterogeneity in spatial processing speed and mental rotation ability were both associated with higher angular differences among group members.

Question 3: Do Group Size and Individual Differences Influence Wayfinding Efficiency?

To examine this question, we conducted an ANCOVA with gender (two levels: male and female) and group size (three levels: individual, dyad, and triad) as between-participant factors, trial set as a within-participant factor (two levels: first five trials and second five trials), and six covariates corresponding to Components 2–7. Note that Component 1 was omitted because it solely concerns LTQ and DP ratings gathered from group sessions (dyads or triads), whereas this analysis considered all three group sizes.

The ANCOVA revealed a main effect of trial set, $F(1, 62) = 30.07$, $p < .001$, and $\eta^2 = .27$, but no significant effect of gender or group size (p's $> .05$). These patterns were qualified by an interaction between trial set and group size, $F(2, 62) = 3.57$, $p < .05$, and $\eta^2 = .06$. As shown in Figure 3.4, during the first five trials, dyads under-performed relative to individuals and triads, but the performance was similar across group sizes within the last five trials. This was confirmed in two paired samples t-tests in Trial Set 1, demonstrating significantly higher path efficiency for individuals versus dyads, $t(43) = 3.21$, $p < .01$, and $d = .48$, and marginally higher path efficiency for triads versus dyads, $t(43) = 1.92$, $p = .06$, and $d = .32$.

The ANCOVA also revealed an interaction between trial set and Component 3 (autism-related traits), $F(1, 65) = 7.13$, $p < .01$, and $\eta^2 = .06$. Comparing Component 3 parameter estimates across the two trial sets demonstrated a negative estimate in Trial Set 1 ($\beta = -.04$ and $p < .01$), but relatively flat estimate in Trial Set 2 ($\beta = -.003$ and $p = .79$). In other words, individuals with higher autism-related traits had lower path efficiency in the first trial set, but this pattern diminished over time.

Finally, the ANCOVA revealed that Component 4 (spatial processing speed) was significantly associated with path efficiency, $F(1, 62) = 9.86$, $p < .01$, and $\eta^2 = .11$; negative parameter estimates demonstrated that slower spatial processing speed was associated with lower path efficiency.

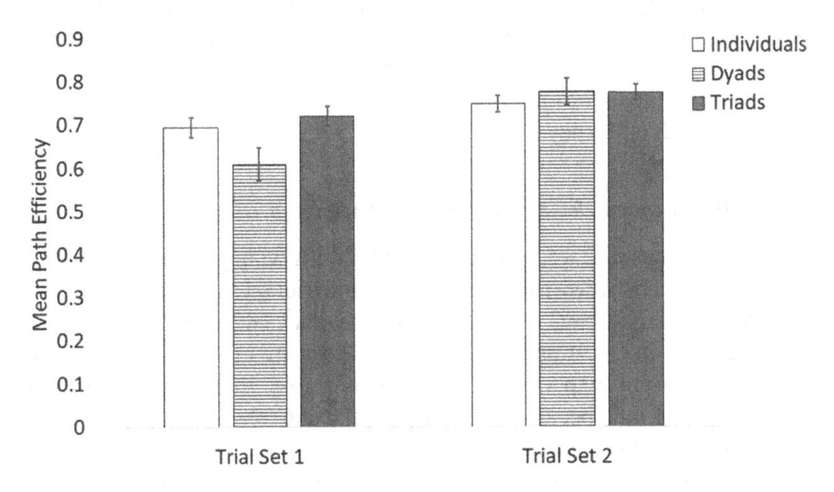

Figure 3.4 Estimated marginal mean and standard error path efficiency as a function of group size (individuals, dyads, and triads) and trial set (first five trials and last five trials).

Question 4: Do Leadership Traits Influence Group Wayfinding Efficiency?

To examine this question, we re-conducted the above (Question 2) analysis while excluding individuals (including only dyads and triads) and including Component 1 (others' perceptions of leadership qualities) along with Components 2–7.

The same overall patterns from the Question 1 analysis were identified, with the interactions between group size and trial set, $F(1, 34) = 5.28$, $p < .05$, and $\eta^2 = .10$, and trial set and Component 3 (autism-related traits), $F(1, 34) = 5.85$, $p < .05$, and $\eta^2 = .11$. However, there was no significant main or interactive effect of the two leadership predictors of interest, Component 1 (p's $> .76$) and Component 2 (p's $> .29$), suggesting that other-perceived or self-perceived leadership traits were not reliably related to path efficiency outcomes, respectively.

Question 5: Does Group Heterogeneity Influence Wayfinding Efficiency?

The same ANCOVA was conducted as in Question 2, except with path efficiency as the outcome. There were no main effects of trial set or interactions (all p's $> .05$). However, there were three significant results related to heterogeneity. First, gender heterogeneity was associated with path efficiency, $F(1, 7) = 6.89$, $p < .05$, and $\eta^2 = .14$; negative parameter estimates demonstrated that mixed-gender groups generally showed lower path efficiency. Second, heterogeneity in Component 4 (spatial processing speed) was associated with path efficiency, $F(1, 7) = 7.07$, $p < .05$, and $\eta^2 = .14$; negative parameter estimates demonstrated that groups with more heterogeneity in spatial processing speed generally showed lower path efficiency. Third, heterogeneity in Component 7 (mental rotation ability) was associated with path efficiency, $F(1, 7) = 8.55$, $p < .05$, and $\eta^2 = .17$; negative parameter estimates demonstrated that groups with more heterogeneity in mental rotation ability generally showed lower path efficiency. A final marginally significant pattern is worth mentioning: Component 3 (autism-related traits) was associated with path efficiency, $F(1, 7) = 4.93$, $p = .06$, and $\eta^2 = .10$; negative parameter estimates suggested that groups with more heterogeneity in autism-related traits generally showed lower path efficiency.

Question 6: Does Group Behavior Influence Wayfinding Efficiency?

To examine this question, we conducted an ANCOVA with group size (two levels: dyad and triad) as a between-participants factor and trial set as a within-participants factor (two levels: Trial Sets 1 and 2). Three covariates detailing group behavior were included. As in Question 2, we included a measure of map sharing behavior (standard deviation of map viewing proportions across group members) and the measures of group member distances and angular differences. The outcome was path efficiency over the course of the two trial sets.

The ANCOVA revealed two specific patterns of interest. Both mean angular heading differences and map possession patterns were associated with path efficiency, $F(1, 12) = 24.09$, $p < .001$, and $\eta^2 = .32$, and $F(1, 12) = 12.94$, $p < .01$, and

$\eta^2 = .16$. Negative parameter estimates demonstrated that lower angular heading differences and lower map sharing among group members were associated with higher path efficiency.

Question 7: Do Group Size and Individual Differences Influence Final Map Memory?

To examine this question, we conducted an ANCOVA with group size (three levels: individual, dyad, and triad) and gender (two levels: male and female) as between-participants factors, learning type as a within-participant factor (three levels: intentional, incidental, and non-learned), and six covariates corresponding to Components 2–7. The outcome measure was mean distance between placed and correct landmark locations. Note that Component 1 was omitted because it solely concerns LTQ and DP ratings gathered from group sessions (dyads or triads), whereas this analysis considered all three group sizes.

The ANCOVA revealed the main effects of learning type, $F(2, 114) = 93.47$, $p < .001$, and $\eta^2 = .52$ and group size, $F(2, 57) = 5.52$, $p < .01$, and $\eta^2 = .14$. These effects were qualified by an interaction between these two factors, $F(4, 114) = 6.44$, $p < .001$, and $\eta^2 = .07$. This interaction is plotted in Figure 3.5. Within intentional learning, triads tended to underperform relative to individuals and dyads, with higher landmark placement distance error. Within incidental learning, dyads tended to underperform relative to individuals and triads. As would be expected, when placing landmarks on the map that were never learned (intentionally or incidentally), the three group sizes showed similar performance. None of the six components was associated with landmark placement performance.

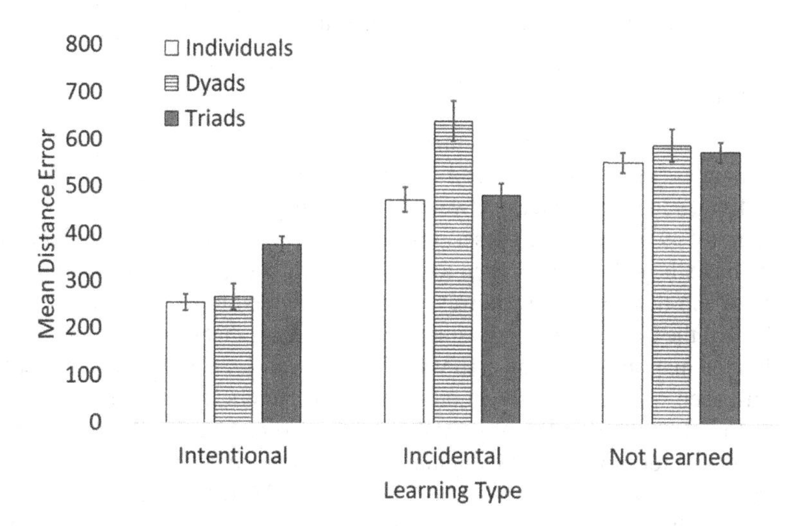

Figure 3.5 Estimated marginal mean distance error (and standard error bars) on the landmark placement task as a function of the three learning types (intentional, incidental, and not learned) and three group sizes (individuals, dyads, and triads).

Question 8: Does Wayfinding Behavior Influence Final Map Memory?

To examine this question, we conducted an ANCOVA with group size (two levels: dyad and triad) and gender (two levels: male and female) as between-participants factors, learning type as a within-participant factor (three levels: intentional, incidental, and non-learned), and three covariates: mean map possession proportions, distance between group members, and angular heading difference between group members corresponding to Components 2–7. The outcome measure was mean distance between placed and correct landmark locations.

The ANOVA revealed a main effect of Learning Type, $F(2, 68) = 3.19$, $p < .05$, and $\eta^2 = .06$, with generally higher distance error across the three learning types (as seen in Figure 3.5). There was no main effect of gender or group size (all p's $> .05$). However, there was one significant interaction of interest (in addition to interactions revealed in Question 5). Learning type interacted with map possession, $F(2, 68) = 3.79$, $p < .05$, and $\eta^2 = .07$; comparing parameter estimates across the three learning types demonstrated that higher map possession proportions were associated with lower distance error during landmark placement, but only when placing intentionally learned landmarks ($\beta = -207.71$ and $p < .001$) and not when placing incidentally learned ($\beta = 56.56$ and $p = .54$) or non-learned ($\beta = -33.65$ and $p = .62$) landmarks.

Discussion

In motivating our study, we posed eight questions related to collective navigation behavior, wayfinding efficiency, and spatial memory outcomes. In answering these questions, our preliminary data showed evidence for several spatial and non-spatial individual differences influencing all three outcomes.

Group Behavior

Navigation behavior was quantified through three primary measures that reflected group dynamics: map possession and sharing, distance between group members, and angular heading differences between group members. When considering map possession, a few (unanalyzed) descriptive statistics are worth noting. Individuals tended to have the map on the screen for most of their session, with the map showing 93% of the time. In dyads, however, the map tended to be on a screen only about 74% of the time, suggesting a small group dynamic that resulted in hesitancy to display the map. In triads, the map was on the screen 86% of the time, closer to the pattern seen with individuals. In our analysis of factors related to map possession, we found that higher levels of spatial anxiety were associated with the map being displayed for more time. Given that higher spatial anxiety is associated with a landmark preference during wayfinding (Schmitz, 1997, 1999), these participants may have a stronger desire to see the labeled landmarks on the map, using them for localization and guiding navigation. In contrast to earlier results, however, this pattern was not predominantly female, as it did not interact with gender (Hund & Minarik, 2009; Lawton & Kallai, 2002). In dyads or triads, allowing individuals with higher spatial anxiety to possess a map may help alleviate anxiety, though it may not help support efficient wayfinding.

In considering the distance and angular heading differences between members of dyads and triads, we found four primary results. First, triads tended to remain closely grouped together over the course of trials, whereas dyads started the session farther apart and then became closer together over time. Second, this pattern may be partially attributed to groups with higher autism-related traits showing higher distance among group members, an effect that also diminished with time. Third, when considering group heterogeneity in individual differences, higher spatial anxiety heterogeneity in groups was related to farther distances among members during navigation. Finally, higher heterogeneity in two spatial individual differences, spatial processing speed and mental rotation ability, was associated with higher angular differences between group members. Together, these results suggest that three spatial attributes and one non-spatial attribute play a role in guiding group dynamics during wayfinding, with high autism-related traits, spatial anxiety, or low spatial skills related to more disaggregated movement patterns. It is likely that heterogeneity among group members challenged group members' communication patterns, planning and coordination, perspective taking, and spatial strategies used to execute the task (Brunyé et al., 2012; Forlizzi et al., 2010; He et al., 2015). In some cases, the team likely adapted to overcome the challenge. For example, there are inherent communication challenges in groups with heterogeneity of autism-related traits, more than groups with consistently high or low levels of these traits (Crompton et al., 2019). The fact that autism-related traits were only related to higher group disaggregation during early wayfinding trials suggests an adaptation of group dynamics that serves to accommodate different styles and needs. It is likely that adapting to group heterogeneity is challenging at times, with the intent of transitioning from weak to strong social wayfinding over time, characterized by reciprocal, bidirectional communication patterns (Dalton et al., 2019). Future research would benefit from recording and scoring communication patterns among group members.

Wayfinding Efficiency

Given that individual differences appear to affect group dynamics during wayfinding, the next logical question is whether these dynamics were associated with wayfinding performance. Two interesting patterns emerged from this analysis. First, lower angular heading differences among group members were associated with higher path efficiency. This pattern suggests that the extent to which group members tended to head in consistent directions, rather than deviating directionally from one another, is related to more efficient movement toward destinations. Interestingly, this did not emerge when examining the distance between group members. It could be the case that shared viewing patterns among group members are conducive to shared spatial awareness, in contrast to relatively uncoordinated looking behavior during navigation. Such a pattern would match research showing the value of shared referential viewing during visual search and language comprehension

(Gergle & Clark, 2011; Richardson et al., 2009; Tanenhaus et al., 1995). Shared heading during collective wayfinding could help group members map conversations onto environmental referents, such as when discussing perceived landmarks or upcoming turns. The second interesting pattern was related to map possession behavior, with lower map sharing related to poorer wayfinding efficiency. Given the fact that those with high spatial anxiety tended to possess the map for longer than low-anxiety peers, this could underlie the low path efficiency outcomes.

The size of the group itself was also associated with path efficiency outcomes, with dyads underperforming relative to triads during early trials. As noted in the group behavior outcomes, dyads tended to disaggregate during early trials more than triads, and they also had more time without a map on the screen. It is likely that these two behavioral patterns led to relatively poor path efficiency outcomes. There were two additional variables related to path efficiency outcomes: Autism-related traits were related to lower path efficiency, but only during early trials, and faster spatial processing speeds were related to higher path efficiency. There were also interesting patterns regarding group heterogeneity. Groups that were more diverse in gender, autism-related traits, spatial processing speed, and mental rotation ability tended to show lower path efficiency. This pattern complements the group behavior results, with higher spatial and non-spatial group heterogeneity associated with higher distances and heading differences among group members. Overall, this pattern suggests that heterogeneity in collectives, whether in personality traits or spatial skills, can alter behavioral dynamics during wayfinding and result in suboptimal performance. This outcome supports and extends recent research finding that higher heterogeneity of spatial skills is associated with relatively poor spatial performance (He et al., 2015). An interesting future direction might concern the potential costs and benefits of the opposite pattern: high group homogeneity, when all participants use similar (and potentially suboptimal Bailenson et al., 2000; Brunyé et al., 2010, 2015b; Hochmair & Karlsson, 2005) strategies. Another future direction might consider the heterogeneity of not only group members but also the information provided to each individual, which could be identical, partially overlapping, or entirely complementary, adding a new layer of complexity in terms of bidirectional information sharing and coordination.

Spatial Memory

An unexpected pattern emerged when navigators were asked to individually reconstruct a map of the navigated environment. Because all landmark identities were included in the landmark placement task, but only 11 landmark identities (goal destinations) were depicted on the mini-map, this gave us the opportunity to compare intentional versus incidental learning of the environment (McLaughlin, 1965; Thorndyke, 1981). When the landmarks served as goal destinations, they were considered intentionally learned; when the landmarks were not goal destinations but were passed during navigations, they were considered incidentally learned. Of course, we cannot guarantee that participants attended to a passed landmark without using eye tracking or a form of knowledge probing during navigation, but it serves as a reasonable proxy for incidental learning in the absence of more advanced measures. These two types of learning were compared to landmarks that did not serve as goal destinations and were not passed, where very high error rates would be expected. With intentional learning, participants who navigated in triads showed markedly higher distance error relative to those who navigated as individuals or in dyads. In contrast, with incidental learning, participants who navigated in dyads showed higher distance error relative to others.

Prior research examining how collaborative groups encode and retrieve memories demonstrates that the memory of individuals who learn in groups tends to be poorer than individuals who learn alone, a phenomenon termed collaborative inhibition (Harris et al., 2008, 2013). For example, when dyads collaboratively learn verbal material, they tend to create less effective cues for later retrieval relative to individual learners (Barber et al., 2010). The same occurs with collaborative triads relative to individuals (Blumen & Rajaram, 2008). This pattern is thought to reflect retrieval inhibition when idiosyncratic encoding strategies are used across group members that subsequently limit effective retrieval among individuals (Barber et al., 2015; Rajaram, 2011). This extant research and theory makes two relevant predictions: first, it predicts larger memory deficits with larger collaborative groups, and second, collaborative inhibition will affect intentional learning more than incidental learning (Marion & Thorley, 2016). However, our results were not entirely consistent with these predictions. As shown in Figure 3.5, if we collapse across intentional and incidental learning, groups generally showed poorer memory relative to individuals ($M = 360.8$), though the means are similar across the two group sizes ($M = 439.6$ and $M = 408.2$). Furthermore, intentional and incidental produced very different patterns when considering group size.

For intentional learning, individuals and dyads outperformed triads. Given how important map exposure was for promoting intentional learning, it could be the case that individual triad members did not have as much map exposure during navigation, leading to inferior memory for the environment. This contrasts individuals who could display the map for the entire duration of navigation and dyads who tended to show higher map sharing across members than triads. Regarding incidental learning, the high error rates seen with dyads may be associated with the fact that dyad members tended to show more distance between one another and lower path efficiency overall. It could be the case that dyads have relatively uncollaborative behavior during navigation (e.g., being relatively far apart or heading in different directions), resulting in fewer shared incidental experiences that might support non-goal location encoding (e.g., "*Look, I've seen that bank before!*"), especially at critical decision points such as intersections (Brunyé, Gardony et al., 2018). Of course, these interpretations are post hoc and speculative, motivating future research.

Overall Findings

Surprisingly, we found no reliable evidence that leadership qualities, whether self-assessed or other-assessed, were associated with navigation behavior or outcomes.

In our review of the leadership literature, we expected to find some evidence that leadership traits would guide group dynamics, with individuals emerging as leaders and serving an organizing and guiding role during wayfinding (Ensari et al., 2011; Zaccaro et al., 2018). Such a pattern would likely manifest itself as tighter coupling of group members, higher resource sharing (map sharing), and ultimately higher efficiency. However, self-assessed leadership traits were not associated with any of these behavioral patterns. Furthermore, when participants were given the chance to rate other group members' leadership qualities at the end of the session, there was no evidence that group dynamics were associated with these subsequent ratings. The only pattern that emerged was when a group member tended to stay closer to others during navigation, and they were subsequently rated as having higher leadership qualities. In a highly specialized domain, such as wayfinding, it could be the case that task-related phenotypic personality traits are more strongly related to group dynamics and performance, outweighing any contribution of leadership-oriented traits (Taylor & Brunyé, 2013). It could also be the case that leadership traits may prove more valuable under relatively protracted circumstances; indeed, clear leadership can take weeks to emerge in newly formed groups (Lemoine et al., 2016). Continuing research in this area may benefit from examining leadership patterns emerging in extended wayfinding settings outside of the laboratory, such as during military land navigation training, or search and rescue training.

Another interesting result was the value of characterizing autism-related traits associated with both navigation behavior and performance outcomes. Autism-related traits in otherwise neurotypical individuals have been termed part of the broader autism phenotype (BAP), which emerges largely due to familial heredity (Gerdts & Bernier, 2011). In general, individuals with autism-related personality traits are characterized as having relatively decreased interest in reciprocal social interactions, lower social communicative skills, heightened special interests, and decreased flexibility and difficulty adjusting to change (Armstrong, 2010; Baron-Cohen et al., 2001; Gerdts & Bernier, 2011). Recent research has demonstrated how challenging heterogeneity in autism-related traits can be for group communication and coordination (Crompton et al., 2019). Our results supported this finding, with heterogeneity in autism-related traits related to higher levels of group disaggregation during navigation and lower path efficiency. Heterogeneity of personality traits and spatial skills can both produce some strain on group cohesion, challenging all group members to accommodate different styles and needs. Showcasing the adaptive nature of our groups, these patterns tended to dissipate with time, leading to more effective teams. This finding also highlights the importance of techniques for relationship building in teams, such as the fast friends methodology, which facilitates interpersonal closeness by having group members take turns answering predetermined (and increasingly personal) questions about one another (Aron et al., 1997).

Limitations

While the present chapter identified some promising patterns of wayfinding behavior as a function of personality and spatial skills, it also lays a foundation for continuing research that can build upon our design. First, this preliminary study used a relatively small sample size, especially given the heterogeneity found in group settings and wayfinding performance. As a pilot study, we found some compelling results that build a preliminary framework for understanding collective navigation, but these results must be considered in the context of our somewhat limited sample. Second, while we collected several measures of individual differences, group behavior, performance, and memory, future research may find benefit in additional measures. For example, an analysis of verbal communications can help elucidate patterns of behavior and identify distinct group strategies (Tenbrink, 2015; Tenbrink et al., 2020). Some wayfinding behavior, such as disaggregation or individual map possession, could arise due to disagreements in groups. Indeed, we informally observed that dyads were more likely to develop competitive disagreements with one another relative to triads, frequently resulting in failure to complete the study as intended; recording communications can lend more insights into the planning processes, push- and pull-of information, and the interpersonal dynamics that unfold over time and shape navigation outcomes.

Research and theory in distributed cognition also points to some additional measures that may prove informative for understanding group wayfinding. In ethnographic research, researchers can observe how individuals and groups perform in a real-world environment, outside of the typical confines of university laboratories. While certainly more complex and challenging to control, engaging participants in real-world navigation tasks provides the opportunity to record and analyze a relatively rich data set that may more readily generalize to other real-world navigation domains. These data can include not only communications and performance outcomes but also information regarding planning and replanning processes and relatively subtle communicative behaviors such as gestures, eye gaze, orienting the body toward another person, and facial expressions (Hutchins & Palen, 1997). For example, sociometric badges have been used in real-world collaborative settings to capture information about collective motion, proximity, and relative orientations among group members (Kim et al., 2012). This research demonstrates that rich data regarding the interpersonal dynamics of group members can reveal new causal links between patterns of collaboration and performance outcomes (Wu et al., 2008).

Third, while we find compelling relationships across levels of analysis, including individual differences, wayfinding behavior, and memory, our large mixed design is not immediately conducive to formal mediation analyses (especially given interactive ANCOVA terms). For example, it could be informative to understand whether wayfinding behavior serves as a mediator between individual differences and performance and memory outcomes of groups, an exciting direction for future research (Meneghetti et al., 2011, 2014). Fourth, the card rotation test tended to show ceiling effects in performance, the road map task is not as commonly used as other tasks (such as the spatial orientation test Guilford & Zimmerman, 1948), and the video game questionnaire was limited in scope, all of which could be improved in future research. Finally, our study used networked virtual environment navigation to elicit and measure group wayfinding through a novel urban environment, and we cannot make any strong conclusions regarding whether our results will translate to real-world collective navigation. Indeed, an ongoing debate in the spatial cognition literature pertains to whether results derived from virtual environment behavior adequately characterize real-world behavior, especially given the lack of idiothetic information during virtual navigation (Chrastil & Warren, 2013; Richardson et al., 1999, 2011). Using head-mounted displays (HUDs) is a step in the right direction, although they introduce more symptoms of nausea relative to desktop virtual environments (Sharples et al., 2008). Future research might benefit from richer self-motion cues offered by ambulatory VR, with the ability to see fellow group members only in the virtual world.

Conclusion

Returning to the challenge problem posed in the introduction, we highlight several important factors that will shape the performance and memory outcomes of groups navigating together through novel environments. In addition to the size of the collective itself, we find that individual differences in personality and spatial skills influence how group behavior manifests during wayfinding. These behaviors are related to the group's success in efficiently finding goal destinations, and individual differences and behavior converge to shape intentional and incidental memory for the navigated space. Many of our patterns point to the inherent complexity of measuring, modeling, and interpreting collective navigation, and the importance of understanding how experience within a collective influences outcomes.

References

Allen, G. L. (1999). Spatial abilities, cognitive maps, and wayfinding: Bases for individual differences in spatial cognition and behavior. In R. G. Golledge (Ed.), *Wayfinding behavior: Cognitive mapping and other spatial processes* (pp. 46–80). Baltimore, MD: Johns Hopkins University Press.

Alper, S., Tjosvold, D., & Law, K. S. (1998). Interdependence and controversy in group decision making: Antecedents to effective self-managing teams. *Organizational Behavior and Human Decision Processes*, *74*, 33–52. https://doi.org/10.1006/obhd.1998.2748

Anacta, V. J. A., & Schwering, A. (2010). Men to the east and women to the right: Wayfinding with verbal route instructions. In C. Holscher, T. F. Shipley, M. O. Belardinelli, J. A. Bateman, & N. S. Newcombe (Eds.), *Spatial cognition VII* (pp. 70–84). Springer. https://doi.org/10.1007/978-3-642-14749-4_9

Armstrong, T. (2010). *Neurodiversity: Discovering the extraordinary gifts of autism, ADHD, dyslexia, and other brain differences*. Cambridge, MA: Da Capo.

Aron, A., Melinat, E., Aron, E. N., Vallone, R. D., & Bator, R. J. (1997). The experimental generation of interpersonal closeness: A procedure and some preliminary findings. *Personality and Social Psychology Bulletin*, *23*, 363–377. https://doi.org/10.1177/0146167297234003

Bailenson, J. N., Shum, M. S., & Uttal, D. H. (2000). The initial segment strategy: A heuristic for route selection. *Memory & Cognition*, *28*, 306–318. https://doi.org/10.3758/BF03213808

Barber, S. J., Harris, C. B., & Rajaram, S. (2015). Why two heads apart are better than two heads together: Multiple mechanisms underlie the collaborative inhibition effect in memory. *Journal of Experimental Psychology: Learning Memory and Cognition, 41*, 559–566. https://doi.org/10.1037/xlm0000037

Barber, S. J., Rajaram, S., & Aron, A. (2010). When two is too many: Collaborative encoding impairs memory. *Memory and Cognition, 38*, 255–264. https://doi.org/10.3758/MC.38.3.255

Baron-Cohen, S., Wheelwright, S., Skinner, R., Martin, J., & Clubley, E. (2001). The autism-spectrum quotient (AQ): Evidence from Asperger syndrome/high-functioning autism, males and females, scientists and mathematicians. *Journal of Autism and Developmental Disorders, 31*, 5–17. https://doi.org/10.1023/A:1005653411471

Bishop, D. V. M., Maybery, M., Wong, D., Maley, A., & Hallmayer, J. (2006). Characteristics of the broader phenotype in autism: A study of siblings using the children's communication checklist-2. *American Journal of Medical Genetics Part B: Neuropsychiatric Genetics, 141B*(2), 117–122. https://doi.org/10.1002/ajmg.b.30267

Blumen, H. M., & Rajaram, S. (2008). Influence of re-exposure and retrieval disruption during group collaboration on later individual recall. *Memory, 16*, 231–244. https://doi.org/10.1080/09658210701804495

Boone, A. P., Gong, X., & Hegarty, M. (2018). Sex differences in navigation strategy and efficiency. *Memory & Cognition, 46*(6), 909–922. https://doi.org/10.3758/s13421-018-0811-y

Bormann, E. G. (1990). *Small group communication: Theory and practice*. New York, NY: Harper Collins College Division.

Britten, K. H. (2008). Mechanisms of self-motion perception. *Annual Review of Neuroscience, 31*, 389–410. https://doi.org/10.1146/annurev.neuro.29.051605.112953

Brown, D. J., & Lord, R. G. (2001). Leadership and perceiver cognition: Moving beyond first order constructs. In M. London (Ed.), *How people evaluate others in organizations* (pp. 181–202). London: Lawrence Erlbaum Associates.

Brown, V. R., & Paulus, P. B. (2002). Making group brainstorming more effective: Recommendations from an associative memory perspective. *Current Directions in Psychological Science, 11*, 208–212. https://doi.org/10.1111/1467-8721.00202

Brunyé, T. T., Collier, Z. A., Cantelon, J., Holmes, A., Wood, M. D., Linkov, I., & Taylor, H. A. (2015a). Strategies for selecting routes through real-world environments: Relative topography, initial route straightness, and cardinal direction. *PLoS One, 10*(5). https://doi.org/10.1371/journal.pone.0124404

Brunyé, T. T., Collier, Z. A., Cantelon, J., Holmes, A., Wood, M. D., Linkov, I., & Taylor, H. A. (2015b). Strategies for selecting routes through real-world environments: Relative topography, initial route straightness, and cardinal direction. *PLoS One, 10*(5), e0124404. https://doi.org/10.1371/journal.pone.0124404

Brunyé, T. T., Ditman, T., Giles, G. E., Mahoney, C. R., Kessler, K., & Taylor, H. A. (2012). Gender and autistic personality traits predict perspective-taking ability in typical adults. *Personality and Individual Differences, 52*(1). https://doi.org/10.1016/j.paid.2011.09.004

Brunyé, T. T., Gardony, A. L., Holmes, A., & Taylor, H. A. (2018). Spatial decision dynamics during wayfinding: Intersections prompt the decision-making process. *Cognitive Research: Principles and Implications, 3*, 13.

Brunyé, T. T., Haga, Z. D., Houck, L. A., & Taylor, H. A. (2017). You look lost: Understanding uncertainty and representational flexibility in navigation. In J. M. Zacks & H. A. Taylor (Eds.), *Representations in mind and world: Essays inspired by Barbara Tversky* (pp. 42–56). Routledge. https://doi.org/10.4324/9781315169781

Brunyé, T. T., Mahoney, C. R., Gardony, A. L., & Taylor, H. A. H. A. (2010). North is up(hill): Route planning heuristics in real-world environments. *Memory & Cognition, 38*(6), 700–712. https://doi.org/10.3758/MC.38.6.700

Brunyé, T. T., Smith, A. M., Hendel, D., Gardony, A. L., Martis, S. B., & Taylor, H. A. (2020). Retrieval practice enhances near but not far transfer of spatial memory. *Journal of Experimental Psychology: Learning Memory and Cognition, 46*, 24–45. https://doi.org/10.1037/xlm0000710

Bryant, K. J. (1982). Personality correlates of sense of direction and geographic orientation. *Journal of Personality and Social Psychology*. https://doi.org/10.1037//0022-3514.43.6.1318

Caron, M. J., Mottron, L., Rainville, C., & Chouinard, S. (2004). Do high functioning persons with autism present superior spatial abilities? *Neuropsychologia, 42*, 467–481. https://doi.org/10.1016/j.neuropsychologia.2003.08.015

Chemers, M. M. (2001). Leadership effectiveness: An integrative review. In M. A. Hogg & R. S. Tindale (Eds.), *Blackwell handbook of social psychology: Group processes* (pp. 376–399). Blackwell. https://doi.org/10.1002/9780470998458.ch16

Cheng, J. T., Tracy, J. L., & Henrich, J. (2010). Pride, personality, and the evolutionary foundations of human social status. *Evolution and Human Behavior, 31*, 334–347. https://doi.org/10.1016/j.evolhumbehav.2010.02.004

Cheng, K., Huttenlocher, J., & Newcombe, N. S. (2013). 25 years of research on the use of geometry in spatial reorientation: A current theoretical perspective. *Psychonomic Bulletin & Review, 20*(6), 1033–1054. https://doi.org/10.3758/s13423-013-0416-1

Chrastil, E. R., & Warren, W. H. (2013). Active and passive spatial learning in human navigation: Acquisition of survey knowledge. *Journal of Experimental Psychology. Learning, Memory, and Cognition, 39*(5), 1520–1537. https://doi.org/10.1037/a0032382

Condon, D. M., Wilt, J., Cohen, C. A., Revelle, W., Hegarty, M., & Uttal, D. H. (2015). Sense of direction: General factor saturation and associations with the big-five traits. *Personality and Individual Differences, 86*, 38–43. https://doi.org/10.1016/j.paid.2015.05.023

Coutrot, A., Silva, R., Manley, E., de Cothi, W., Sami, S., Bohbot, V. D., Wiener, J. M., Hölscher, C., Dalton, R. C., Hornberger, M., & Spiers, H. J. (2018). Global determinants of navigation ability. *Current Biology*. https://doi.org/10.1016/j.cub.2018.06.009

Crompton, C. J., Ropar, D., Vans-Williams, C. V. M., Flynn, E. G., & Fletcher-Watson, S. (2019). Autistic peer to peer information transfer is highly effective. *Open Science Framework Preprints*. https://doi.org/10.31219/OSF.IO/J4KNX

Dabbs, J. M., Chang, E. L., Strong, R. A., & Milun, R. (1998). Spatial ability, navigation strategy, and geographic knowledge among men and women. *Evolution and Human Behavior, 19*, 89–98. https://doi.org/10.1016/S1090-5138(97)00107-4

Dalton, R. C. (2003). The secret is to follow your nose: Route path selection and angularity. *Environment & Behavior, 35*(1), 107–131. https://doi.org/10.1177/0013916502238867

Dalton, R. C., Hölscher, C., & Montello, D. R. (2019). Wayfinding as a social activity. *Frontiers in Psychology, 10*, 142. https://doi.org/10.3389/fpsyg.2019.00142

De Lisi, R., & Wolford, J. L. (2002). Improving children's mental rotation accuracy with computer game playing. *Journal of Genetic Psychology, 163*, 272–282. https://doi.org/10.1080/00221320209598683

Drew, T., & Marigold, D. S. (2015). Taking the next step: Cortical contributions to the control of locomotion. *Current Opinion in Neurobiology, 33*, 25–33. https://doi.org/10.1016/j.conb.2015.01.011

Ekstrom, R. B., French, J. J. W., Harman, H. H., & Dermen, D. (1976). Manual for kit of factor-referenced cognitive tests. In *Educational testing service*. Educational Testing Services. https://doi.org/10.1073/pnas.0506897102

Ensari, N., Riggio, R. E., Christian, J., & Carslaw, G. (2011). Who emerges as a leader? Meta-analyses of individual differences as predictors of leadership emergence. *Personality and Individual Differences*, *51*, 532–536. https://doi.org/10.1016/j.paid.2011.05.017

Forlizzi, J., Barley, W. C., & Seder, T. (2010). *Where should I turn? Moving from individual to collaborative navigation strategies to inform the interaction design of future navigation systems*. Proceedings of the Conference on Human Factors in Computing Systems, pp. 1261–1270. https://doi.org/10.1145/1753326.1753516

Foti, R. J., Fraser, S. L., & Lord, R. G. (1982). Effects of leadership labels and prototypes on perceptions of political leaders. *Journal of Applied Psychology*, *67*, 326–333. https://doi.org/10.1037/0021-9010.67.3.326

Frith, U. (2001). Mind blindness and the brain in autism. *Neuron*, *32*(6), 969–979. https://doi.org/10.1016/S0896-6273(01)00552-9

Gächter, S., Nosenzo, D., Renner, E., & Sefton, M. (2012). Who makes a good leader? Cooperativeness, optimism, and leading-by-example. *Economic Inquiry*, *50*, 953–967. https://doi.org/10.1111/j.1465-7295.2010.00295.x

Galea, L. A. M., & Kimura, D. (1993). Sex differences in route-learning. *Personality and Individual Differences*, *14*, 53–65. https://doi.org/10.1016/0191-8869(93)90174-2

Gärling, T., Book, A., & Lindberg, E. (1984). Cognitive mapping of large-scale environments: The interrelationship of action plans, acquisition, and orientation. *Environment and Behavior*, *16*(1), 3–34. https://doi.org/10.1177/0013916584161001

Gärling, T., & Gärling, E. (1988). Distance minimization in downtown pedestrian shopping. *Environment and Planning A*, *20*(4), 547–554. https://doi.org/10.1068/a200547

Gerdts, J., & Bernier, R. (2011). The broader autism phenotype and its implications on the etiology and treatment of autism spectrum disorders. *Autism Research and Treatment*, *2011*, 1–19. https://doi.org/10.1155/2011/545901

Gergle, D., & Clark, A. T. (2011). *See what i'm saying? Using dyadic mobile eye tracking to study collaborative reference*. Proceedings of the ACM Conference on Computer Supported Cooperative Work, CSCW, pp. 435–444. https://doi.org/10.1145/1958824.1958892

Giannopoulos, I., Kiefer, P., Raubal, M., Richter, K. F., & Thrash, T. (2014). Wayfinding decision situations: A conceptual model and evaluation. *Lecture Notes in Computer Science (Including Subseries Lecture Notes in Artificial Intelligence and Lecture Notes in Bioinformatics)*, 221–234.

Ginzburg, S. B., Schwartz, J., Gerber, R., Deutsch, S., Elkowitz, D. E., Ventura-Dipersia, C., Lim, Y. S., & Lucito, R. (2018). Assessment of medical students' leadership traits in a problem/case-based learning program. *Medical Education Online*, *23*, 1–8. https://doi.org/10.1080/10872981.2018.1542923

Goldberg, L. R. (1993). The structure of phenotypic personality traits. *American Psychologist*, *48*, 26–34. https://doi.org/10.1037/0003-066X.48.1.26

Goleman, D. (1998). What makes a leader? *Harvard Business Review*, *76*, 93–102.

Golledge, R. G. (1999). *Wayfinding behavior: Cognitive mapping and other spatial processes*. Baltimore, MD: Johns Hopkins University Press.

Greer, L. L., & Chu, C. (2020). Power struggles: When and why the benefits of power for individuals paradoxically harm groups. *Current Opinion in Psychology*, *33*, 162–166. https://doi.org/10.1016/j.copsyc.2019.07.040

Guilford, J. P., & Zimmerman, W. S. (1948). The Guilford-Zimmerman aptitude survey. *Journal of Applied Psychology*, *32*(1), 24–34. https://doi.org/10.1037/h0063610

Haghani, M., & Sarvi, M. (2017). Following the crowd or avoiding it? Empirical investigation of imitative behaviour in emergency escape of human crowds. *Animal Behaviour*, *124*, 47–56. https://doi.org/10.1016/j.anbehav.2016.11.024

Halpern, D. F. (2004). A cognitive-process taxonomy for sex differences in cognitive abilities. In *Current directions in psychological science*. https://doi.org/10.1111/j.0963-7214.2004.00292.x

Halpern, D. F., & Collaer, M. L. (2009). Sex differences in visuospatial abilities: More than meets the eye. In P. Shah & A. Miyake (Eds.), *The Cambridge handbook of visuospatial thinking* (pp. 170–212). Cambridge University Press. https://doi.org/10.1017/cbo9780511610448.006

Harris, C. B., Barnier, A. J., & Sutton, J. (2013). Shared encoding and the costs and benefits of collaborative recall. *Journal of Experimental Psychology: Learning Memory and Cognition*, *39*, 183–195. https://doi.org/10.1037/a0028906

Harris, C. B., Paterson, H. M., & Kemp, R. I. (2008). Collaborative recall and collective memory: What happens when we remember together? *Memory*, *16*, 213–230. https://doi.org/10.1080/09658210701811862

Hart, P. E., Nilsson, N. J., & Raphael, B. (1968). A formal basis for the heuristic determination of minimum cost paths. *IEEE Transactions on Systems Science and Cybernetics*, *4*(2), 100–107. https://doi.org/10.1109/TSSC.1968.300136

He, G., Ishikawa, T., & Takemiya, M. (2015). Collaborative navigation in an unfamiliar environment with people having different spatial aptitudes. *Spatial Cognition and Computation*, *15*, 285–307. https://doi.org/10.1080/13875868.2015.1072537

Hegarty, M. A., Richardson, A. E., Montello, D. R., Lovelace, K., & Subbiah, I. (2002). Development of a self-report measure of environmental spatial ability. *Intelligence*, *30*(5), 427–447.

Hegarty, M. A., & Waller, D. (2004). A dissociation between mental rotation and perspective-taking spatial abilities. *Intelligence*, *32*(2), 175–191. https://doi.org/10.1016/j.intell.2003.12.001

Helbing, D., Farkas, I., & Vicsek, T. (2000). Simulating dynamical features of escape panic. *Nature*, *407*(6803), 487–490. https://doi.org/10.1038/35035023

Hirtle, S. C. (2017). Wayfinding and orientation: Cognitive aspects of human navigation. In D. R. Montello (Ed.), *Handbook of behavioral and cognitive geography* (pp. 141–153). Cheltenham: Edward Elgar.

Hochmair, H. H. (1995). Towards a classification of route selection criteria for route planning tools. In P. Fisher (Ed.), *Developments in spatial data handling* (pp. 481–492). Springer-Verlag. https://doi.org/10.1007/b138045.

Hochmair, H. H., & Karlsson, V. (2005). Investigation of preference between the least-angle strategy and the initial segment strategy for route selection in unknown environments. In C. Freksa, M. Knauff, B. Krieg-Brckner, B. Nebel & T. Barkowsky (Eds.), *Spatial cognition IV: Reasoning, action, interaction* (Vol. 3343, pp. 79–97). Berlin: Springer.

Hölscher, C., Meilinger, T., Vrachliotis, G., Brösamle, M., & Knauff, M. (2006). Up the down staircase: Wayfinding strategies in multi-level buildings. *Journal of Environmental Psychology*, *26*(4), 284–299. https://doi.org/10.1016/j.jenvp.2006.09.002

Hund, A. M., & Minarik, J. L. (2009). Getting from here to there: Spatial anxiety, wayfinding strategies, direction type, and wayfinding efficiency. *Spatial Cognition and Computation: An Interdisciplinary Journal*, *6*(6:3), 179–201. https://doi.org/10.1207/s15427633scc0603_1

Hund, A. M., & Padgitt, A. J. (2010). Direction giving and following in the service of wayfinding in a complex indoor environment. *Journal of Environmental Psychology*. https://doi.org/10.1016/j.jenvp.2010.01.002

Hurlebaus, R., Basten, K., Mallot, H. A., & Wiener, J. M. (2008). Route learning strategies in a virtual cluttered environment. *Lecture Notes in Computer Science*, 104–120. https://doi.org/10.1007/978-3-540-87601-4_10

Hutchins, E. (1995a). *Cognition in the wild*. Cambridge, MA: The MIT Press.

Hutchins, E. (1995b). How a cockpit remembers its speeds. *Cognitive Science*, *19*(3), 265–288. https://doi.org/10.1207/s15516709cog1903_1

Hutchins, E. (2006). The distributed cognition perspective on human interaction. In *Roots of human sociality: Culture, cognition and interaction* (pp. 375–398). Oxford: Berg.

Hutchins, E. (2014). The technology of team navigation. In *Intellectual teamwork: Social and technological foundations of cooperative work* (pp. 191–220). London: Psychology Press.

Hutchins, E., & Palen, L. (1997). Constructing meaning from space, gesture, and speech. In L. B. Resnick, R. Säljö, C. Pontecorvo, & B. Burge (Eds.), *Discourse, tools and reasoning: Essays on situated cognition* (pp. 23–40). Springer. https://doi.org/10.1007/978-3-662-03362-3_2

Iaria, G., Petrides, M., Dagher, A., Pike, B., & Bohbot, V. D. (2003). Cognitive strategies dependent on the hippocampus and caudate nucleus in human navigation: Variability and change with practice. *The Journal of Neuroscience: The Official Journal of the Society for Neuroscience*, *23*, 5945–5952.

Ishikawa, T., & Montello, D. R. (2006). Spatial knowledge acquisition from direct experience in the environment: Individual differences in the development of metric knowledge and the integration of separately learned places. *Cognitive Psychology*, *52*(2), 93–129. https://doi.org/10.1016/j.cogpsych.2005.08.003

Janis, I. L. (1982). *Groupthink* (2nd ed.). Boston, MA: Houghton Mifflin.

John, O. P., & Srivastava, S. (1999). The big five trait taxonomy: History, measurement, and theoretical perspectives. In *Handbook of personality: Theory and research* (2nd ed., pp. 102–138). https://doi.org/citeulike-article-id:3488537

Johnson, S. D., & Bechler, C. (1998). Examining the relationship between listening effectiveness and leadership emergence: Perceptions, behaviors, and recall. *Small Group Research*, *29*, 452–471. https://doi.org/10.1177/1046496498294003

Joietz, D., & Kiefer, P. (2017). Uncertainty in wayfinding: A conceptual framework and agent-based model. In E. Clementini, M. Donnelly, M. Yuan, C. Kray, P. Fogliaroni, & A. Ballatore (Eds.), *13th International conference on spatial information theory* (pp. 1–15). Wadern: Schloss Dagstuhl-Leibniz-Zentrum fuer Informatic.

Karau, S. J., & Williams, K. D. (1993). Social loafing: A meta-analytic review and theoretical integration. *Journal of Personality and Social Psychology*, *65*, 681–706. https://doi.org/10.1037/0022-3514.65.4.681

Kim, T., McFee, E., Olguin, D. O., Waber, B., & Pentland, A. "Sandy." (2012). Sociometric badges: Using sensor technology to capture new forms of collaboration. *Journal of Organizational Behavior*, *33*(3), 412–427. https://doi.org/10.1002/job.1776

Klippel, A., & Winter, S. (2005). Structural salience of landmarks for route directions. In A. G. Cohn & D. M. Mark (Eds.), *Proceedings of the international conference, COSIT* (pp. 347–362). Berlin and Heidelberg: Springer.

Kozhevnikov, M., Motes, M. A., Rasch, B., & Blajenkova, O. (2006). Perspective-taking vs. Mental rotation transformations and how they predict spatial navigation performance. *Applied Cognitive Psychology*, *20*(3), 397–417. https://doi.org/10.1002/acp.1192

Kuipers, B. (1978). Modeling spatial knowledge. *Cognitive Science*, *2*(2), 129–153. https://doi.org/10.1016/S0364-0213(78)80003-2

Kuipers, B. (2000). Spatial semantic hierarchy. *Artificial Intelligence*, *119*, 191–233. https://doi.org/10.1016/S0004-3702(00)00017-5

Kuipers, B., Tecuci, D. G., & Stankiewicz, B. J. (2003). The skeleton in the cognitive map: A computational and empirical exploration. *Environment & Behavior*, *35*(1), 81–106. https://doi.org/10.1177/0013916502238866

Lawton, C. A. (1994). Gender differences in way-finding strategies: Relationship to spatial ability and spatial anxiety. *Sex Roles*, *30*(11–12), 765–779. https://doi.org/10.1007/BF01544230

Lawton, C. A. (1996). Strategies for indoor wayfinding: The role of orientation. *Journal of Environmental Psychology*, *16*(2), 137–145. https://doi.org/10.1006/jevp.1996.0011

Lawton, C. A. (2001). Gender and regional differences in spatial referents used in direction giving. *Sex Roles*, *44*(5–6), 321–337. https://doi.org/10.1023/A:1010981616842

Lawton, C. A., & Kallai, J. (2002). Gender differences in wayfinding strategies and anxiety about wayfinding: A cross-cultural comparison. *Sex Roles*, *47*(9–10), 389–401. https://doi.org/10.1023/A:1021668724970

Lemoine, G. J., Aggarwal, I., & Steed, L. B. (2016). When women emerge as leaders: Effects of extraversion and gender composition in groups. *Leadership Quarterly*, *27*, 470–486. https://doi.org/10.1016/j.leaqua.2015.12.008

Maguire, E. A., Burgess, N., Donnett, J. G., Frackowiak, R. S., Frith, C. D., & O'Keefe, J. (1998). Knowing where and getting there: A human navigation network. *Science*, *280*, 921–924. https://doi.org/10.1126/science.280.5365.921

Maguire, E. A., Intraub, H., & Mullally, S. L. L. (2015). Scenes, spaces, and memory traces: What does the hippocampus do? *The Neuroscientist*. https://doi.org/10.1177/1073858415600389

Marion, S. B., & Thorley, C. (2016). A meta-analytic review of collaborative inhibition and postcollaborative memory: Testing the predictions of the retrieval strategy disruption hypothesis. *Psychological Bulletin*, *142*, 1141–1164. https://doi.org/10.1037/bul0000071

Matthews, G., Deary, I. J., & Whiteman, M. C. (2009). Personality traits, Third edition. In *Personality traits* (3rd ed.). https://doi.org/10.1017/CBO9780511812743

McLaughlin, B. (1965). "Intentional" and "incidental" learning in human subjects: The role of instructions to learn and motivation. *Psychological Bulletin*. https://doi.org/10.1037/h0021759

Meneghetti, C., De Beni, R., Pazzaglia, F., & Gyselinck, V. (2011). The role of visuo-spatial abilities in recall of spatial descriptions: A mediation model. *Learning and Individual Differences*, *21*, 719–723. https://doi.org/10.1016/j.lindif.2011.07.015

Meneghetti, C., Ronconi, L., Pazzaglia, F., & De Beni, R. (2014). Spatial mental representations derived from spatial descriptions: The predicting and mediating roles of spatial preferences, strategies, and abilities. *British Journal of Psychology*, *105*, 295–315. https://doi.org/10.1111/bjop.12038

Money, J., Alexander, D., & Walker, H. T. (1965). *A standardized road-map test of direction sense*. Baltimore, MD: Johns Hopkins University Press.

Montello, D. R. (1998). A new framework for understanding the acquisition of spatial knowledge in large-scale environments. In M. J. Egenhofer & R. G. Golledge (Eds.), *Spatial and temporal reasoning in geographic information systems* (pp. 143–154). Oxford University Press. https://doi.org/10.1088/1748-6041/6/2/025001

Montello, D. R. (2005). Navigation. In P. Shah & A. Miyake (Eds.), *The Cambridge handbook of visuospatial thinking* (pp. 257–294). Cambridge: Cambridge University Press.

Montello, D. R., Lovelace, K. L., Golledge, R. G., & Self, C. M. (1999). Sex-related differences and similarities in geographic and environmental spatial abilities. *Annals of the Association of American Geographers*, *89*, 515–534. https://doi.org/10.1111/0004-5608.00160

Morrison, K. E., DeBrabander, K. M., Jones, D. R., Faso, D. J., Ackerman, R. A., & Sasson, N. J. (2019). Outcomes of real-world social interaction for autistic adults paired with autistic compared to typically developing partners. *Autism*. https://doi.org/10.1177/1362361319892701

Mullen, B., Johnson, C., & Salas, E. (1991). Productivity loss in brainstorming groups: A meta-analytic integration. *Basic and Applied Social Psychology*, *12*, 3–24. https://doi.org/10.1207/s15324834basp1201_1

Nazareth, A., Huang, X., Voyer, D., & Newcombe, N. (2019). A meta-analysis of sex differences in human navigation skills. *Psychonomic Bulletin & Review*, *26*(5), 1503–1528. https://doi.org/10.3758/s13423-019-01633-6

Northouse, P. G. (2019). *Introduction to leadership: Concepts and practice*. Newbury Park, CA: Sage Publications.

Pailhous, J. (1984). The representation of urban space: Its development and its role in the organisation of journeys. In R. Farr & S. Moscovici (Eds.), *Social representations* (pp. 311–327). Cambridge: Cambridge University Press.

Passini, R. (1981). Wayfinding: A conceptual framework. *Urban Ecology*, *5*(1), 17–31. https://doi.org/10.1016/0304-4009(81)90018-8

Passini, R. (1984). Spatial representations, a wayfinding perspective. *Journal of Environmental Psychology*, *4*(2), 153–164. https://doi.org/10.1016/S0272-4944(84)80031-6

Passini, R. (1992). *Wayfinding in architecture*. New York, NY: Van Nostrand Reinhold.

Pearson, A., Marsh, L., Hamilton, A., & Ropar, D. (2014). Spatial transformations of bodies and objects in adults with autism spectrum disorder. *Journal of Autism and Developmental Disorders*, *44*(9), 2277–2289. https://doi.org/10.1007/s10803-014-2098-6

Prestopnik, J. L., & Roskos-Ewoldsen, B. (2000). The relations among wayfinding strategy use, sense of direction, sex, familiarity, and wayfinding ability. *Journal of Environmental Psychology*, *20*(2), 177–191. https://doi.org/10.1006/jevp.1999.0160

Quaiser-Pohl, C., Geiser, C., & Lehmann, W. (2006). The relationship between computer-game preference, gender, and mental-rotation ability. *Personality and Individual Differences*, *40*, 609–619. https://doi.org/10.1016/j.paid.2005.07.015

Rajaram, S. (2011). Collaboration both hurts and helps memory: A cognitive perspective. *Current Directions in Psychological Science*, *20*, 76–81. https://doi.org/10.1177/0963721411403251

Raubal, M., & Worboys, M. (1999). A formal model of the process of wayfinding in built environments. *Spatial Information Theory: Cognitive and Computational Foundations of Geographic Information Science (COSIT 99)*, *1661*, 381–399. https://doi.org/10.1007/3-540-48384-5_25

Reilly, D., Mackay, B., Watters, C., & Inkpen, K. (2009). Planners, navigators, and pragmatists: Collaborative wayfinding using a single mobile phone. *Personal and Ubiquitous Computing*, *13*, 321–329. https://doi.org/10.1007/s00779-008-0207-2

Richardson, A. E., Dale, R., & Tomlinson, J. M. (2009). Conversation, gaze coordination, and beliefs about visual context. *Cognitive Science*, *33*, 1468–1482. https://doi.org/10.1111/j.1551-6709.2009.01057.x

Richardson, A. E., Montello, D. R., & Hegarty, M. (1999). Spatial knowledge acquisition from maps and from navigation in real and virtual environments. *Memory & Cognition*, *27*(4), 741–750. https://doi.org/10.3758/BF03211566

Richardson, A. E., Powers, M. E., & Bousquet, L. G. (2011). Video game experience predicts virtual, but not real navigation performance. *Computers in Human Behavior*, *27*(1), 552–560. https://doi.org/10.1016/j.chb.2010.10.003

Saloman, G. (1993). *Distributed cognitions: Psychological and educational considerations. Learning in doing: Social, cognitive, and computational perspectives*. Cambridge: Cambridge University Press.

Sandstrom, N. J., Kaufman, J., & Huettel, S. A. (1998). Males and females use different distal cues in a virtual environment navigation task. *Brain Research. Cognitive Brain Research*, *6*(4), 351–360. https://doi.org/10.1016/s0926-6410(98)00002-0

Saucier, D. M., Green, S. M., Leason, J., MacFadden, A., Bell, S., & Elias, L. J. (2002). Are sex differences in navigation caused by sexually dimorphic strategies or by differences in the ability to use the strategies? *Behavioral Neuroscience*. https://doi.org/10.1037/0735-7044.116.3.403

Schmitz, S. (1997). Gender-related strategies in environmental development: Effects of anxiety on wayfinding in and representation of a three-dimensional maze. *Journal of Environmental Psychology*, *17*, 215–228. https://doi.org/10.1006/jevp.1997.0056

Schmitz, S. (1999). Gender differences in acquisition of environmental knowledge related to wayfinding behavior, spatial anxiety and self-estimated environmental competencies. *Sex Roles*, *41*, 71–93. https://doi.org/10.1023/A:1018837808724

Sharples, S., Cobb, S., Moody, A., & Wilson, J. R. (2008). Virtual reality induced symptoms and effects (VRISE): Comparison of head mounted display (HMD), desktop and projection display systems. *Displays*, *29*, 58–69. https://doi.org/10.1016/j.displa.2007.09.005

Siegel, A. W., & White, S. H. (1975). The development of spatial representations of large-scale environments. *Advances in Child Development and Behavior*, *10*(C), 9–55. https://doi.org/10.1016/S0065-2407(08)60007-5

Spence, I., & Feng, J. (2010). Video games and spatial cognition. *Review of General Psychology*, *14*(2), 92–104. https://doi.org/10.1037/a0019491

Spiers, H. J., & Maguire, E. A. (2006). Thoughts, behaviour, and brain dynamics during navigation in the real world. *Neuro Image*, *31*(4), 1826–1840. https://doi.org/10.1016/j.neuroimage.2006.01.037

Stroebe, W., & Diehl, M. (1994). Why groups are less effective than their members: On productivity losses in idea-generating groups. *European Review of Social Psychology*, *5*, 271–303. https://doi.org/10.1080/14792779543000084

Tanenhaus, M. K., Spivey-Knowlton, M. J., Eberhard, K. M., & Sedivy, J. C. (1995). Integration of visual and linguistic information in spoken language comprehension. *Science*, *268*, 1632–1634. https://doi.org/10.1126/science.7777863

Taylor, H. A. A., & Brunyé, T. T. (2013). The cognition of spatial cognition: Domain-general within domain-specific. In B. Ross (Ed.), *The psychology of learning and motivation* (Vol. 58, pp. 77–116). Academic Press. https://doi.org/10.1016/B978-0-12-407237-4.00003-7

Tenbrink, T. (2015). Cognitive discourse analysis: Accessing cognitive representations and processes through language data. *Language and Cognition*, *7*, 98–137. https://doi.org/10.1017/langcog.2014.19

Tenbrink, T., Bergmann, E., & Konieczny, L. (2011). *Wayfinding and description strategies in an unfamiliar complex building*. Proceedings of the 33rd Annual Conference of the Cognitive Science Society.

Tenbrink, T., Taylor, H. A., Brunyé, T. T., Gagnon, S. A., & Gardony, A. L. (2020). Cognitive focus affects spatial decisions under conditions of uncertainty. *Cognitive Processing*. https://doi.org/10.1007/s10339-020-00952-0

Terlecki, M. S., & Newcombe, N. S. (2005). How important is the digital divide? The relation of computer and videogame usage to gender differences in mental rotation ability. *Sex Roles*, *53*, 433–441. https://doi.org/10.1007/s11199-005-6765-0

Thorndyke, P. W. (1981). Distance estimation from cognitive maps. *Cognitive Psychology*, *13*(4), 526–550. https://doi.org/10.1016/0010-0285(81)90019-0

Thorndyke, P. W., & Hayes-Roth, B. (1982). Differences in spatial knowledge acquired from maps and navigation. *Cognitive Psychology*, *14*(4), 560–589.

Toader, A. F., & Kessler, T. (2018). Team mental models, team goal orientations, and information elaboration, predicting team creative performance. *Creativity Research Journal*. https://doi.org/10.1080/10400419.2018.1530912

Walkowiak, S., Foulsham, T., & Eardley, A. F. (2015). Individual differences and personality correlates of navigational performance in the virtual route learning task. *Computers in Human Behavior*. https://doi.org/10.1016/j.chb.2014.12.041

Weisberg, S. M., Schinazi, V. R., Newcombe, N. S., Shipley, T. F., & Epstein, R. A. (2014). Variations in cognitive maps: Understanding individual differences in navigation. *Journal of Experimental Psychology. Learning, Memory, and Cognition*, *40*(3), 669–682. https://doi.org/10.1037/a0035261

Wiener, J. M., Büchner, S. J., & Hölscher, C. (2009). Taxonomy of human wayfinding tasks: A knowledge-based approach. *Spatial Cognition & Computation*, *9*(2), 152–165. https://doi.org/10.1080/13875860902906496

Wiener, J. M., & Mallot, H. A. (2003). "Fine-to-coarse" route panning and navigation in regionalized environments. *Spatial Cognition and Computation*, *3*(4), 331–358. https://doi.org/10.1207/s15427633scc0304_5

Wiener, J. M., Schnee, A., & Mallot, H. A. (2004). Use and interaction of navigation strategies in regionalized environments. *Journal of Environmental Psychology*, *24*(4), 475–493. https://doi.org/10.1016/j.jenvp.2004.09.006

Wilson, D. S., Timmel, J. J., & Miller, R. R. (2004). Cognitive cooperation: When the going gets tough, think as a group. *Human Nature*, *15*(3), 225–250. https://doi.org/10.1007/s12110-004-1007-7

Wolbers, T., & Hegarty, M. A. (2010). What determines our navigational abilities? *Trends in Cognitive Sciences*, *14*(3), 138–146. https://doi.org/10.1016/j.tics.2010.01.001

Wu, L., Waber, B. N., Aral, S., Brynjolfsson, E., & Pentland, A. (2008). *Mining face-to-face interaction networks using sociometric badges: Predicting productivity in an IT configuration task* (SSRN Scholarly Paper ID 1130251). Social Science Research Network. https://doi.org/10.2139/ssrn.1130251

Xia, J., Packer, D., & Dong, J. (2009). *Individual differences and tourist wayfinding behaviors*. Proceedings of the 18th World IMACS Congress and MODSIM09 International Congress on Modelling and Simulation, pp. 1272–1278.

Zaccaro, S. J., Dubrow, S., & Kolze, M. (2018). Leader traits and attributes. In *The nature of leadership* (pp. 29–55). Newbury Park, CA: Sage Publications.

Zacks, J. M., Mires, J. O. N., Tversky, B., & Hazeltine, E. (2000). Mental spatial transformations of objects and perspective. *Spatial Cognition and Computation*, *2*(4), 315–332. https://doi.org/10.1023/A:1015584100204

Zhang, H., Zherdeva, K., & Ekstrom, A. D. (2014). Different "routes" to a cognitive map: Dissociable forms of spatial knowledge derived from route and cartographic map learning. *Memory and Cognition*, *42*(7), 1106–1117. https://doi.org/10.3758/s13421-014-0418-x

4 Facilitating Collective Cognition During Group Wayfinding Through the Human Eye

Ioannis Giannopoulos and Daniel R. Montello

Introduction

Wayfinding requires individuals to make a series of spatial decisions. These decisions can be of recreational nature (e.g., finding the way to a restaurant while exploring Rome for the first time) or essential for survival (e.g., finding an escape route in case you get lost while exploring the Roman catacombs). In many cases, these decisions may involve more than one person and function at different temporal scales. For instance, tourists exploring Rome can afford to make mistakes and explore the city, while those lost in the catacombs may wish to locate an exit as soon as possible. Despite the differences in the two scenarios described earlier, spatial decision-making in unfamiliar environments is often supported by navigation aids.

Unlike individual navigation, the group decision-making process can be hampered by the consensus often required among group members. Every group member brings a different expertise and knowledge level relevant for the task at hand. Leveraging this group member knowledge can improve decision-making (Bonner & Bolinger, 2013). Unfortunately, however, group decision-making can suffer from overconfident members without much knowledge who cannot easily be distinguished (Littlepage & Mueller, 1997).

There is considerable research on group navigational decision-making in the animal kingdom, showing roughly similar functional characteristics to humans (Conradt & Roper, 2005; Couzin & Krause, 2003; Sumpter, 2006) but also following different patterns, which can be very inspiring and useful for designing assistance systems. For instance, ants use their numbers to collectively expand their coverage area (Gelblum et al., 2020), adapting their movement in response to a chemical stimulus, that is, chemotaxis. Ants modify their environment in order to communicate and coordinate their individual activities, by following paths with higher concentrations of pheromones (Theraulaz & Bonabeau, 1999). In nature, we often observe a very synchronized behavior of animals, giving the impression of the group following a leader (Couzin & Krause, 2003; Selous, 1932) or even the false impression of the existence of telepathic properties. We know that this behavior is feasible through sensory modalities such as sound, pressure, and vision, when animals align their behavior based on their peers in close proximity (Couzin,

DOI: 10.4324/9781003202738-6

2007). The individuals in such groups differ concerning their information state, that is, their own informational status and their awareness of the informational state of others (Couzin, 2007).

Consensus decisions are common in social species, allowing them to resolve significant conflicts of interest and jointly act to a solution (Conradt & Roper, 2005). According to Sasaki and Pratt (2012), highly integrated groups have a larger cognitive capacity, which can help overcome individual limitations by reducing cognitive load during decision-making. Furthermore, groups are typically able to proceed with complex problems more efficiently than individuals with limited access to information (Kearns et al., 2006), thus enjoying the benefits of pooling information and correcting errors; this shows group superiority over individuals (Hill, 1982).

This chapter will discuss, from a technologically driven perspective, how observing the human eyes can be utilized to investigate and facilitate collective cognition during group wayfinding, serving as a mediator for consensus decisions. We first discuss the wayfinding process and assistance systems at an individual level and then at the group level.

The Process of Wayfinding

Navigation is coordinated and goal-directed travel through the environment (Montello, 2005); it includes both wayfinding and locomotion. When navigating, individuals may acquire different types of spatial knowledge about their environment. Indeed, an important branch of wayfinding research is dedicated to the investigation of the cognitive processes involved in the acquisition of spatial knowledge in both familiar and unfamiliar environments (Kuipers, 1982). Here, researchers have proposed that spatial knowledge can be acquired in discrete stages (Siegel & White, 1975) characterized by a strict progression from landmark knowledge to route knowledge to configurational knowledge of their environment or in a continuous manner, in which landmark, route, and survey knowledge are acquired simultaneously with experience (Montello, 1998; Schinazi & Epstein, 2010; Schinazi et al., 2013).

We utilize these abilities and knowledge during decision-making, which is one of the main foci in wayfinding research (Passini, 1981). Decision-making, execution, and information processing are the three main wayfinding processes, according to Arthur and Passini (1992). Downs and Stea (1977) broke the wayfinding process into four steps, namely orientation, route choice, monitoring the track, and recognition of the destination. Several factors can influence wayfinding performance and the maintenance of orientation during travel in environments, such as visual access and the degree of differentiation of appearance (Weisman, 1981).

There have been many empirical studies investigating the process of wayfinding, spatial knowledge acquisition, and the impact of assistance systems. Most of these studies utilize some combination of think-aloud protocols, performance measures, and error metrics in order to study human navigation. Eye tracking is also used frequently in recent years as a research tool for investigating human behavior during navigation, especially focusing on understanding the cognitive processes of navigation and learning (Wiener et al., 2011).

Eye tracking as a research tool has mostly been utilized for research in controlled laboratory settings, providing valuable insights for many aspects of wayfinding and learning. Wiener et al. (2011) demonstrated that gaze behavior reflects decision-making during wayfinding; Emo (2012) highlighted the role of spatial configuration during wayfinding. This tool has also been used to investigate interaction with geospatial assistance systems (e.g., Kiefer & Giannopoulos, 2012), for evaluating visual strategies and the use of interactive maps (Çöltekin et al., 2009, 2010), and for evaluating map perception (Opach & Nossum, 2011).

Wayfinding Assistance

Next, to research investigating the wayfinding processes, there is a large research corpus on wayfinding assistance. Cognitive work refers to demanding mental processing during decision-making; assistance systems can reduce complexity by off-loading cognitive work (Clark, 1997; Scaife & Rogers, 1996). In its most traditional form, cartographic maps have been utilized in order to off-load cognitive work. Cognitive load should be considered when designing assistance systems. "Cognitive load theory" (Bunch & Lloyd, 2006; Chandler & Sweller, 1991) allows assessing critical components during the design process, which depend on several factors, including the cognitive abilities of individuals, the nature of the current task, the surrounding environment, and the interactions involved. According to Allen (1999), the spatial abilities of the individual are crucial for successful wayfinding (see Ishikawa & Montello, 2006); they may vary according to gender, age, the capacity of working memory and reasoning strategies, and other factors (Wolbers & Hegarty, 2010). Thus, it is important that assistance systems are designed to be able to adapt according to the individual's abilities and the current task. A large number of factors can be considered important for adaptive systems, such as the complexity of the surrounding environment (Giannopoulos et al., 2014) and the mode of transportation (Reichenbacher, 2007). Raubal and Panov (2009) introduced a model for map adaptation on mobile phones, which considers the current context, the user, and the task.

The adaptive nature of such systems is not relevant only for background processes, that is, the logic of the system, but also for foreground processes, that is, the graphical user interface, which serves as a means for communicating with the user by enabling interaction dialogues that obtain and manipulate provided information. Such adaptive interfaces focus on reducing the cognitive load during interaction with such systems.

Adaptive interfaces support users in achieving their goals without requiring them to learn and adapt to the system. Hutchins et al. (1985) consider an effective interface as one in which the user has to interact only with elements relevant to their current task. According to Rouse (1988), "There are three general methods for aiding a user: (1) an aid can make a task easier, (2) an aid can perform part of a task, and (3) an aid can completely perform a task" (p. 439). Individual differences are very important for adaptive interfaces, especially individual differences in cognition and personality (Raubal, 2009). Benyon (1993) suggested that an interface should adapt on the basis of the needs of a group of users.

Nowadays, it is easy to utilize an assistance aid for almost anything. Many of us ask YouTube how to fix the doorbell, we ask Siri whether we should take an umbrella with us, and we rely on Google Maps to find and navigate to a nice restaurant. As an example, a trio of foreign tourists in an unfamiliar environment would experience difficulties in communicating with locals and would coordinate and prioritize their preferences, getting a city experience on foot without a digital assistance system. In some cases, they could be successful with a typical cartographic paper map. Many assistance systems today know our preferences and what we are currently doing (Kiefer et al., 2013) and can adapt in real time based on behavioral changes (e.g., detecting interest through eye movements and eye gaze or even our cognitive load [Duchowski et al., 2018]). Furthermore, such a system can compute a nice route through the city in accordance with our preferences, it can help us communicate with locals, it can collaborate with other assistance systems (i.e., handling the collective spatial cognition of a group by retrieving the context and knowledge states of members), and it can help us learn the surrounding environment in an implicit way (Schwering et al., 2017). An assistance system can adapt its supportive behavior based on the collective strategy of the trio or even based purely on the strategy of an individual. Gaze-based interaction (among other types of interactions, such as mobility patterns) of tourists with the surrounding environment could eventually be used to compute and inform the assistance system about the spatial knowledge acquired by the individuals and thus by the whole team. In return, the assistance system could increase or decrease the level of assistance, changing the spatial behavior of the tourists, and helping to bolster collective cognition.

A (collective) assistance system could handle individual differences at an implicit level (i.e., without explicit user input), providing a different level of assistance to each user. Several human factors could be considered, such as spatial abilities (Hegarty et al., 2002) and the current cognitive load of the tourists, providing them with a common base for making decisions as a team and reducing confusion and uncertainty related to individual differences among group members.

Eye Tracking for Research and Application

Eye tracking is a valuable tool utilized for research and application in multiple disciplines. It can be utilized as a remote technology placed under a screen monitor, usually in laboratory settings, as well as a mobile system, facilitating research in outdoor spaces and, in general, in scenarios where human mobility is required. Research with remote eye-tracking technology has yielded considerable research insights, but since this chapter is about outdoor wayfinding, we focus on the potential of mobile systems.

Mobile eye-tracking devices look almost like regular correction glasses or sunglasses. They are equipped with several tiny cameras in order to track the user's eye movements as well as the surrounding environment through the field view camera. The captured eye movements can further be processed as eye events and overlaid on a video recording (i.e., the field of view recording) to visualize the point of gaze in the surrounding environment of the user. The most relevant eye events are the so-called *fixations* and *saccades*. Fixations occur when our eyes remain relatively still for a certain period of time and can indicate the regions in our surroundings that are currently subject to cognitive processing (i.e., the "eye-mind assumption" [Just & Carpenter, 1976]). Saccades are the rapid eye movements between fixations, often utilized in machine learning methods as features for predictions.

In the research domain of human–computer interaction (HCI), mobile eye trackers are often used as an input method, enabling gaze-based interaction with mobile devices. The users' gaze can be utilized for typing purposes and information selection (Majaranta et al., 2009), for interface adaptation purposes (Kiefer et al., 2016), and to provide orientation cues during mobile phone interaction (Giannopoulos et al., 2012), among other uses. In combination with mobile phone devices, mobile eye trackers have also been employed to enable users to interact with their surrounding environment (Anagnostopoulos et al., 2017; Giannopoulos et al., 2015).

In the fields of spatial cognition and geographic information science, mobile eye trackers are usually utilized in order to investigate how users interact with their surrounding environment, focusing on understanding human behavior during wayfinding, among other uses. For instance, Kiefer et al. (2014) investigated the process of self-localization in unfamiliar environments, recording eye movements to reveal and analyze different strategies between successful and unsuccessful travelers. Wenczel et al. (2017) investigated visual attention to landmarks, while travelers followed a memorized route. Kiefer et al. (2017) presented an overview of eye tracking for spatial research and discussed research challenges and opportunities for eye tracking. Schwarzkopf et al. (2017) introduced *gaze angle analysis*, demonstrating that partners during collaborative wayfinding tend to shift their gaze toward their peers, contradicting the idea that there is a "spatial division-of-labor" to different areas of the environment during group search.

Research utilizing eye tracking as a means to interact and as an analysis tool has revealed the potential of this technology to help us understand better how humans acquire and use knowledge concerning their surrounding environment. Furthermore, assistance systems can employ this technology to allow for explicit and implicit interaction with a device and with the surrounding environment. Such a system could support the user in a multitude of ways and furthermore, also multiple users through integrative methods, thus being beneficial for collective spatial cognition. Some of the most relevant capabilities of such a gaze-based assistance system include the detection of the spatio-temporal context and task of the user as well as the associated cognitive load.

Gaze-Based Assistance and Human Behavior

The processing and output of a gaze-based assistance system can be based on *content-dependent* information, that is, the elements in the environment gazed at, as well as on content-independent information, that is, the eye movements of the user. Employing content-dependent information, which is a very challenging computational and retrieval task (see the Discussion), may allow the system to detect user interests by analyzing the interactions with the device as well as with elements within the surrounding environment, providing targeted support. *Content-independent* information, that is, eye movements, is easy to retrieve and has been a valuable source for automatic detection of human activities and prediction. Already since the 1960s, we know that the gaze behavior of humans during scene perception is task-dependent (Yarbus, 1967), which is also the case for map perception (Steinke, 1987). For instance, our eye movements exhibit different behavior when we are asked to plan a route between two locations on a cartographic map compared to a task where we are asked to find a specific location. During the route task, we will probably look at the start and end points and afterward inspect different possible routes in-between these points, resulting in saccades between these two locations. In contrast, during the search task, we will explore the map, resulting in more unstructured gaze behavior, having a gaze sequence distributed all over the map. In both cases, the output of such an assistance system can be explicit or implicit. In the first case, the system reacts to the explicit input of the user. In the second case, the system reacts based on a context or state chance.

Assistance systems utilizing eye movements have revealed promising results over the years. Kiefer et al. (2013) demonstrated that certain map activities (e.g., route planning and search) can be successfully predicted using only eye-movement data. They extracted features from collected eye-movement data and trained a classifier to predict the current cartographic task of the user within only a few seconds of interaction with the map. In addition to gaze fixations and saccades, in recent years, the human pupil is getting increased attention. Research in pupillometry has yield promising results concerning automatic detection of cognitive load. Duchowski et al. (2018) showed that they were able to measure changes in cognitive load under controlled conditions. They proposed a metric termed the "index of pupillary activity." During controlled experiments, participants were instructed to fixate a central target and, at the same time, solve mental arithmetic tasks of varying complexity. The pupil diameter oscillation was successfully used as an indicator of cognitive load. Although pupillometry studies show promising results in controlled environments, the development of algorithmic filters that would facilitate the use of these findings in outdoor environments is still an open challenge.

In the last decade, eye tracking has also been successfully utilized in order to enable interaction with the surrounding environment of a user, focusing on content-dependent support. Anagnostopoulos et al. (2017) demonstrated the feasibility and importance of gaze-informed location-based services. Their work provides a method that allows mapping the user's gaze in real time to surrounding objects (e.g., buildings). This approach provides the ability for users to directly interact with their surroundings, for example, receiving information for an object of interest by looking at it for a defined period of time (e.g., interest threshold). Giannopoulos et al. (2015) followed a similar approach and presented a gaze-based navigation assistance aid named "GazeNav," which could also be extended to serve as a mediator for collective spatial cognition. Users can interact through GazeNav with the surrounding environment and receive navigation instructions. These instructions are limited to simple mobile phone vibrations. The initialization process of this assistance system is very similar to that of the typical map-based turn-by-turn system. Users have to type in their destination, and instead of keeping their mobile phone in their hand, which requires them to constantly look at it in order to consume relevant information, GazeNav allows the user to just put the mobile phone in their pocket (i.e., a non-visual navigation approach). While users walk and interact with their surrounding environment, once they gaze at the correct street segment (i.e., the one that leads toward their desired destination), the mobile phone in their pocket starts vibrating. This vibration indicates to the user that the gazed street is the one that has to be followed in order to reach the destination.

Group Wayfinding

Wayfinding can be influenced by the past and current actions of other people in the surroundings, and thus, to a certain degree, it can also be classified as a social activity. The focus of this chapter is on group wayfinding, of the type Dalton et al. (2019) termed "strong synchronous" social wayfinding. The member makeup of wayfinding groups may vary in terms of gender, spatial ability, personality, social status, and more. Since group wayfinding generally requires extensive coordination and communication, this collaboration can be quite difficult, even in dyads (He et al., 2015). Below, we discuss how recent research and technological innovations of the last few years can be leveraged in order to assist in the investigation and support of group wayfinding, in group sizes of two or beyond.

Group Wayfinding Assistance

Of course, like individual travelers, groups engaged in wayfinding will typically employ wayfinding assistance systems. In fact, as Dalton et al. (2019) identified, such systems themselves reflect social and cultural influences of system designers, existing cultural practices, and so on. They themselves thus constitute expressions of the social or the collective in spatial cognition.

But here, we focus on the active use of wayfinding assistance systems during strong synchronous wayfinding by groups of travelers. Even when wayfinding activities are outsourced to an assistance system, decision-making by navigating groups relies on consensus, from route planning to following a route and recognizing the destination. This consensus decision-making is important in order to keep the group together (Conradt & Roper, 2005). Consensus decision-making requires successful communication among group members, which in turn requires a constantly updated common ground concerning their knowledge status (Clark & Brennan, 1991). Typical assistance systems do not incorporate the context of peer wayfinders, operating at an individualized and personalized level.

Research on gaze-based assistance systems can be valuable in supporting group wayfinding, acting as a mediator between the involved wayfinders. In the case of content-dependent eye tracking, the elements wayfinders are looking at in the environment, for how long and how often, can help an assistance system build an internal representation of the spatial knowledge of each group member and, thus, compose a more holistic representation for the whole group (see Figure 4.1). Based on this representation, which can be seen as the mental representation of the environment for the whole group, an assistance system can act as a mediator at both the individual and group levels. Figure 4.1 shows the use case of a group mental representation of the environment. It remains a challenging research question as how to approximate the mental representation of a user based on tracked gaze events. In post-analyses, typical methods for this approximation rely on verbal protocols and sketch map analyses (e.g., Schwering et al., 2014). Under the assumption that the elements in the environment that are part of the mental representation of a human, for example, signs, landmarks, and street segments, have to be gazed at, content-dependent gaze data might be a valuable source for automatic approximation. If the gaze of the individual wayfinder in the surrounding environment can be mapped with adequate accuracy, it may be possible to approximate the actual mental representation first at an individual and then at a group level. Achieving this goal will facilitate research in the area of human behavior as well as in the area of HCI, allowing us to better understand the process of group wayfinding and collective cognition, among other processes.

Group wayfinding can range from recreational activities to disaster management. An example of the latter is the scenario of firefighters working in a disaster area by exploring the ruins for survivors. An assistance system incorporating the

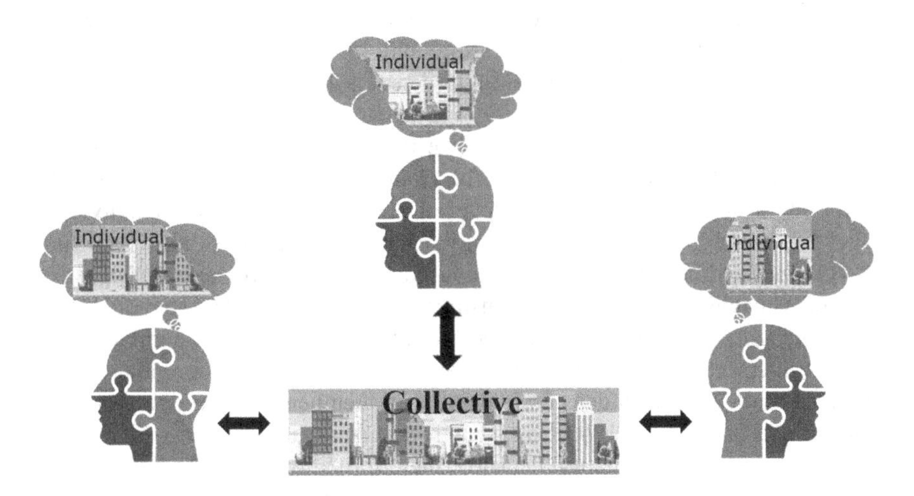

Figure 4.1 Parts of the mental representations of individuals could be detected and organized by an assistance system, which uses eye-tracking technology, but this would hold only if eye-tracking data are sufficient to understand which environmental information is being consumed and mentally organized by each user (the figure was created using CC0 1.0 images from rawpixel.com).

abilities we discuss earlier would know which areas have been already explored by the collection of firefighters through content-dependent gaze analyses, that is, the individual's ideas about the environment could be combined into a holistic mental representation of the group. In turn, such an assistance system would guide its user to unexplored areas.

Systems such as GazeNav (Giannopoulos et al., 2015) can support group wayfinding by providing non-visual feedback to the wayfinders based on what they and their peers are looking at, and resolving ambiguities and uncertainty due to miscommunication (Bae & Montello, 2019). Such a system can also adapt the modality of assistance based on the type of environment, for example, high-density city, ruins, and wilderness. For instance, it can provide instructions in a form that suits the pace of the wayfinding task, for example, when decision-making must be carried out under time pressure (Zheng, 2012).

Recording the wayfinders' gazes can reveal their interest in specific elements in the environment as well as in elements displayed on a visual assistance system. For instance, a system might detect that the majority of the wayfinders in the group showed interest in a specific landmark (Anagnostopoulos et al., 2017), thus providing wayfinding instructions that incorporate this information, for example, *"turn left after the church" or "this area is still unexplored."* This type of assistance could potentially facilitate consensus decisions by providing instructions that are grounded on a common knowledge base. In cases of travel for recreational purposes, wherein groups may want to take the most scenic route instead of the shortest one, the assistance system could guide the wayfinders through a route that best reflects the visual interests of the group.

Of course, for centuries, wayfinders have often used maps in order to reach destinations. Prior research on eye movements and artificial intelligence has shown that activity recognition on cartographic maps is feasible. Kiefer et al. (2013) demonstrated that it is possible to detect at least six activities on cartographic maps, such as route planning and place search. An assistance system having access to such a model can utilize this information in order to understand the context of each wayfinder in the group; furthermore, by analyzing their pupils, the system can identify the level of cognitive load for each individual (Duchowski et al., 2018). Artificial intelligence approaches have also been successfully utilized in order to predict the turn activity of individuals during wayfinding. Alinaghi et al. (2021) demonstrated that it is possible to predict from eye movements whether a wayfinder will turn left or right or go straight ahead at the next intersection. This information can be valuable for a system that acts as a mediator, allowing it to detect whether group members are about to make the same turn decision or not. If such a system recognizes that all members of a group are about to make the same turn decision, the information exchange between the system and the group members could be kept to a minimum in order to avoid distraction. On the other hand, if the system detects a mismatch, it could adapt in order to support a consensus by prioritizing the context of the most relevant group members.

Understanding Group Wayfinding

Above, we considered applying the measurement of human gaze to two scenarios, one involving time pressure and one involving a recreational tour. Gaze measurement can also be utilized as a tool in order to understand group wayfinding and collective cognition. When investigating collective cognition, it is important to understand the interaction between the individual abilities of humans, the knowledge they share, as well as the structure of their communications. Content-dependent and content-independent gaze data can facilitate this understanding in multiple ways.

Having a holistic mental representation of the environment can facilitate the study of the group, understanding the structure of their communication based on their shared knowledge, which can be inferred from patterns of gaze. The main advance that this represents over current research is assessment of the dynamic nature of the approximated mental representations of the wayfinders, allowing us to observe the development of the mental representation over time (instead of just at the end of an experiment). This allows us to associate specific activities and the current context of the user with the elements of the representation being developed.

During wayfinding, we carry out multiple reoccurring cognitive activities, including route planning, self-localization and monitoring, and recognition of the destination (Downs & Stea, 1977). Activity recognition models could be trained and utilized in order to create activity sequences over time, which would allow us to understand the process of wayfinding—when, for how long, and how often does a specific activity occur. They would further allow us to make inferences about the development of mental representations and collective cognition over time. Such models could further support the study of collective cognition by investigating the turning behavior and cognitive load of wayfinders in familiar and unfamiliar environments, allowing us to explore environmental factors that affect collective cognition.

Discussion and Outlook

This chapter highlights the possibilities of eye-tracking technology as an assistance aid for collective wayfinding, as well as a research tool for investigating collective cognition. The application scenarios we presented demonstrate how the current level of technology can help enable interaction among wayfinders and between the environment, facilitating collective spatial cognition, and serving as a mediator for the acquisition, organization, exchange, and utilization of spatial data and knowledge, supporting consensus decision-making.

The gaze-based content-dependent scenarios we introduced rely on the annotation of gaze fixations during travel in real environments. For interactive scenarios, this will have to be automatic and in real-time. In non-interactive research on human behavior, this does not necessarily need to be in real time, but annotations should still be computed automatically in order to avoid excessive manual coding, typically a substantial bottleneck to research progress. This is a very challenging task, but current approaches in eye tracking, computer vision, and deep learning encourage the possibility that this will become reality in the next few years (Deane et al., 2022; Müller & Mann, 2021).

An assistance aid is no more than what it is designed and programed to be, and we can programmatically change it, and limit or increase its capabilities. For instance, we can implement an assistance system to recognize bad navigation strategies and teach us a better, more efficient and effective, strategy in an implicit way. An example of such implicit assistance is suggested by the example of group wayfinding in the animal kingdom we discussed in our introduction: the pheromones used by ants that signal the way for other ants. By analogy, the human gaze may be utilized by the system similar to the pheromones, somewhat like information system "bread crumbs" (Gloor, 1997) (ironically, this term for a design feature that helps maintain orientation in cyberspaces was originally borrowed metaphorically from Hansel and Gretel's orientation in real environmental spaces, which are the focus of the wayfinding systems we discuss in this chapter). Like wayfinding pheromones, the distribution and concentration of eye movements and gaze in the environment could play a significant role for wayfinding assistance systems. Strategies like those used by ants can inspire feasible design features in human assistance systems. That said, it cannot be denied that relying on an assistance aid blindly can, under some circumstances, be catastrophic (e.g., the battery dies, devices are lost or broken, and localization does not work properly while escaping a battle field or evacuation area).

A lot of research is still necessary in this area, and we are currently among the researchers investigating the feasibility and impact of such assistance systems. We are currently conducting or planning experiments to investigate the impact of such an assistance system on the collective spatial cognition. Besides developing such tools, it is important to understand how such systems affect our spatial cognition as individuals, as well as collectives. Understanding how and to what extent we change our competences when offloading tasks to such a system is a question that continues to receive attention (e.g., Ishikawa, 2019). Researchers are also investigating the possibility that such assistance systems can be designed to support our wayfinding and, at the same time, increase our wayfinding competences and spatial knowledge acquisition (e.g., Gramann et al., 2017). It would be valuable and interesting to extend these efforts specifically to scenarios involving collective wayfinding. Another worthy research question relates to the communication and exchange of relevant information within navigating groups using assistance systems. What types of interaction dialogues and visualizations can best communicate relevant information among group members? Finally, is it possible to evaluate the quality of spatial knowledge obtained during travel from eye tracking alone? One of the most challenging tasks we currently face is to identify which of the gazed elements in the environment are actually part of the mental representations of individual wayfinders and how these individual representations can be optimally merged into a collective representation.

References

Alinaghi, N., Kattenbeck, M., Golab, A., & Giannopoulos, I. (2021). Will you take this turn? Gaze-based turning activity recognition during navigation. In K. Janowicz & J. A. Verstegen (Eds.), *11th International conference on geographic information science (GIScience 2021) Part II* (Vol. 208, pp. 5:1–5:16). Dagstuhl, Germany: Schloss Dagstuhl–Leibniz-Zentrum fu¨r Informatik.

Allen, A. L. (1999). Spatial abilities, cognitive maps, and wayfinding: Bases for individual differences in spatial cognition and behavior. In R. G. Golledge (Ed.), *Wayfinding behavior: Cognitive mapping and other spatial processes* (pp. 46–80). Baltimore, MD: The Johns Hopkins University Press.

Anagnostopoulos, V., Havlena, M., Kiefer, P., Giannopoulos, I., Schindler, K., & Raubal, M. (2017). Gaze-informed location-based services. *International Journal of Geographical Information Science*, *31*(9). doi:10.1080/13658816.2017.1334896

Arthur, P., & Passini, R. (1992). *Wayfinding: People, signs, and architecture*. Toronto: McGraw-Hill Ryerson.

Bae, C. J., & Montello, D. R. (2019). Dyadic route planning and navigation in collaborative wayfinding. In S. Timpf, C. Schlieder, M. Kattenbeck, B. Ludwig, & K. Stewart (Eds.), *14th International conference on spatial information theory* (pp. 24:1–24:20). Article No. 24, Proceedings of COSIT '19. doi:10.4230/LIPIcs.COSIT.2019.24

Benyon, D. (1993). Accommodating individual differences through an adaptive user interface. *Human Factors in Information Technology*, *10*, 149–149.

Bonner, B. L., & Bolinger, A. R. (2013). Separating the confident from the correct: Leveraging member knowledge in groups to improve decision making and performance. *Organizational Behavior and Human Decision Processes*, *122*(2), 214–221. doi:10.1016/j.obhdp.2013.07.005

Bunch, R. L., & Lloyd, R. E. (2006). The cognitive load of geographic information. *The Professional Geographer*, *58*(2), 209–220. doi:10.1111/j.14679272.2006.00527.x

Chandler, P., & Sweller, J. (1991). Cognitive load theory and the format of instruction. *Cognition and Instruction*, *8*(4), 293–332. doi:10.1207/s1532690xci0804_2

Clark, A. (1997). *Being there: Putting brain, body, and world together again*. Cambridge, MA: MIT Press.

Clark, H., & Brennan, S. E. (1991). Grounding in communication. In L. B. Resnick, J. M. Levine, & S. D. Teasley (Eds.), *Perspectives on socially shared cognition* (pp. 127–149). American Psychological Association. doi:10.1037/10096-006

Çöltekin, A., Fabrikant, S. I., & Lacayo, M. (2010). Exploring the efficiency of users' visual analytics strategies based on sequence analysis of eye movement recordings. *International Journal of Geographical Information Systems*, *24*(10), 1559–1575. doi:10.1080/13658816.2010.511718

Çöltekin, A., Heil, B., Garlandini, S., & Fabrikant, S. I. (2009). Evaluating the effectiveness of interactive map interface designs: A case study integrating usability metrics with eye-movement analysis. *Cartography and Geographic Information Science*, *36*, 5–17. doi:10.1559/152304009787340197

Conradt, L., & Roper, T. J. (2005). Consensus decision making in animals. *Trends in Ecology and Evolution*, *20*(8), 449–456. doi:10.1016/j.tree.2005.05.008

Couzin, I. D. (2007). Collective minds. *Nature*, *445*(7129), 715. doi:10.1038/445715a

Couzin, I. D., & Krause, J. (2003). Self-organization and collective behavior in vertebrates. *Advances in the Study of Behavior*, *32*, 1–75. doi:10.1016/S0065-3454(03)01001-5

Dalton, R. C., H¨olscher, C., & Montello, D. R. (2019, February). Wayfinding as a social activity. *Frontiers in Psychology*, *10*, 1–14. doi:10.3389/fpsyg.2019.00142

Deane, O., Toth, E., & Yeo, S. H. (2022). Deep-SAGA: A deep-learning-based system for automatic gaze annotation from eye-tracking data. *Behavior Research Methods*, 1–20.

Downs, R. M., & Stea, D. (1977). *Maps in minds: Reflections on cognitive mapping*. New York: Harper & Row.

Duchowski, A. T., Krejtz, K., Krejtz, I., Biele, C., Niedzielska, A., Kiefer, P., . . . Giannopoulos, I. (2018). The index of pupillary activity: Measuring cognitive load vis-à-vis task difficulty with pupil oscillation. In *Proceedings of the 2018 CHI conference on human factors in computing systems* (pp. 1–13). doi:10.1145/3173574.3173856

Emo, B. (2012). Wayfinding in real cities: Experiments at street corners. In C. Stachniss, K. Schill, & D. Uttal (Eds.), *Spatial cognition VII* (pp. 461–491). Berlin and Heidelberg: Springer. doi:10.1007/978-3-642-32732-2$_3$0

Gelblum, A., Fonio, E., Rodeh, Y., Korman, A., & Feinerman, O. (2020). Ant collective cognition allows for efficient navigation through disordered environments. *eLife*, *9*, 1–47. doi:10.7554/eLife.55195

Giannopoulos, I., Kiefer, P., & Raubal, M. (2012). Geogazemarks: Providing gaze history for the orientation on small display maps. In *Proceedings of the 14th ACM international conference on multimodal interaction* (pp. 165–172). doi:10.1145/2388676.2388711

Giannopoulos, I., Kiefer, P., & Raubal, M. (2015). GazeNav: Gaze-based pedestrian navigation. In *Proceedings of the 17th international conference on human-computer interaction with mobile devices and services* (pp. 337–346). doi:10.1145/2785830.2785873

Giannopoulos, I., Kiefer, P., Raubal, M., Richter, K.-F., & Thrash, T. (2014). *Wayfinding decision situations: A conceptual model and evaluation*. In *International conference on geographic information science* (pp. 221–234). Cham: Springer. doi:10.1007/978-3-319-11593-1_15

Gloor, P. (1997). *Elements of hypermedia design: Techniques for navigation & visualization in cyberspace*. New York: Springer Science + Business Media.

Gramann, K., Hoepner, P., & Karrer-Gauss, K. (2017). Modified navigation instructions for spatial navigation assistance systems lead to incidental spatial learning. *Frontiers in Psychology*, *8*, 193.

He, G., Ishikawa, T., & Takemiya, M. (2015). Collaborative navigation in an unfamiliar environment with people having different spatial aptitudes. *Spatial Cognition and Computation*, *15*(4), 285–307. doi:10.1080/13875868.2015.1072537

Hegarty, M., Richardson, A. E., Montello, D. R., Lovelace, K., & Subbiah, I. (2002). Development of a self-report measure of environmental spatial ability. *Intelligence*, *30*(5), 425–447. doi:10.1016/S0160-2896(02)00116-2

Hill, G. W. (1982). Group versus individual performance: Are N + 1 heads better than one? *Psychological Bulletin*, *91*(3), 517–539. doi:10.1037/00332909.91.3.517

Hutchins, E., Hollan, J., & Norman, D. (1985). Direct manipulation interfaces. *Human-Computer Interaction*, *1*(4), 311–338. doi:10.1207/s15327051hci0104$_2$

Ishikawa, T. (2019). Satellite navigation and geospatial awareness: Long-term effects of using navigation tools on wayfinding and spatial orientation. *The Professional Geographer*, *71*(2), 197–209. doi:10.1080/00330124.2018.1479970

Ishikawa, T., & Montello, D. R. (2006). Spatial knowledge acquisition from direct experience in the environment: Individual differences in the development of metric knowledge and the integration of separately learned places. *Cognitive psychology*, *52*(2), 93–129. doi:10.1016/j.cogpsych.2005.08.003

Just, M. A., & Carpenter, P. A. (1976). The role of eye-fixation research in cognitive psychology. *Behavior Research Methods Instrumentation*, *8*(2), 139–143. doi:10.3758/BF03201761

Kearns, M., Suri, S., & Montfort, N. (2006). An experimental study of the coloring problem on human subject networks. *Science*, *313*(5788), 824–827. doi:10.1126/science.1127207

Kiefer, P., & Giannopoulos, I. (2012). Gaze map matching: Mapping eye tracking data to geographic vector features. In *Proceedings of the 20th international conference on advances in geographic information systems* (pp. 359–368). doi:10.1145/2424321.2424367

Kiefer, P., Giannopoulos, I., & Raubal, M. (2013). Using eye movements to recognize activities on cartographic maps. In *Proceedings of the 21th Singspiel international conference on advances in geographic information systems* (pp. 498–501). doi:10.1145/2525314.2525467

Kiefer, P., Giannopoulos, I., & Raubal, M. (2014). Where am I? Investigating map matching during self-localization with mobile eye tracking in an urban environment. *Transactions in GIS*, *18*(5), 660–686. doi:10.1111/tgis.12067

Kiefer, P., Giannopoulos, I., Raubal, M., & Duchowski, A. (2017). Eye tracking for spatial research: Cognition, computation, challenges. *Spatial Cognition and Computation*, *17*(1–2). doi:10.1080/13875868.2016.1254634

Kuipers, B. (1982). The "map in the head" metaphor. *Environment and Behavior*, *14*(2), 202–220. doi:10.1177/0013916584142005

Littlepage, G. E., & Mueller, A. L. (1997). Recognition and utilization of expertise in problem-solving groups: Expert characteristics and behavior. *Group Dynamics: Theory, Research, and Practice*, *1*(4), 324–328. doi:10.1037/1089-2699.1.4.324

Majaranta, P., Ahola, U.-K., & Špakov, O. (2009). Fast gaze typing with an adjustable dwell time. In *Proceedings of the SIGCHI conference on human factors in computing systems* (pp. 357–360). doi:10.1145/1518701.1518758

Montello, D. R. (1998). A new framework for understanding the acquisition of spatial knowledge in large-scale environments. In M. J. Egenhofer & R. G. Golledge (Eds.), *Spatial and temporal reasoning in geographic information systems* (pp. 143–154). New York, NY: Oxford University Press.

Montello, D. R. (2005). Navigation. In P. Shah & A. Miyake (Eds.), *The Cambridge handbook of visuospatial thinking* (pp. 257–294). Cambridge University Press. doi:10.1017/CBO9780511610448.008

Müller, D., & Mann, D. (2021, May). Algorithmic gaze classification for mobile eye-tracking. In *ACM symposium on eye tracking research and applications* (pp. 1–4). doi:10.1145/3450341.3458886

Opach, T., & Nossum, A. (2011). Evaluating the usability of cartographic animations with eye-movement analysis. In *25th International cartographic conference* (p. 11). Paris: International Cartographic Association.

Passini, R. (1981). Wayfinding: A conceptual framework. *Urban Ecology*, *5*(1), 17–31. doi:10.1016/0304-4009(81)90018-8

Raubal, M. (2009). Cognitive engineering for geographic information science. *Geography Compass*, *3*(3), 1087–1104. doi:10.1111/j.1749-8198.2009.00224.x

Raubal, M., & Panov, I. (2009). A formal model for mobile map adaptation. In G. Gartner & K. Rehrl (Eds.), *Location based services and tele cartography II* (pp. 11–34). Berlin and Heidelberg: Springer. doi:10.1007/978-3-540-87393-82

Reichenbacher, T. (2007). Adaptation in mobile and ubiquitous cartography. In W. Cartwright, M. P. Peterson, & G. Gartner (Eds.), *Multimedia cartography* (pp. 383–397). Berlin and Heidelberg: Springer. doi:10.1007/978-3-540-36651-5_27

Rouse, W. (1988). Adaptive aiding for human/computer control. *Human Factors*, *30*(4), 431–443. doi:10.1177/001872088803000405

Sasaki, T., & Pratt, S. C. (2012). Groups have a larger cognitive capacity than individuals. *Current Biology*, *22*(19), R827–R829. doi:10.1016/j.cub.2012.07.058

Scaife, M., & Rogers, Y. (1996). External cognition: How do graphical representations work? *International Journal of Human-Computer Studies*, *45*(2), 185–213. doi:10.1006/ijhc.1996.0048

Schinazi, V. R., & Epstein, R. A. (2010). Neural correlates of real-world route learning. *Neuroimage*, *53*(2), 725–735. doi:10.1016/j.neuroimage.2010.06.065

Schinazi, V. R., Nardi, D., Newcombe, N. S., Shipley, T. F., & Epstein, R. A. (2013). Hippocampal size predicts rapid learning of a cognitive map in humans. *Hippocampus*, *23*(6), 515–528. doi:10.1002/hipo.22111

Schwarzkopf, S., Büchner, S. J., Hölscher, C., & Konieczny, L. (2017). Perspective tracking in the real world: Gaze angle analysis in a collaborative wayfinding task. *Spatial Cognition & Computation, 17*(1–2), 143–162. doi:10.1080/13875868.2016.1226841

Schwering, A., Krukar, J., Li, R., Anacta, V. J., & Fuest, S. (2017). Wayfinding through orientation. *Spatial Cognition and Computation, 17*(4), 273–303. doi:10.1080/13875868.2017.1322597

Schwering, A., Wang, J., Chipofya, M., Jan, S., Li, R., & Broelemann, K. (2014). SketchMapia: Qualitative representations for the alignment of sketch and metric maps. *Spatial Cognition and Computation, 14*(3), 220–254. doi:10.1080/13875868.2014.917378

Selous, E. (1932). Thought transference (or what?) in birds. *Nature*, *129*(3251), 263. doi:10.1038/129263c0

Siegel, A. W., & White, S. H. (1975). The development of spatial representations of large-scale environments. *Advances in Child Development and Behavior*, *10*, 9–55. doi:10.1016/S0065-2407(08)60007-5

Steinke, T. R. (1987). Eye movement studies in cartography and related fields. *Cartographica: The International Journal for Geographic Information and Geovisualization*, *24*(2), 40–73. doi:10.3138/J166-635U-7R56-X2L1

Sumpter, D. J. (2006). The principles of collective animal behaviour. *Philosophical Transactions of the Royal Society B: Biological Sciences*, *361*(1465), 5–22. doi:10.1098/rstb.2005.1733

Theraulaz, G., & Bonabeau, E. (1999). A brief history of stigmergy. *Artificial Life*, *5*(2), 97–116. doi:10.1162/106454699568700

Weisman, J. (1981). Evaluating architectural legibility: Way-finding in the built environment. *Environment and Behavior*, *13*, 189–204. doi:10.1177/0013916581132004

Wenczel, F., Hepperle, L., & von Stu¨lpnagel, R. (2017). Gaze behavior during incidental and intentional navigation in an outdoor environment. *Spatial Cognition and Computation*, *17*(1–2), 121–142. doi:10.1080/13875868.2016.1226838

Wiener, J. M., H¨olscher, C., Bu¨chner, S., & Konieczny, L. (2011). Gaze behaviour during space perception and spatial decision making. *Psychological Research*, *76*(6), 1–17. doi:10.1007/s00426-011-0397-5

Wolbers, T., & Hegarty, M. (2010). What determines our navigational abilities? *Trends in Cognitive Sciences*, *14*(3), 138–146. doi:10.1016/j.tics.2010.01.001

Yarbus, A. L. (1967). *Eye movements and vision*. Springer. doi:10.1007/978-1-4899-5379-7

Zheng, M.-C. (2012, December). Time constraints in emergencies affecting the use of information signs in wayfinding behavior. *Procedia Social and Behavioral Sciences*, *35*, 440–448. http://dx.doi.org/10.1016/j.sbspro.2012.02.109 doi:10.1016/j.sbspro.2012.02.109

5 Virtual Humans and Their Influence on Navigation

Peter Khooshabehadeh, Kimberly A. Pollard, Ashley H. Oiknine, Benjamin T. Files, Bianca Dalangin, Anne M. Sinatra, Steven D. Fleming, and Tiffany R. Raber

Introduction

Military missions often involve entering buildings and pursuing high-value targets. In the near future, technological advancements in remote sensing, machine vision, and computer graphics could make it possible to capture the interior layout of a building, albeit with some relative uncertainty. With improvements in those innovations on the horizon, warfighters will be able to train virtually to navigate the interior space of buildings, including the important objects within them, that they may never have physically been inside prior to a mission. While viewing floor plans and even accessing them from a third-person perspective beforehand can provide knowledge benefits, the added value from training in an immersive, virtual environment can result in faster action and more accurate problem-solving (Gruchalla, 2004). Our recent research (Pollard et al., 2020) examined how a relatively brief experience of a virtual environment (VE) under different levels of immersion leads to an accurate spatial understanding of the environment and the objects within it.

Wayfinding, an example of spatial problem-solving, is the process of orienting one's position to physical space while navigating toward a goal location in the environment. One overlooked but common real-world consideration of wayfinding during navigation is the co-presence, and possible social influence, of other human characters (Dalton et al., 2019). While landmarks, such as unique structures and signs, are common aids in wayfinding, the presence of humans can also be used as social cues, directly or indirectly, to guide navigation (Bönsch et al., 2021).

Given this, we envision that spatial cognition might be enhanced by including social presence via virtual humans (VHs); we also propose that this would be moderated by the social fidelity of the VHs. Here, we propose research that focuses on how the collective social context provided by one or more VHs affects the spatial memory of building interiors. We implement social cues through advanced VH technology. As participants in our experiment navigate within various virtually rendered interior building spaces, they will encounter a VH that communicates through vignettes about objects in the environment in which they are embedded. The collective aspect of this work consists of the presence of additional human characters in the environment during wayfinding, but it is also embedded within the VH vignettes that mention additional people not present in the environment. As the

DOI: 10.4324/9781003202738-7

participant performs a simulated scavenger hunt, the VHs appear in a subset of the environmental locations to provide social vignettes for certain objects. Each social vignette will be carefully crafted to include a first- and third-person mention to help prime the collective social.

Our study would be the first to examine how the social fidelity of VHs affects participants' spatial analysis of the information conveyed by the VH and the environment. We will also look at the mediating relationship of individual differences among participants, such as *need for cognition*, a trait that captures the extent to which people have the propensity to enjoy deep, effortful thought, or whether they just process information more on a shallow level (e.g., heuristic processing) (Cacioppo & Petty, 1982; Oiknine et al., 2021). Other traits such as personality, social skills, and spatial ability will be measured as well. We are interested in understanding which traits, including those critical for motivation, most affect participants' response to VHs of varying social fidelity, for example, different levels of quality in their animated gestures and speech production. Understanding how the impact of social fidelity on spatial knowledge varies across individuals will enable more effective modeling of trainees and, ultimately, more effective navigation interventions and training programs.

Next, we provide an overview of prior work on the role of social fidelity in VH learning interactions. We then sketch the methods and motivation of our proposed social wayfinding study with VHs that prime a collective social context.

Prior Work on the Role of Social Agents in Training Contexts

Social Fidelity

A VH may be considered to have high social fidelity if it accurately reproduces socially realistic human behavior. Such behaviors may include gaze direction, vocal prosody, personalized language, and many others (Sinatra et al., 2021). A VH with high social fidelity can increase the feeling of social presence experienced by users (Kasap & Magnenat-Thalmann, 2012; Lee & Nass, 2005; Moreno & Mayer, 2004). Social presence in turn may increase engagement in virtual environment training (Alexander et al., 2005). Social fidelity has also been found to enhance non-social forms of presence and attention (Moreno & Mayer, 2004), which may further improve learning outcomes or motivation (Saerbeck et al., 2010).

In real-world social interactions, humans speak in a personalized manner; that is, they use the first person ("I" and "my") when speaking about themselves. A virtual agent that speaks about itself and objects related to itself in this manner is more socially accurate than one that uses a more detached narrative point of view. Moreno and Mayer's (2004) *personalization hypothesis* suggested that the use of personalized language should increase social presence and learning outcomes. Moreno and Mayer (2000) examined the personalization hypothesis by training students using narrated animations about lightning, with either personalized or non-personalized language, in text or spoken aloud. Students scored better on knowledge-transfer tests when using the personalized (high social fidelity) system, but more straightforward low-level knowledge retention did not differ. Moreno and Mayer (2000) similarly trained students in botany using an animated pedagogical agent. Students scored better on knowledge transfer and retention in the personalized conditions. Moreno and Mayer (2004) again trained students in botany using a virtual agent, finding again that personalized messages improve both retention and transfer. The authors argue that the increased social presence caused students to pay more attention to the lesson. The higher social-fidelity intervention also yielded greater reported levels of physical presence, suggesting that social fidelity may be a promising method to improve learning outcomes in physical or spatial training.

In addition to personalized language, expressive agents that display socially supportive behavior have a positive impact on learning performance (Saerbeck et al., 2010). This includes familiar dialog and positive interactions, which ultimately serve as the primary motivation for learning. Saerbeck et al. (2010) examined the effects of socially supportive behavior of a robot tutor providing feedback either in a neutral, machine-like tone, or with the enthusiastic sound of a life-like, extroverted personality, in addition to non-verbal feedback such as gestures. Results displayed significantly higher test scores in the socially supportive condition and even a decrease in the distance the participants kept from their robot tutor. This further supports that a noticeable improvement in socialness was perceived by the participants during the experiment and suggests that engaging with a robot that displays socially supportive behavior in language and/or gestures is more effective than impassive knowledge transfer alone.

Humans use gestures when speaking and interacting with one another. Inclusion of physical gestures is thus another way of increasing a VH's social fidelity, which may improve learning outcomes. A few studies have examined this, but the results are mixed. For example, Craig et al. (2002) did not find improved performance on recall or transfer test scores when including gestures in an animated lesson on lightning formation. Frechette and Moreno (2010) also found no benefit of gestures or facial expressions in their use of a virtual teaching agent. In contrast, Baylor et al. (2003) found that including gestures and emotional expression in virtual agents marginally improved information recall and learning transfer in the training of teaching skills. Saerbeck et al. (2010) also found that the display of empathy and non-verbal feedback such as nodding or shaking of the head attributed to a positive effect on the participants' learning performance.

A study by Randhavane et al. (2019) experimented with nonverbal movement characteristics of a virtual agent, such as waving, nodding, and gazing to improve perceived friendliness and social presence. By having participants complete tasks involving the agent's awareness of the environment and influence on it, Randhavane et al. determined that non-verbal gestures did improve the perceived friendliness of the agent. This increased social presence, specifically the spatial presence and social richness of the environment, for the user.

Direct personal and supportive verbal and non-verbal communications are characteristics of high social fidelity in VHs. A VH's high social fidelity can increase personal presence during training or the feeling of being "immersed" in the virtual environment. Heeter (1992) asserted that social presence can add to the feeling of the user's subjective personal presence simply by increasing the number of VHs in a virtual world, whether users are interacting with them or not, because it mimics real-world experiences. A VH's fidelity impacts both the user's social and personal presence inside a virtual environment, which directly impacts engagement and potentially learning outcomes in training scenarios.

Trust

There is a lack of literature on factors that make VHs able to influence user decisions in a virtual environment, specifically in situations involving decision-making in navigational training scenarios. One factor that may determine influence is *co-presence*, the sense that the virtual agent and human are working together (Pimentel & Vinkers, 2021). In a study by Kang and Gratch (2011), human users were interviewed by VH counselors. Users reported more co-presence and attraction to the VHs that first disclosed highly personal information and stories about themselves, encouraging the human participants to disclose more intimate details in deeper interactions with the VHs. While this is an example that persuades users to share more about themselves, few studies have explored how VHs can influence action in goal-oriented scenarios.

The co-presence of VHs can affect how participants view and make decisions in an environment during wayfinding (Yassin et al., 2021). The idea that individuals determine what to do by observing other people—the *principle of social proof*—may also be applied to the influence of VHs in virtual worlds (Cialdini, 2007). In some cases, carrying out tasks in virtual environments with VHs co-present impacted users' decisions more than signage, with users indicating that crowds of VHs moving against the guidance of a sign made them question that sign's validity (Yassin et al., 2021). Souza de Araújo et al. (2010) suggested that VHs can build confidence for users navigating in the environment with supportive guidance. In their study, set in a virtual art gallery with many exhibition options, VHs with complimentary suggestions to the participants' initial preferences encouraged most users to follow the VH and accept the VHs suggestions in later interactions. Souza de Araújo et al. also noted that when a VH made a suggestion and disappeared from the gallery, nobody accepted its suggestion, implying that the loss of physical presence resulted in diminished ability to influence decisions. VHs actually generated mistrust among users when making suggestions contrary to the navigator's intended decisions and lost the confidence of users and the ability to make influential recommendations.

For an intelligent agent to make acceptable decisions, its behavior must be predictable and trustworthy, with an underlying reasoning process that is understandable to its human colleagues. This is especially critical in route planning, which involves spatial awareness of the route and important objects relied on for accurate planning. The core issue for human versus agent approaches to planning routes exists in the different approaches to making spatial decisions, specifically the extent to which they consider information in making decisions; this can result in the breakdown of trust (Perelman et al., 2020). In a route-planning task, Perelman et al. suggested that there is no single way humans solve spatial problems, but they do tend to develop similar spatial mental models. To succeed, intelligent agents on these teams should consider individual differences among human planners, in addition to weighing goals, risks, and the priorities of a mission. Perelman et al. concluded that viewing an intelligent agent's planned route can influence human spatial mental models, but trust ultimately determines whether the route is accepted or rejected.

Social Cues in the Scavenger Hunt Task

For the purposes of our social wayfinding scavenger hunt, we designed a VH that had personalized language and gaze following in all conditions, as VHs without these traits were found to be disconcerting. However, we varied the fidelity of speech reproduction and gesture accuracy. The VH is interactive in that participants are able to engage it to receive (one-way) information by approaching the VH and triggering the interaction with joystick controls. See Figure 5.1 for images of a VH in the virtual environment.

Our study will explore ways in which VHs can exert "synchronous weak" influences on human wayfinding decision-making (Dalton et al., 2019). We manipulate social fidelity across three different levels of how the VH renders its speech and gesture behavior. In addition to social fidelity, these three levels also vary in terms of the cost effectiveness of implementing such levels in training. Therefore, we can assess if and when enhanced levels of social fidelity are worthwhile in improving training versus when they are not cost justified. The low social-fidelity condition uses the native text-to-speech (TTS) Zira female voice in Windows 10 and does not perform any gestures. The middle social-fidelity condition uses a custom TTS from the Amazon Polly female voice and automated nonverbal behavior (gestural) generation (NVBG) (Lee & Marsella, 2006). Finally, the highest level of fidelity is

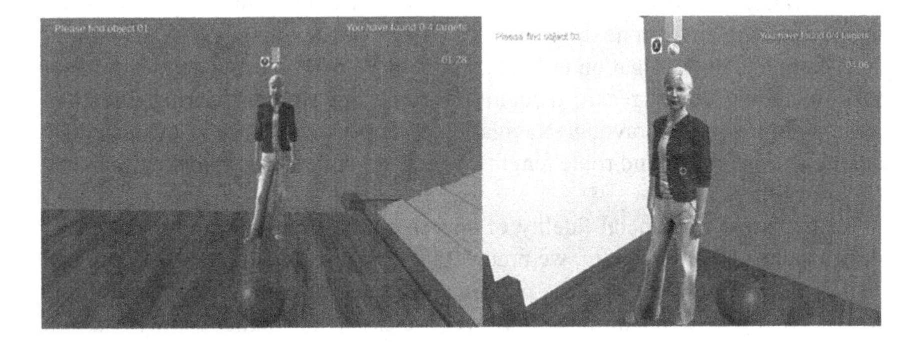

Figure 5.1 Left: virtual human embedded in the scavenger hunt navigation task. Participants have to get within a certain distance to engage the VH for her to speak the vignette. Right: close-up view that shows the participant's perspective when they approach to hear the vignette.

a prerecorded female voice of one of our staff members. The prerecorded voice is then lip-synched to the virtual human character (Khooshabeh et al., 2017) and the NVBG is customized with particular beat gestures and other gesture additions to make the characters' gestures appear most authentic.

Prior to engaging in the main environments of the experimental task, participants will complete a battery of individual difference questionnaires, including spatial ability, cognitive style, need for cognition, anthropomorphic tendency (a person's propensity to see human characteristics in nonhuman and inanimate objects), immersive tendency (a person's capacity to feel present or inside of a technologically mediated environment), and others. Next, they will complete a navigational task similar to the one they will do in the main test environment in a smaller (mini) environment that is intended to demonstrate to the participant the interaction technique and control mechanics for the task. During both the mini and main navigation tasks, participants will explore different virtual environments (built in Unity 3D) that are modeled after various themes (e.g., home, school, and office). These VEs will all be of equal complexity, including similar shapes with equivalent numbers of rooms, doorways, dimensions, and so on (Files et al., 2019). To ensure that all participants sufficiently explore each aspect of the VEs, they will be instructed to complete a scavenger hunt for eight objects in serial order. To avoid conflating semantics with experiential memory of environment objects, target objects will be marked via numbered flags and not mentioned by name. Throughout the scavenger hunt, participants will encounter a VH who provides spoken vignettes that are contextually specific to an object in each location. For example, when the participant enters the house gardening room, the VH will appear beside the watering can and, when clicked, will say, "This watering can is a little big for my needs, but I'm attached to it because it belonged to my father." The social collective here is thus three-way and, in this case, encompasses the participant, the (present) VH character, and the (non-present) VH character's father. Although the utterance is not directly linked to navigation (e.g., no distances or directions), we predict the collective social prime paired with the presence of the VH may affect wayfinding and the construction of interior memory representations by placing environment objects in a collective social context. After participants complete the navigation task, spatial memory will be evaluated via transfer tasks, including object recall, recognition, route description, bearing direction, and topological map drawing. Navigational success will also be evaluated by looking at time points and route lengths, which are collected automatically by the VE software.

To the extent that social fidelity of speech and gesture are important factors for enhancing spatial memory, we predict that the prerecorded human speech and naturalistic gestures will increase performance on transfer tasks compared to the two TTS and gesture conditions. Moreover, we expect the higher quality Amazon Polly TTS condition to lead to better performance compared to the lower quality Microsoft TTS Zira voice. In addition to our hypotheses regarding the role of social fidelity on the memorability of the vignettes about objects in the VE and thus memorability of the spatial environment overall, we predict that certain individual differences will mediate performance (e.g., Wang et al., 2008). For example, we predict that the need for cognition may predict which participants are more affected by social fidelity (Wang et al., 2008, 2013).

In summary, our vignettes attach a social narrative involving multiple characters to certain features of the environment by means of social interaction (storytelling). We are interested to know whether attaching collective social meaning to objects in an environment enhances spatial learning, reasoning, and recall of that environment and its features. Of special interest is whether the enhanced spatial understanding only applies to objects with attached collective social elements or whether that attachment affects other objects, either positively or negatively. Broader implications of this research include the design of improved next-generation navigation technologies and training in novel environments that can be personalized to individual traits.

References

Alexander, A. L., Brunyé, T., Sidman, J., & Weil, S. A. (2005). From gaming to training: A review of studies on fidelity, immersion, presence, and buy-in and their effects on transfer in PC-based simulations and games. *DARWARS Training Impact Group, 5*, 1–14.

Baylor, A. L., Ryu, J., & Shen, E. (2003). The effects of pedagogical agent voice and animation on learning, motivation, and perceived persona. In *EdMedia+ innovate learning* (pp. 452–458). Waynesville, NC: Association for the Advancement of Computing in Education (AACE).

Bönsch, A., Güths, K., Ehret, J., & Kuhlen, T. W. (2021). Indirect user guidance by pedestrians in virtual environments. In J. Maiero, M. Weier, and D. Zielasko (Eds.), *ICAT-EGVE 2021—International Conference on Artificial Reality and Telexistence and Eurographics Symposium on Virtual Environments—Posters and Demos*. The Eurographics Association. https://doi.org/10.2312/EGVE.20211336

Cacioppo, J. T., & Petty, R. E. (1982). The need for cognition. *Journal of Personality and Social Psychology, 42*, 116–131.

Cialdini, R. B. (2007). *Influence: The psychology of persuasion*. New York: HarperCollins.

Craig, S. D., Gholson, B., & Driscoll, D. M. (2002). Animated pedagogical agents in multimedia educational environments: Effects of agent properties, picture features, and redundancy. *Journal of Educational Psychology, 94*(2), 428–434.

Dalton, R. C., Hölscher, C., & Montello, D. R. (2019). Wayfinding as a social activity. *Frontiers in Psychology, 10*, 142.

Files, B. T., Oiknine, A. H., Thomas, J., Khooshabeh, P., Sinatra, A. M., & Pollard, K. A. (2019, July). Same task, different place: Developing novel simulation environments with equivalent task difficulties. In *International conference on applied human factors and ergonomics* (pp. 108–119). Cham: Springer.

Frechette, C., & Moreno, R. (2010). The roles of animated pedagogical agents' presence and nonverbal communication in multimedia learning environments. *Journal of Media Psychology, 22*, 61–72.

Gruchalla, K. (2004). Immersive well-path editing: Investigating the added value of immersion. *IEEE Virtual Reality 2004*, 157–164.

Heeter, C. (1992). Being there: The subjective experience of presence. *Presence: Teleoperators and Virtual Environments, 1*(2), 262–271.

Kang, S.-H., & Gratch, J. (2011). People like virtual counselors that highly-disclose about themselves. *Studies in Health Technology and Informatics, 167*, 143–148.

Kasap, Z., & Magnenat-Thalmann, N. (2012). Building long-term relationships with virtual and robotic characters: The role of remembering. *The Visual Computer, 28*(1), 87–97.

Khooshabeh, P., Dehghani, M., Nazarian, A., & Gratch, J. (2017). The cultural influence model: When accented natural language spoken by virtual characters matters. *AI & Society, 32*(1), 9–16.

Lee, J., & Marsella, S. (2006, August). Nonverbal behavior generator for embodied conversational agents. In *International workshop on intelligent virtual agents* (pp. 243–255). Berlin and Heidelberg: Springer.

Lee, K. M., & Nass, C. (2005). Social-psychological origins of feelings of presence: Creating social presence with machine-generated voices. *Media Psychology, 7*(1), 31–45.

Moreno, R, & Mayer, R. E. (2000). Engaging students in active learning: The case for personalized multimedia messages. *Journal of Educational Psychology, 92*(4), 724–733.

Moreno, R., & Mayer, R. E. (2004). Personalized messages that promote science learning in virtual environments. *Journal of Educational Psychology, 96*(1), 165–173.

Oiknine, A. H., Pollard, K. A., Khooshabeh, P., & Files, B. T. (2021). Need for cognition is positively related to promotion focus and negatively related to prevention focus. *Frontiers in Psychology, 12*.

Perelman, B. S., Evans III, A. W., & Schaefer, K. E. (2020). Where do you think you're going? Characterizing spatial mental models from planned routes. *ACM Transactions on Human-Robot Interaction, 9*(4), 1–55.

Pimentel, D., & Vinkers, C. (2021). Copresence with virtual humans in mixed reality: The impact of contextual responsiveness on social perceptions. *Frontiers in Robotics and AI, 8*.

Pollard, K. A., Oiknine, A. H., Files, B. T., Sinatra, A. M., Patton, D., Ericson, M., & Khooshabeh, P. (2020). Level of immersion affects spatial learning in virtual environments: Results of a three-condition within-subjects study with long intersession intervals. *Virtual Reality, 24*(4), 783–796.

Randhavane, T., Bera, A., Kapsaskis, K., Gray, K., & Manocha, D. (2019). FVA: Modeling perceived friendliness of virtual agents using movement characteristics. *IEEE Transactions on Visualization and Computer Graphics, 25*(11), 3135–3145.

Saerbeck, M., Schut, T., Bartneck, C., & Janse, M. D. (2010). Expressive robots in education: varying the degree of social supportive behavior of a robotic tutor. In *Proceedings of the SIGCHI conference on human factors in computing systems* (pp. 1613–1622). https://doi.org/10.1145/1753326.1753567

Sinatra, A. M., Pollard, K. A., Files, B. T., Oiknine, A. H., Ericson, M., & Khooshabeh, P. (2021). Social fidelity in virtual agents: Impacts on presence and learning. *Computers in Human Behavior, 114*, 106562.

Souza de Araújo, A., Vidal de Carvalho, L. A., & Moreira da Costa, R. M. E. (2010). The influence of intelligent characters on users' navigation through a three-dimensional virtual environment. *Presence: Teleoperators and Virtual Environments, 19*(3), 253–264.

Wang, N., Johnson, W. L., Mayer, R. E., Rizzo, P., Shaw, E., & Collins, H. (2008). The politeness effect: Pedagogical agents and learning outcomes. *International Journal of Human-Computer Studies, 66*, 98–112.

Wang, Y., Khooshabeh, P., & Gratch, J. (2013, August). Looking real and making mistakes. In *International workshop on intelligent virtual agents* (pp. 339–348). Berlin and Heidelberg: Springer.

Yassin, M., El Antably, A., & Abou El-Ela, M. A. (2021). The others know the way: A study of the impact of co-presence on wayfinding decisions in an interior virtual environment. *Automation in Construction*, *128*, 103782.

Knowledge Acquisition and Reasoning

6 Adverse Consequences of Collaboration on Spatial Problem-Solving

Jessica Andrews-Todd and David N. Rapp

Introduction

Much of what we know about how people comprehend, make decisions about, and remember information comes from studying individuals working in isolation. But human activity usually occurs in social environments, with people working interactively, either cooperatively or competitively, as they think about and act on the world (e.g., Andrews & Rapp, 2015; Springer et al., 1999). Collaborations can prove beneficial as individuals work together, relying on their own and each other's understandings and skill sets. In the best of situations, these collaborations result in behavioral products that are more effective than if any one member of the group worked alone. For example, when a dyad is asked to learn words in a list or to build an argument to support a particular position, they could each bring to bear their overlapping and complementary understandings to complete the task. These understandings should include more information than if they relied on what each knows individually and separately without the other person's knowledge and assistance. Existing work corroborates these expectations, showing that collaborative activities support comprehension, knowledge elaboration, encoding, and retrieval (e.g., Rajaram, 2011; van Boxtel et al., 2000).

These collaborative activities are relevant when considering diverse kinds of behaviors and activities, including spatial cognition. Spatial thinking necessarily involves encoding, retrieval, elaboration, inferencing, and comprehension, often with other processes. Spatial cognition is therefore a particularly relevant area for thinking about the consequences of collaboration as people routinely work together to identify, navigate, explore, and discuss locations, landmarks, and geographic systems. For example, a pair might work together to learn locations in an environment, with each producing both overlapping and complementary understandings to support their activities in that space. Similarly, members in a group might work together to generate a subsequent spatial visualization of that environment, whether it involves a map or a virtual tour, based on their memories for what they explored. Their combined memories should offer more fodder for problem-solving than would memories encoded and retrieved by only a single member. In fact, the collaborating members of the group might trigger particular memories or even construct new understandings that conceivably could go beyond what any single

DOI: 10.4324/9781003202738-9

member would produce on their own. Activities of this type regularly take place in hiking and climbing groups, represent crucial considerations for military teams, lead to arguments among travel companions, and are often at the forefront of instructional designers' plans and products.

The types of collaborative activities described earlier, involving combinations of spatial and verbal considerations as people interact to build understandings and enact behaviors, usually foregrounds the ostensible benefits of working in groups. Unfortunately, a variety of projects have called into question such an optimistic outlook by demonstrating the challenges associated with collaboration. One obvious challenge emerges when group members do not pull equal weight, leading to less than collaborative activities and the potential diffusion of responsibility. These kinds of issues are well-documented in a variety of research works (e.g., Andrews & Rapp, 2015; Karau & Williams, 1993; Simms & Nichols, 2014; Suleiman & Watson, 2008). Recent work has also identified another set of concerns focused on instances in which partners cooperatively support each other but unfortunately provide contributions that are unrelated or inaccurate during their group activity. A core challenge for such cases is that these irrelevant or inaccurate partner contributions can potentially contaminate the memories and understandings retained by other group members, influencing their decisions and judgments after the group experience has concluded.

Prior work has examined the consequences of these kinds of problematic contributions, revealing that the inaccuracies produced by one member of a group can be encoded by other group members and used to complete subsequent tasks. For example, imagine a pair of students working together to understand different topics in anticipation of a course exam. When the pair discusses those topics, inaccuracies produced by one partner can inform the understandings of the other, potentially being available for use on that exam. As another example, consider a team collaborating to sketch out the most scenic as compared to the most efficient routes available for exploring new terrain. As they draw out the map in front of them, one member might place a landmark in an incorrect location on their team-built map. This inaccurate information, if not noticed and corrected during discussion like the previously mentioned test topic, is now available for encoding by the other team members, to potentially be used as they head out into the environment. The phenomenon of individuals taking up potentially inaccurate information provided by others during collaborative work has been referred to as *social contagion of memory* (Roediger et al., 2001). It refers to people's understandings being informed and influenced by others' contributions, both when they are aware of it (e.g., involving strategic attempts to think about and use information provided by someone else) and when they are unaware of it (e.g., previously shared activity informing and biasing subsequent thoughts and behaviors without the intention to recollect and rely upon those interactions). Clearly, social contagion can be beneficial when information is correct; in the best of possible worlds, the contributions that others provide in their group are well-planned, well-stated, and useful. But social contagion is problematic when the productions that group partners provide are inaccurate. If members of a group fail to adequately evaluate their partners' contributions, inaccurate offerings can be taken up by other members as useful and relevant for future work.

Previous research examining the negative consequences of social contagion has tended to focus on instances involving verbal tasks, such as with groups of people learning verbal information (e.g., instructions to try to memorize all of the words on a list or in a category and then to collaborate to recall those items) or picture memorization tasks (e.g., instructions to try to memorize all of the items appearing in a photo and then to collaborate to recall those items), after which participants are tasked with completing individual recalls (e.g., Meade et al., 2016; Muller & Hirst, 2014; Park et al., 2016; Rush & Clark, 2013). The degree to which information provided in the group activity spills over into the individual tasks provides an indication of social contagion in these verbal and image-based studies. In the current project, we explore the extent to which social contagion also emerges in a task that involves both verbal and spatial considerations, and involving problems for which group members should have relevant prior knowledge.

Background

Social Contagion of Memory

Research has identified social contagion as a ubiquitous phenomenon, persisting across a variety of conditions (Andrews & Rapp, 2014; Carol et al., 2013; French et al., 2008; Mudd & Govern, 2004; Peker & Tekcan, 2009). As previously mentioned, past studies have often focused on memorization tasks, either involving word lists or depictions of objects. Additionally, most previous studies have applied specific methods intended to experimentally induce inaccurate suggestions among group members. The prototypical manipulation involves reliance on the naïveté of participants as they unknowingly work with a confederate as their collaborative partner (e.g., Allan et al., 2012; Allan & Gabbert, 2008; Davis & Meade, 2013; Goodwin et al., 2013). For example, Roediger et al. (2001) had participants independently study pictures of household scenes (e.g., bathroom and bedroom) and subsequently complete a collaborative recall of items from the scenes with a partner in a turn-taking manner. Unbeknownst to the participants, their partner was a confederate who, during the collaborative recall, purposefully made mistakes by recalling items that had not appeared in the scenes. On a subsequent individual memory test, participants tended to recall a significant proportion of the erroneous items that had been suggested by the confederate partner during the collaborative activity, despite never having actually seen the items in the pictures, and in contrast to when those items had not been mentioned by the confederate at all.

The kinds of information that participants might be asked to think about and remember has included word lists and pictorial presentations, but across these cases, they have involved memorizing information presented during the experience to be recalled in a list-like fashion and/or serial fashion (e.g., French et al., 2011; Gabbert et al., 2006; Garry et al., 2008; Hope et al., 2008). As an example, Gabbert et al. (2003) asked pairs of participants to separately view videos of a criminal event. Participants were led to believe that they were viewing the same video as their partner, but in actuality, each member of the pair viewed unique presentations of the video that offered discrepant segments of information. For instance, among other discrepancies, one member's video showed someone committing an opportunistic crime, while the other member's video did not show it taking place. In the experimental condition, the pairs discussed the video before completing an individual recall recounting the events in the video. In a control condition, the pairs did not discuss the video before completing their individual recalls. Experimental pairs who had discussed the videos were more likely to incorporate discrepant, unseen events into their recalls than were control pairs who did not discuss their memories of the videos. As with the confederate case discussed previously, but this time involving pairs of actual participants, information provided by a partner was relied upon even if it had not initially been personally experienced during viewing of the video.

These tasks have usefully applied standard cognitive psychological methodologies involving list learning, picture studying, and event presentations to assess the influence of a partner's inaccurate contributions. To date, we are unaware of any studies that have incorporated spatial considerations into the elements to be studied and understood or that would influence the ways in which participants might organize their understandings or provide their recalls at test. Additionally, previous projects have provided people with entirely new information to study, which, while potentially connecting with things they already know (e.g., tokens on a categorized word list, such as "birds"), did not rely on prior knowledge for supporting retrieval. The current project attempted to apply these considerations, specifically for spatially organized understandings of geographical environments with which participants might be familiar, to examine the potential negative impacts of social contagion.

The Current Study

Previous psychological and educational studies utilizing methods intended to experimentally manipulate the presence of erroneous information have relied on stimuli such as word lists and pictures that offer new information to participants, and for good reason. These projects are designed to evaluate what people learn from new information and whether sharing that information can have positive and/or negative consequences on subsequent performance. We asked whether the same memorial effects would occur in collaborative contexts more akin to educational settings and contexts in which no experimental manipulations are present and no confederates are involved. This approach was applied to help us to determine whether social contagion emerges when collaborative partners discuss information for which they possess useful prior knowledge. In the current study, we asked participant dyads to collaborate to recall and locate the states and capitals of the United States highlighted on a blank map of the country. The recall of geographical information in such a task requires spatial knowledge, which here we measure as the ability to recall information that requires memory for both name and location in an organized space (Fotheringham & Curtis, 1999). This differs from tasks in which participants are only asked to remember words from a list, items from a picture, or events in a scene in any order or organization they wish. The spatial organization associated with the items to be recalled in the map task constitutes a very different type of activity: Specifically, participants need to recall verbal information associated with the name of the states and capitals and also visual or spatial information associated with where each state and capital is located on the map.

In this experiment, we analyzed the discussions of participant dyads to examine whether inaccurate contributions from a partner would result in social contagion. This involved testing whether a partner's inaccurate productions would be taken up by the other member of the dyad when they were later asked to individually fill in a map of the U.S. states and capitals. Examining the sorts of naturalistic productions each partner produced in the task allowed us to investigate particular discourse factors that could potentially influence the likelihood of falling victim to or avoiding social contagion. The data offer exploratory insight as to whether and how social contagion can emerge on collaborative tasks involving verbal and spatial considerations. They also offer a preliminary examination of how individuals may spontaneously protect themselves from and signal the presence of inaccurate information in collaborative situations, which has not been the focus of descriptive projects intended to demonstrate the consequences of social contagion and make up much of the extant literature to date in the area.

Method

Participants

Twenty-two undergraduates (11 acquainted dyads) from Northwestern University completed the study in pairs for monetary compensation. There were seven males and 15 females who when paired created three male–male dyads, one male–female dyad, and seven female–female dyads. We included this convenience sample both because we had access to recruit these participants and also because they would likely be familiar with the map contents.

Spatial Information

A blank map of the United States was modified so that 30 states were grayed out to highlight 20 states left in white (see Figure 6.1). The 20 highlighted states were selected at random and included Washington, California, Utah, Wyoming, Colorado, Texas, Kansas, Missouri, Minnesota, Louisiana, Alabama, Illinois, Florida, Virginia, Delaware, New York, Vermont, Massachusetts, Rhode Island, and Hawaii. We elected to use 20 states for the task to provide a number we believed participants could complete in a timely manner. We selected this map as our stimulus as it resembles the kinds of map-drawing and identification activities that people might do on their own (e.g., when contemplating places to visit, building representations of places they have already seen, or learning a map) and emulates decisions people

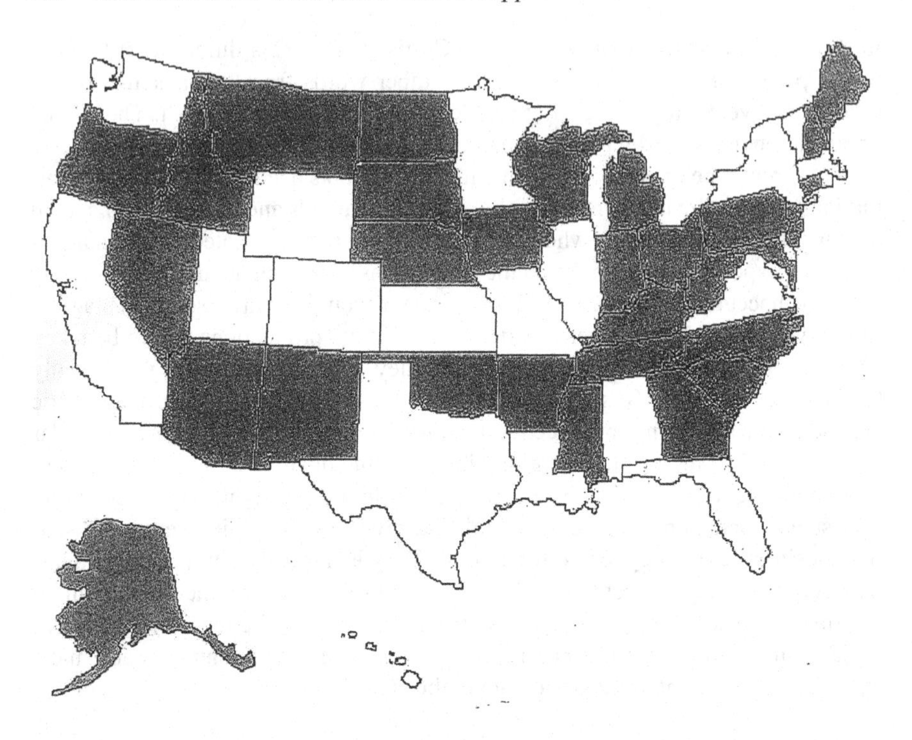

Figure 6.1 Blank United States map used for the map task.

might make in groups about locations and landmarks as they contemplate maps, displays, and locations.

Procedure

Participant pairs were seated together at a table and received one blank map of the United States and two pens. Pairs were asked to work together to fill in the names of the states and capitals for the 20 highlighted states on the map to the best of their ability. Participants were asked to place the name of the state and capital within the correct state but were not instructed to place the capital city in the exact location within the state. When the pairs agreed that they were done with the task, they were asked to let the experimenter know. Participants were next given four minutes to individually complete multiplication problems at their own pace to reduce potential rehearsal strategies. Each participant was then placed on opposite sides of the room, given an identical blank map of the United States, and asked to independently fill in the names of the states and capitals for the highlighted states. The individual map recall was used to assess the extent to which participants incorporated incorrect information from their partner into their maps. Pairs spent anywhere from 6.5 minutes to 18.5 minutes (average 12.5 minutes) to complete the collaborative map task and up to 15 minutes (range and average unavailable) to complete the individual map task. The individual map task was used to examine whether and how the previous collaborative activity influenced subsequent individual performance. After completing the individual task, participants were debriefed as to the purposes of the study and thanked for their participation.

Coding

Each pair's discussions were video recorded and all participants' productions audio recorded using lapel microphones. The discussions were transcribed and coded for each instance of a state or capital mentioned as a response for one of the highlighted states. Using the videos, audio recordings, and the transcriptions of each pair's discussions, two raters coded each mention of a state or capital as accurate or inaccurate and identified who mentioned the item. Specifically, the raters viewed the videos along with the transcriptions to ascertain which locations were being named during pair discussions, where participants were pointing, and whether state and capital location placements were accurate or inaccurate. For capital placements, raters examined whether participants placed the correct capital name within the correct state rather than whether the capital label was placed in the exact location within the state.[1]

Additional codes were utilized to identify the certainty with which the item was mentioned, how the partner responded to the mentioned item, and whether the item was used on the final individual test. Statements and mentions were specifically coded as certain, equivocal, or uncertain. Codings of *certain* statements included confident assertions of a particular state or capital. (e.g., "I know it's Des Moines," "This is Illinois," "That has to be Colorado," or "It's definitely Tallahassee"). These assertions included explicit discourse markers signaling certainty such as *know*, *definitely*, *must*, *I remember*, or *I'm sure* (Kärkkäinen, 2003). Codings of *equivocal* statements were applied when the certainty was sometimes unclear, with statements including discourse markers of questionable certainty (e.g., "This is Colorado, right?" "I think this is Utah," "This is probably Delaware," or "I feel like the one on top is Iowa"). Discourse markers signaling such indecisive hedges included items such as *seems*, *maybe*, *probably*, *I think*, *perhaps*, *might*, and *could*. Codings of *uncertain* statements included explicit markers of definite uncertainty and/or queries about the names of states or capitals (e.g., "What is the capital of Illinois?" "Providence. I actually don't know, I'm just guessing," or "Is this New York?"). Discourse markers of definite uncertainty included items such as *I don't remember* or *I don't know*. The interrater reliability for the coding indicated good agreement ($\kappa = 0.83$). Response type to these productions was coded using five response options: explicit accept (e.g., "Yeah," "Mmhm," or "Okay"), implicit accept (e.g., naming the capital of a state that a partner named in the previous turn or not responding verbally but writing down the state or capital that a partner named), uncertain response (e.g., "I have no idea what that is," "I don't know," or "Is it? Maybe"), no response (e.g., an individual ignores their partner's statement and moves on to another state or capital)[2], and rejection (e.g., "No, Massachusetts is that one," "I thought it wasn't New York," or "Rhode Island is definitely on the other side"). The interrater reliability for response type also indicated good agreement ($\kappa = 0.82$).

As an additional exploratory metric of spatial judgments, we noted distance for state and capital map placements with respect to the testing location for the experiment, in Evanston, IL (on Northwestern University's campus). We categorized state placements as being near or far from Illinois and categorized capital placements as being near or far from the city of Evanston. This distance category was operationalized using a median split. States that were less than or equal to 994 miles from Illinois (the median) were categorized as "near," and states that were greater than 994 miles from Illinois were categorized as "far." Capitals that were less than or equal to 976 miles from Evanston (the median) were categorized as "near," and capitals that were greater than 976 miles from Evanston were categorized as "far."[3]

Results

There were a total of 502 mentions of a state or capital during all of the pair discussions as a response for a highlighted state. Of those 502 mentions, 214 were inaccurately mentioned (43%, i.e., inaccurate productions). The majority of these inaccurate productions involved inaccurate mentions of capitals (55%). When inaccuracies were suggested during collaboration, they were subsequently provided as an answer on the individual test 30% of the time. Inaccuracies were mentioned with certainty 43% of the time, were equivocal 17% of the time, and were mentioned with uncertainty 40% of the time during collaborative discussion. Additionally, during collaboration, inaccurate information was explicitly accepted by the partner 34% of the time, implicitly accepted 8% of the time, given an uncertain response 4% of the time, given no response 10% of the time, and rejected 44% of the time.

To examine whether these discourse factors would predict whether the inaccurate information produced during the collaborative task was used on the subsequent individual test, mixed effects logistic regression was used. The models included certainty (certain, equivocal, and uncertain), response type (explicit, implicit, uncertain, no response, and reject), who mentioned the inaccurate information (self, i.e., the person originally giving the inaccuracy during the discussion; or other, i.e., the person receiving the inaccuracy), and their interactions as fixed factors, with participants individually and participants nested in pairs as random effects. The dependent measure was the presence or absence of inaccurate information on the final test. To determine *p* values, a model comparison approach was utilized in which a likelihood-ratio chi-square test compared the goodness of fit for the full model to models in which each fixed factor was removed. A significant change in the goodness of fit indicates that the effect in question had a significant impact on the goodness of fit in the full model.

Analyses of inaccurate information used at test revealed non-significant effects for certainty ($\chi^2(2) = 0.98$ and $p = .61$), who mentioned the inaccurate information during the discussion ($\chi^2(7) = 1.63$ and $p = .98$), and all interactions (all $ps > .05$). However, the effect for response type was significant ($\chi^2(4) = 140.40$ and $p < .001$), indicating that rejecting inaccuracies during the task was beneficial. Contrasts within response type revealed that the likelihood of using inaccurate information on the individual test was significantly greater if it had been explicitly accepted ($\beta = 3.44$, SE = 0.55, and $p < .001$), implicitly accepted ($\beta = 3.16$, SE = 0.73, and $p < .001$), or received no response ($\beta = 1.84$, SE = 0.71, and $p = .009$) during the collaboration, compared to when it had been rejected. Simply put, inaccurate information was less likely to be used by a partner if it had earlier been discounted as useful than if it had been accepted or undiscussed. The contrast between the use of inaccuracies for uncertain responses and rejected responses was not significantly different ($\beta = 1.52$, SE = 0.96, and $p = .11$).

When inaccurate information was suggested during collaboration, it was not always used in the test. As noted previously, the inaccurate information was used on average 30% of the time by either the person receiving the inaccuracies or the person producing the inaccuracies. But what other kinds of responses did participants provide on the test when they did not use the suggested inaccuracies? Table 6.1 displays the proportion of the different test response types for self (i.e., instances in which a participant produced the inaccuracy during discussion) and other (i.e., instances in which a participant received an inaccuracy during discussion). Specifically, if participants produced an inaccuracy during the pair discussion (i.e., self), they used the inaccuracy 30% of the time, provided an incorrect answer (i.e., a wrong answer that was not the suggested inaccuracy) 17% of the time, left the answer blank 13% of the time, and produced the correct answer 40% of the time. If participants received inaccurate information during the pair discussion (i.e., other), they used that inaccuracy 27% of the time, gave an incorrect but different answer 22% of the time, left the answer blank 15% of the time, or produced the correct answer 36% of the time. A chi-square analysis revealed no differences in test response type as a function of whether a participant was producing or receiving the inaccurate information, $\chi^2(3, N = 428) = 2.66$ and $p = .45$. These results demonstrate instances of recovery in which participants were able to sometimes provide the correct answer at test even after inaccurate content was made available during discussion. However, more often than not, when inaccurate information was suggested or received during discussion, subsequent test responses were plainly incorrect (i.e., participants used inaccuracies, got the answer incorrect, or left the answer blank).

Table 6.1 Frequency and Proportion of Inaccurate Information by Self and Other as a Function of Test Response Type

Test Response Type	*Self*		*Other*	
	N	*(%)*	*N*	*(%)*
Correct Answer	86	40	78	36
Used the Inaccuracy	64	30	57	27
Incorrect Answer	36	17	48	22
Left Blank	28	13	31	15
Total	214	100	214	100

Though certainty did not significantly predict the influence of inaccurate productions at test, it is possible that certainty is an important discourse marker that individuals use to signal, intentionally or not, the utility and accuracy of their contributions for others. If so, individuals should show more uncertain statements for contributions that are inaccurate as opposed to accurate and more certain statements for accurate than inaccurate contributions. A chi-square analysis revealed a significant association between the accuracy of the suggestion and the certainty with which it was mentioned, $\chi^2(2, N = 502) = 18.68$ and $p < .001$. Data from this analysis appear in Table 6.2. Specifically, there were more uncertain statements for inaccurate (54%) than accurate (46%) contributions. Equivocal statements were similar for inaccurate (49%) and accurate contributions (51%). There were fewer certain statements for inaccurate (34%) than accurate contributions (66%).

The results from the mixed effects logistic regression suggest that participants, unfortunately, may *not* have used these markers of uncertainty as a warning to pay more careful attention to the accuracy of their partner's suggestions. To investigate this further, a chi-square analysis was used to determine whether participants were responding differently based on the certainty of their partner's contributions. There was a significant association between certainty and response type, $\chi^2(8, N = 502) = 80.20$ and $p < .001$. For certain statements, participants were more likely to accept, either explicitly (43%) or implicitly (26%), those statements than to give an uncertain response (2%), no response (14%), or a rejection (15%). For equivocal statements, participants showed more explicit acceptances (55%) than implicit acceptances (10%), uncertain responses (3%), no responses (8%), and rejections (25%). Interestingly, for uncertain statements, participants were just as likely to explicitly accept (39%) those statements as they were to reject them (40%). Uncertain statements were also implicitly accepted 4% of the time, given an uncertain response 11% of the time, and given no response 6% of the time.

Given that the accuracy of state and capital placements could be related in some ways to familiarity, it is possible that inaccurate placements would be more likely for locations further from where students were living at the time of the test, which is also where they took the test (i.e., Evanston, IL). Thus, as an additional exploratory analysis, we examined whether the accuracy of state and capital placements was related to the distance from the testing location. For state placements, we conducted a chi-square analysis with accuracy of state placement at test (correct or incorrect) and distance of placement from Illinois (near or far) as variables. The

Table 6.2 Frequency and Proportion of Accurate and Inaccurate Information for Uncertain, Equivocal, and Certain Statements

Contribution Accuracy	*Uncertain*		*Equivocal*		*Certain*	
	N	*(%)*	*N*	*(%)*	*N*	*(%)*
Accurate Information	72	46	37	51	179	66
Inaccurate Information	86	54	36	49	92	34
Total	158	100	73	100	271	100

association between state accuracy and distance was marginally significant, $\chi^2(1, N = 371) = 3.51$ and $p = .06$, but patterns of results were in expected directions. Incorrect state placements were more likely to occur for states far from Illinois (60%) than near Illinois (40%). Conversely, correct state placements were more likely to occur for states near Illinois (53%) than far from Illinois (47%). We conducted a similar analysis for capital placements, with capital placements at test (correct or incorrect) and distance of placements from Evanston (near or far) as variables. The association between capital accuracy and distance was statistically significant, but not in the expected directions, $\chi^2(1, N = 267) = 11.43$ and $p = .001$. Incorrect capital placements were actually more likely to occur for capitals near Evanston (64%) than far from Evanston (36%). On the other hand, correct capital placements were more likely to occur for capitals far from Evanston (59%) than near Evanston (41%).

Discussion

This project offers an exploratory examination as to whether social contagion occurs in collaborative contexts that involve thinking about familiar information to be used in combined spatial and verbal tasks. The activity we asked participants to complete, filling in a familiar map, is akin to activities that can occur in educational settings and informal contexts, unlike the relatively decontextualized, memorization-based tasks that have been evaluated in most studies of social contagion of memory. We also assayed the kinds of discourse factors present in collaborative discussions that might make individuals more or less susceptible to encoding and relying upon their partner's and their own inaccurate contributions. In a single experiment, participants were tasked with working in pairs to fill in a map of the United States. This spatial information should be familiar to participants, although remembering the names of capitals and where they are located in the geographical context of the map was clearly (and unsurprisingly) more difficult than was remembering the names and locations of the states. We specifically examined instances in which participants produced inaccurate suggestions for the names and spatial locations of the states and capitals, with an eye toward examining the naturalistic kinds of errors that can emerge and the likelihood those errors have downstream effects on participants' subsequent attempts to complete the same activities on their own. This tested whether the inaccuracies that participants contemplated in the task would have negative repercussions later.

The results revealed patterns that replicate previous demonstrations of social contagion, showing that people often reproduce the inaccurate contents that their conversational and task partners have earlier provided in a collaborative task (e.g., Andrews & Rapp, 2014; Meade & Roediger, 2002; Wright & Villalba, 2012). In those experiments, with a procedure requiring memorizing lists of items in a category, followed by a collaborative recall with a confederate partner who occasionally produced items never appearing in the lists, participants consistently exhibit use of those inaccuracies on a final individual recall task. The current project differed in a variety of ways, by not involving a confederate, by focusing on materials that participants likely were familiar with before the task given real-world interests and concerns, and in requiring both verbal retrieval for names and spatial coordination for completing the task. In line with previous projects, the current findings indicated clear problematic influences of exposure to inaccurate information.

As pairs collaborated to recall and locate the states and capitals on a map of the United States, members of each pair often produced inaccurate information. These inaccurate suggestions were fodder for reproduction on a subsequent individual test, revealing clear cases of social contagion with a collaborative spatial activity involving free-flowing discussion and information for which pairs likely possessed relevant prior knowledge. These findings indicate that social contagion may be a more general phenomenon linked to comprehension, discourse, and group activity, signifying a potential negative outcome from collaboration. Thankfully, collaboration need not always result in such problematic outcomes. Analyses of the conversational productions by the pairs indicated that the manner in which an individual responded to an inaccurate suggestion from their partner predicted whether it would be used on their subsequent individual recall. Specifically, accepting any suggested inaccuracies, explicitly or implicitly, greatly increased the odds that those inaccuracies would be provided on a subsequent individual test, compared to when the inaccuracies were rejected during the collaboration. These results are consistent with prior work demonstrating that disputes or disagreements can reduce an individual's susceptibility to misinformation (French et al., 2008; Garry et al., 2008; Muller & Hirst, 2014). They also highlight the importance of ongoing discussion and evaluation during the course of any collaborative experience, to ensure that participants might usefully tag inaccurate information as inappropriate for further consideration and to help avoid encoding those inaccuracies into memory. (See Rapp, Hinze et al., 2014 for a discussion of these activities in the service of successful comprehension.)

The association between response type and people's use of inaccurate productions emerged regardless of who produced the inaccurate information and regardless of the certainty with which the inaccuracy was offered during collaboration. While the certainty of people's inaccurate suggestions showed no association with their use of the inaccuracies, we did find that individuals used discourse markers associated with certainty as a cue to the accuracy of their own statements to potentially warn partners about their confidence. Unfortunately, partners failed to take up such markers as a warning or cue for the need to evaluate the accuracy of those statements more closely. They were just as likely to explicitly accept those contributions as they were to reject them.

These findings provide insight into potential challenges associated with collaborative spatial tasks and for problem-solving tasks more generally. Similar to the results from the previous experiments, individuals do not seem to routinely evaluate the veracity of their partner's contributions or their own contributions. This demonstrates the need for interventions, strategies, and design supports that can promote critical evaluation of the information encountered during collaboration. Such evaluation, if developed appropriately, would help people identify and repair potential inaccuracies and misunderstandings while also strengthening accurate conceptions as they defend and explain their ideas. This is clearly not an easy thing to do, nor an easy thing to design for, but there are emerging findings and suggestions to consider in any such developments. To promote critical evaluation, warnings regarding the potential for inaccuracies from others may be provided prior to group work (Chambers & Zaragoza, 2001; Eakin et al., 2003; Echterhoff et al., 2005; Ecker et al., 2010). Effective warnings might include instructions for attending to particular behaviors exhibited by partners that provide a signal as to the accuracy of their contributions or to the confidence with which a partner provides a contribution. For example, in the current study, participants often used discourse markers exemplifying uncertainty to indicate inaccurate productions. Awareness of these tendencies could prove informative and instructional.

Furthermore, warnings could encourage individuals to attend to the source of each contribution, both as coming from particular collaborative partners and as derived from some evidentiary basis associated with the discourse experience. Andrews and Rapp (2014) demonstrated the utility of attending to source information, showing that when sources are perceived as problematic (e.g., less competent or less confident), individuals more readily engage in careful source monitoring of information, helping to protect them against attending to and acquiring erroneous information. Individuals may also be provided with explicit instruction or supporting materials concerning how to engage in beneficial argumentation (Asterhan & Schwarz, 2007) or groups may be strategically assembled in a such a way that group members hold differing views about the topic being covered (Howe et al., 1992). These sorts of rhetorical trainings and strategies can be beneficial in drawing attention to individuals' accurate knowledge as well as their knowledge gaps, to facilitate knowledge acquisition and updating (Andrews & Rapp, 2015; Rapp, Jacovina et al., 2014). Such interventions, however, are not guaranteed to be successful or beneficial, as previous work has also shown that warnings and information about credibility, at times, seem to go ignored in a variety of necessary and seemingly obvious circumstances (Andrews-Todd et al., 2021; Eslick et al., 2011; Fazio et al., 2013; Rapp, 2008). One challenge is that these consequences can arise out of the routine operation of processes associated with memory, comprehension, reading, and problem-solving (Marsh et al., 2016; Rapp, 2016; Rapp & Donovan, 2017; Rapp & Salovich, 2018). That is, people tend to encode information they experience precisely because it is usually useful. It is entirely likely, although offered here as conjecture, that the routine processes underlying spatial cognition and associated experiences may also allow for an influence of inaccurate contributions to contaminate people's understandings.

People's understandings and performance on tasks like the one in this chapter may be related to other factors as well. Through an exploratory analysis, we examined whether one of those factors could be the distance of the states and capitals from the location in which students were living and participating. While states near Illinois were more likely to have been placed correctly on the individual test maps than were states far from Illinois, capitals near the testing location were actually less likely to have been correctly placed on the individual test maps than were capitals far from the testing location. The surprising results for capitals could partly be due to the fact that capital names are generally less well-known than are state names and capitals for states farther away from Illinois may be more well-known (e.g., California, New York) than capitals for the middle states surrounding Illinois (e.g., Kansas, Missouri). (The population of students from whom our participants were sampled includes many people who did not grow up in the midwestern United States, which likely influences their familiarity with state and capital names.) To flesh out the potential hypotheses raised by these contradictory data, a more complete analysis is needed that could involve examinations of where people take the test, familiarity with locations, locations where people grew up, and so on. These demographic data were not part of the current experiment.

The obtained results contribute to our understanding of the effect of social influences on the acquisition of erroneous knowledge and are relevant for a range of collaborative situations, including various types of group structures, social settings, and subject matter. However, the results are more applicable to certain types of tasks—specifically tasks that involve individuals working together to retrieve information. For example, in an educational context, relevant tasks may include several classmates gathering together to discuss information from a previous U.S. history lecture or a small group of students collaborating to create a concept map depicting the relationships between the components of photosynthesis. In each of these situations, students are retrieving previously learned information in the presence of others.

Similarly, a variety of spatially focused tasks (e.g., navigating, environmental search, orienteering, and managing perspectives) often require interactions with collaborative partners or take place in competitive spatial environments that involve opposing groups (e.g., military action and sports). The dynamic interactions that occur in such cases, involving moment-by-moment verbal and gestural language use, considerations of location and movement, evaluations of environmental contingencies, inferences about future events and experiences, and so on, emerge through collaborative discourse opportunities. This allows for useful productions that can inform and support people's thoughts and behaviors, for which we might hope that contagion would occur precisely because what other people have said has the potential to be quite useful. However, the often time-delimited, stressful nature of many kinds of spatial interactions might allow for even more mistakes to pollute people's conversations and subsequent understandings, increasing the likelihood that false information might propagate between group members. The current project represents only a first step in identifying the potentially problematic consequences of contagion for a task involving spatial considerations. Future work should further interrogate these issues with even more naturalistic and dynamic spatial activities that go beyond the retrieval of spatial understandings, including unfolding spatial decisions that occur in real time with real consequences. Future work could also explore situations in which correct information produced by the dyad is carried over into individual performance. Our previous findings have indicated that people take up accurate information as readily as inaccurate information in collaborative settings and when they read (Andrews & Rapp, 2014; Andrews-Todd et al., 2021;

Rapp, 2016); however, we have yet to explore such effects in tasks with spatial considerations.

The experiment here used a U.S. map as materials to examine performance on a spatial identification and recall task, and this is clearly a spatial object that differs from the kinds of word lists and non-map images that have been used in many learning and recall tasks. The map materials afforded the opportunity to begin examining interesting questions about familiarity and space, as reflected in our analyses examining participants' location and their state and capital responses. These spatially focused materials and analyses offer useful extensions to the constrained sets of materials and tasks that have traditionally been used to study social contagion. But there are many additional means of exploiting spatial analyses in future projects. For example, topological measures (e.g., adjacency or containment) could be applied to interrogate response accuracy with regard to participants' designations on the map. Other measures, focused on area and distance, could be leveraged to assess people's beliefs and understandings about various characteristics of geographical locations, including the size of spatial features and/or the separation among them. The rich and growing field of spatial statistics could also be applied to usefully describe and measure subject performance. As examples, quadrat analyses (to determine spatial arrangements), mean nearest neighbor, and other statistical considerations could be used to determine whether participants' responses are more or less clustered around a correct location (i.e., more accurate) when the study and retrieval tasks are performed individually compared to when they are performed as a team. These examples are by no means exhaustive but are intended to help illustrate the numerous possibilities available for further integrating spatiality into team, group, and collaborative cognition experiments. The project here represents an initial examination, and one we hope others will follow up to better understand social interaction in spatially important and diverse settings.[4] Our suggestions and intended extensions discussed thus far, as associated with varied kinds of ambiguous and difficult materials, and involving various kinds of tasks and differing levels of participant familiarity and even comfort, represent exciting directions for future work.

Synthesis for Collective Spatial Cognition

This project has shown that during group retrieval activities, information offered by group members can be potentially inaccurate, and when this occurs, the inaccurate information becomes available for encoding by everyone else in the group. The ease with which such inaccurate information is encoded by others and used for subsequent tasks has been repeatedly demonstrated in previous experiments. In the current study, it occurred as participants were afforded the opportunity to produce naturalistic discourse and during which inaccurate productions spontaneously emerged. Interventions have the potential to provide useful tools and trainings that can help people overcome such collaborative problems. But interventions like these are not easy to develop, and they have largely been focused on verbal learning experiences. The current results offer a demonstration that spatially relevant tasks, even simple ones associated with retrieving names and placing them in their appropriate geophysical locations, and for which people likely have relevant prior knowledge, can reveal problematic examples of social contagion. Future work should continue to test the situations during which collaborative activities prove beneficial while also maintaining a critical eye toward the unintended negative consequences that occur as people work together in groups to learn about, navigate through, and report on the world.

Acknowledgments

This material is based upon work supported by the National Science Foundation under Grant DGE-0824162 awarded to the first author. The opinions expressed are those of the authors and do not necessarily represent views of the National Science Foundation. We thank Bruce Sherin and William S. Horton for their help with this research.

Notes

1 Future work could look at precise placement of the capitals to determine location accuracy within the states as an additional analytic consideration. If so, we advise using large maps to support such calculations.
2 Given that the lack of a response could be construed as an implicit accept in some circumstances, additional analyses were completed in which these two response types were collapsed. The results remained largely the same.
3 We note that we did not collect data with respect to where participants were from, the location of their permanent home, or their familiarity with locations in the United States.
4 We thank Kevin M. Curtin for offering suggestions and language to push ideas forward as discussed in this paragraph.

References

Allan, K., & Gabbert, F. (2008). I still think it was a banana: Memorable 'lies' and forgettable 'truths.' *Acta Psychologica*, *127*(2), 299–308.

Allan, K., Midjord, J. P., Martin, D., & Gabbert, F. (2012). Memory conformity and the perceived accuracy of self versus other. *Memory & Cognition*, *40*(2), 280–286.

Andrews, J. J., & Rapp, D. N. (2014). Partner characteristics and social contagion: Does group composition matter? *Applied Cognitive Psychology*, *28*(4), 505–517.

Andrews, J. J., & Rapp, D. N. (2015). Benefits, costs, and challenges of collaboration for learning and memory. *Translational Issues in Psychological Science*, *1*, 182–191.

Andrews-Todd, J., Rapp, D. N., & Salovich, N. A. (2021). Differential effects of pressure on social contagion of memory. *Journal of Experimental Psychology: Applied*, Advance online publication. https://doi.org/10.1037/xap0000346

Asterhan, C. S., & Schwarz, B. B. (2007). The effects of monological and dialogical argumentation on concept learning in evolutionary theory. *Journal of Educational Psychology*, *99*(3), 626–639.

Carol, R. N., Carlucci, M. E., Eaton, A. A., & Wright, D. B. (2013). The power of a co-witness: When more power leads to more conformity. *Applied Cognitive Psychology*, *27*(3), 344–351.

Chambers, K. L., & Zaragoza, M. S. (2001). Intended and unintended effects of explicit warnings on eyewitness suggestibility: Evidence from source identification tests. *Memory & Cognition, 29*(8), 1120–1129.

Davis, S. D., & Meade, M. L. (2013). Both young and older adults discount suggestions from older adults on a social memory test. *Psychonomic Bulletin & Review, 20*(4), 760–765.

Eakin, D. K., Schreiber, T. A., & Sergent-Marshall, S. (2003). Misinformation effects in eyewitness memory: The presence and absence of memory impairment as a function of warning and misinformation accessibility. *Journal of Experimental Psychology: Learning, Memory, & Cognition, 29*, 813–825.

Echterhoff, G., Hirst, W., & Hussy, W. (2005). How eyewitnesses resist misinformation: Social postwarnings and the monitoring of memory characteristics. *Memory & Cognition, 33*(5), 770–782.

Ecker, U. K., Lewandowsky, S., & Tang, D. T. (2010). Explicit warnings reduce but do not eliminate the continued influence of misinformation. *Memory & Cognition, 38*(8), 1087–1100.

Eslick, A. N., Fazio, L. K., & Marsh, E. J. (2011). Ironic effects of drawing attention to story errors. *Memory, 19*, 184–191.

Fazio, L. K., Barber, S. J., Rajaram, S., Ornstein, P., & Marsh, E. J. (2013). Creating illusions of knowledge: Learning errors that contradict prior knowledge. *Journal of Experimental Psychology: General, 142*, 1–5.

Fotheringham, A. S., & Curtis, A. (1999). Regularities in spatial information processing: Implications for modeling destination choice. *The Professional Geographer, 51*(2), 227–239.

French, L., Garry, M., & Mori, K. (2008). You say tomato? Collaborative remembering leads to more false memories for intimate couples than for strangers. *Memory, 16*(3), 262–273.

French, L., Garry, M., & Mori, K. (2011). Relative–not absolute–judgments of credibility affect susceptibility to misinformation conveyed during discussion. *Acta Psychological, 136*(1), 119–128.

Gabbert, F., Memon, A., & Allan, K. (2003). Memory conformity: Can eyewitnesses influence each other's memories for an event? *Applied Cognitive Psychology, 17*(5), 533–543.

Gabbert, F., Memon, A., & Wright, D. B. (2006). Memory conformity: Disentangling the steps toward influence during a discussion. *Psychonomic Bulletin & Review, 13*(3), 480–485.

Garry, M., French, L., Kinzett, T., & Mori, K. (2008). Eyewitness memory following discussion: Using the MORI technique with a Western sample. *Applied Cognitive Psychology, 22*(4), 431–439.

Goodwin, K. A., Kukucka, J. P., & Hawks, I. M. (2013). Co-witness confidence, conformity, and eyewitness memory: An examination of normative and informational social influences. *Applied Cognitive Psychology, 27*(1), 91–100.

Hope, L., Ost, J., Gabbert, F., Healey, S., & Lenton, E. (2008). "With a little help from my friends. . .": The role of co-witness relationship in susceptibility to misinformation. *Acta Psychological, 127*(2), 476–484.

Howe, C., Tolmie, A., & Rodgers, C. (1992). The acquisition of conceptual knowledge in science by primary school children: Group interaction and the understanding of motion down an incline. *British Journal of Developmental Psychology, 10*(2), 113–130.

Karau, S. J., & Williams, K. D. (1993). Social loafing: A meta-analytic review and theoretical integration. *Journal of Personality and Social Psychology, 65*(4), 681–706.

Kärkkäinen, E. (2003). *Epistemic stance in English conversation: A description of its interactional functions, with a focus on I think*. Amsterdam/Philadelphia, PA: John Benjamins.
Marsh, E. J., Cantor, A. D., & Brashier, N. M. (2016). Believing that humans swallow spiders in their sleep: False beliefs as side effects of the processes that support accurate knowledge. *Psychology of Learning and Motivation, 64*, 93–132.
Meade, M. L., McNabb, J. C., Lindeman, M. I. H., & Smith, J. L. (2016). Discounting input from older adults: The role of age salience on partner age effects in the social contagion of memory. *Memory, 25*(5), 704–716. https://doi.org/10.1080/09658211.2016.1207783
Meade, M. L., & Roediger, H. L. (2002). Explorations in the social contagion of memory. *Memory & Cognition, 30*(7), 995–1009.
Mudd, K., & Govern, J. M. (2004). Conformity to misinformation and time delay negatively affect eyewitness confidence and accuracy. *North American Journal of Psychology, 6*, 227–238.
Muller, F., & Hirst, W. (2014). Remembering stories together: Social contagion and the moderating influence of disagreements in conversations. *Journal of Applied Research in Memory and Cognition, 3*(1), 7–11.
Park, S. H., Son, L. K., & Kim, M.-S. (2016). Social contagion in competitors versus cooperators. *Applied Cognitive Psychology, 30*, 305–313.
Peker, M., & Tekcan, A. İ. (2009). The role of familiarity among group members in collaborative inhibition and social contagion. *Social Psychology, 40*(3), 111–118.
Rajaram, S. (2011). Collaboration both hurts and helps memory: A cognitive perspective. *Current Directions in Psychological Science, 20*, 76–81.
Rapp, D. N. (2008). How do readers handle incorrect information during reading? *Memory & Cognition, 36*(3), 688–701.
Rapp, D. N. (2016). The consequences of reading inaccurate information. *Current Directions in Psychological Science, 25*, 281–285.
Rapp, D. N., & Donovan, A. M. (2017). Routine processes of cognition result in routine influences of inaccurate content. *Journal of Applied Research in Memory and Cognition, 6*(4), 409–413.
Rapp, D. N., Hinze, S. R., Kohlhepp, K., & Ryskin, R. A. (2014). Reducing reliance on inaccurate information. *Memory & Cognition, 42*(1), 11–26.
Rapp, D. N., Jacovina, M. E., & Andrews, J. J. (2014). Mechanisms of problematic knowledge acquisition. In D. N. Rapp & J. L. G. Braasch (Eds.), *Processing inaccurate information: Theoretical and applied perspectives from cognitive science and the educational sciences* (pp. 181–202). Cambridge, MA: MIT Press.
Rapp, D. N., & Salovich, N. A. (2018). Can't we just disregard fake news? The consequences of exposure to inaccurate information. *Policy Insights from the Behavioral and Brain Sciences, 5*(2), 232–239.
Roediger, H. L., Meade, M. L., & Bergman, E. T. (2001). Social contagion of memory. *Psychonomic Bulletin & Review, 8*(2), 365–371.
Rush, R. A., & Clark, S. E. (2013). Social contagion of correct and incorrect information in memory. *Memory, Ahead-of-Print*, 1–12. https://doi.org/10.1080/09658211.2013.859268
Simms, A., & Nichols, T. (2014). Social loafing: A review of the literature. *Journal of Management Policy and Practice, 15*(1), 58.
Springer, L., Stanne, M. E., & Donovan, S. S. (1999). Effects of small-group learning on undergraduates in science, mathematics, engineering, and technology: A meta-analysis. *Review of Educational Research, 69*(1), 21–51.

Suleiman, J., & Watson, R. T. (2008). Social loafing in technology-supported teams. *Computer Supported Cooperative Work (CSCW)*, *17*(4), 291–309.
van Boxtel, C., van der Linden, J., & Kanselaar, G. (2000). Collaborative learning tasks and the elaboration of conceptual knowledge. *Learning and Instruction*, *10*(4), 311–330.
Wright, D. B., & Villalba, D. K. (2012). Memory conformity affects inaccurate memories more than accurate memories. *Memory*, *20*(3), 254–265.

7 Central Coordination and Integration of Diverse Information to Form a Single Map

Elizabeth R. Chrastil and You (Lily) Cheng

Introduction

A platoon of soldiers explores a city in which all previous maps and information sources have been rendered obsolete due to the destruction of war. The soldiers must communicate information about the city back to a central location and then move to a new part of the city as quickly as possible. The task is a high-stakes version of the game Telephone, with information passed back and forth, changing hands along the way. The potential for distortion and misinterpretation is high, even more so due to the stressful situation. Therefore, a large number of factors could affect the communication of spatial information and integration into a common map.

Because of the complexity of this task, we will focus on the specific challenge of the leader's integration of information they receive from different scouting groups, which we term "central coordination." Although the scout and squad team are also critical to understanding collective spatial cognition, here we focus on the leader to provide a tighter examination of the problem. This challenge requires integrating spatial information that comes from multiple sources, all derived from ground-level perspectives of the environment. Furthermore, that information must be communicated remotely through verbal means—without visual aids—which could complicate this integration. The stress and urgency of the situation compound these factors even more.

This chapter largely focuses on specifying the problem space and how to study it, which is a critical first step in the road to establishing guidelines for optimizing behavior during this challenge. First, we will delve into possible factors that could affect the outcomes of this scenario, as well as central coordination of spatial information more broadly. Then, we will propose three methods for testing these factors, each with several opportunities for full lines of research. Simply understanding the richness of this problem space will be a major undertaking for years to come.

Assessment of the factors involved

To complete this task successfully, there are several factors that must be considered, ranging from individual abilities to the overall assessment of risk of the situation.

DOI: 10.4324/9781003202738-10

The spatial abilities of each person or squad

The spatial abilities of the individuals and squads are a major factor for successful task completion. The scout must be able to learn the spatial information about the city from the ground level. Individuals differ in their abilities to learn the spatial arrangement of complex environments, with knowledge ranging from simple landmarks to a complete metrically accurate understanding of the city. Although this problem of individual abilities is not fully resolved, there has been a reasonable amount of research on this topic (Condon et al., 2015; Hegarty et al., 2002; Ishikawa & Montello, 2006; Marchette et al., 2011; Ngo et al., 2016; Weisberg et al., 2014; Wolbers & Hegarty, 2010). These endeavors have made strides in categorizing people into different navigational types (Hegarty et al., 2002; Ishikawa & Montello, 2006; Marchette et al., 2011; Weisberg et al., 2014) or correlating navigational ability with different cognitive factors (Condon et al., 2015; Ngo et al., 2016; Wolbers & Hegarty, 2010).

Although an in-depth examination of individual navigational abilities is outside the scope of this chapter, it is vital that the leader is aware of the abilities of the members of the platoon. The leader needs to determine how much each person or squad can be tasked with. If everyone's navigation ability is similar, the leader may simply split up the navigation responsibility among soldiers. More consideration needs to be given when there is an imbalance in navigation ability. Members with low spatial abilities might be sent to explore areas that are less likely to be complex or that are closer to the base or might be given non-spatial tasks. The leader might not send individuals with low abilities out on their own but rather group them with people who have higher spatial abilities. Research suggests that spatial abilities might not be a singular entity (Wolbers & Hegarty, 2010). Thus, it is possible that some scouts can best contribute to the exploration by converting first-person information into a map, while others might be better suited to remembering the series of turns they took during exploration. This grouping and splitting up of tasks could increase the safety of the team members (Elgar, 1989; Pulliam, 1973) and lead to more reliable spatial information (Brodbeck et al., 2007; Gaertner et al., 1990) sent back to the command center.

Finally, the platoon leader needs to have fairly high spatial abilities. It would be quite difficult for a person with low spatial abilities to integrate all the various spatial information stemming from multiple sources. The leader must have self-awareness of their own abilities and limitations in order to effectively lead the group and to derive a common map of the environment (Davidson, 1994; Markessini, 1996; Marshall-Mies et al., 2000).

Communication of spatial information

Once the members of the platoon obtain spatial information about the surrounding area, they must communicate it back to the leader. The ground-level soldier must determine the best way to verbally communicate the spatial information they have gleaned. Optimization of spatial descriptions is needed here (e.g., Aylett & Turk,

2004; Jaeger & Levy, 2007), but the best spatial descriptions may depend on how that information would be used. If a survey-like representation of the environment is ultimately needed, then the platoon members would likely need to provide information about the distances and directions between locations. In contrast, if the platoon is looking for a safe route through a neighborhood, then landmark information and the connections between different streets might be more important. The leader must interpret that communication to develop an understanding of the region, likely in the form of a physical map at headquarters. Route and survey instructions have been studied in some detail (Brunyé et al., 2008; Denis et al., 1999; Lawton, 2001; Lovelace et al., 1999; Meneghetti et al., 2011; Taylor & Tversky, 1992; Vogel & Jurafsky, 2010), including optimization, the resulting mental maps, and their interactions with individual navigational abilities.

The leader may change the type or amount of communication based on the spatial ability of the person sending in the information (Bell & Kozlowski, 2002; Yukl & Mahsud, 2010). The leader might ask more probing questions from people with lower abilities or ask them to fill in specific information, considering that unambiguous spatial information is easier to remember (e.g., Mani & Johnson-Laird, 1982). The leader might give those with better spatial abilities more latitude to communicate how they wish. On the other hand, the leader might ask for a standardized set of information or request that the information arrives in a standard format, in order to better integrate across multiple sources. The leader's knowledge of each individual's or group's spatial ability could also affect their judgment of the validity of the incoming reports (Stadtler & Bromme, 2014). Information coming from people with low spatial ability—rendering it potentially less reliable—might be given less weight than information coming from people with high spatial ability.

Critically, the scouts themselves cannot see the common map while it is being developed. Thus, they cannot give feedback to the leader as to whether the map matches what they are seeing on the ground. Consider a task in which you must direct someone how to draw a picture of an alien lifeform that you imagined. It would be difficult enough to get an approximation of your imaginings when you can see the drawing and are able to instruct the drawer how to correct their errors. Without being able to see the drawing, extremely precise language is needed to achieve anything close to the original. Likewise, sending spatial information into essentially a black hole without any feedback could compound errors and lead to major distortions (Butler et al., 2008; Butler & Roediger, 2008).

Coordination of multiple sources of potentially conflicting information

Communication with the headquarters originates from multiple soldiers, both in terms of multiple people within each scout group and multiple scout groups. Within each scout group, the members need to determine whether it is better to communicate uncertainty and disagreements within the group to headquarters or come to a single mutual solution. Different members of the group might have doubts about some information; the doubts themselves could potentially be useful to the leader (Ellsberg, 1961, 2015). The members of each scout group could also have unique information to communicate. For example, one scout could focus on safety hazards, while another measures distances between locations, and a third tracks the landmarks in each street. Together, they need to communicate a single coherent message to the leader for central coordination. Soldiers who are trained in communication can facilitate the process of information integration.

The coordination of information across multiple scout groups brings a different set of challenges, and the leader needs to make decisions about how to best deploy their scouts. Is it better to have multiple accounts of the same region to improve accuracy or to make a wider sweep of the area but with a greater risk of incorrect accounts (e.g., Gagnon et al., 2018)? The leader needs to command the units to explore the gaps or be willing to leave some areas unexplored. The leader must weigh the benefits of each approach. Overlapping information could facilitate the process of creating a high-quality common map. If the sources converge, the leader can feel more confident in the final outcome (Little, 1961). If the sources disagree, then the leader must weigh the reliability of these conflicting accounts, based on either what they know about the abilities of the unit or the number of diverging accounts they get. Approaches to weighing reliability include using probability theory and Bayesian approaches, examining constraints on the system, or developing heuristics to simplify complex problems (Brown & Duren, 1986; Plous, 1993; Zhao et al., 2012). The leader could also probe for more information by sending additional people to explore the area of uncertainty (Gwet, 2014) or by asking clarifying questions of the scouts in the area (Altman & Bland, 1994; Pulford et al., 2014). Leaders could be trained to have expertise in this information coordination process.

Furthermore, several factors could contribute to the effectiveness of information coordination in the leader–follower dynamics. This includes psychological capacities (e.g., self-confidence, optimism, and secure-attachment styles) of military leaders (Mayseless, 2010; Popper et al., 2004), an individual's affective state (e.g., sad vs. happy) in information processing (Forgas, 1999), attitudes toward conflict (Erb et al., 2002; Mackie, 1987), as well as the group members' level of affiliation to each other (Burger et al., 2004, 2001; Cialdini & Trost, 1998). In addition, it is also worth mentioning that effective information coordination takes time to develop (Gilbert, 1991).

Finally, it is important to note that in some situations, the information could be coming from technology, such as drone information or satellite imagery. This information could conflict with reports on the ground. Although information stemming from technology might seem more reliable than information coming from humans, other factors could limit their usefulness. For example, these sources often take an overhead perspective, rather than a first-person view, which could differ in the type of information acquired and how the information is used. A full discussion of the impact of integrating technology into the coordination is beyond the scope of this chapter, but it is important to keep this factor in mind as emerging technologies become more prevalent.

Perception of risk

In this platoon scenario, risk comes from the potential military dangers in the unknown territory and during the deployment to the target zone (e.g., snipers and improvised explosive devices), but also from getting lost. The leader must weigh the benefits of gaining information about a particular region against the potential risks to the safety of the unit (Mehlhorn et al., 2015). Is leaving holes in knowledge with just a "here be monsters" warning sufficient if that area is deemed too dangerous to enter? If the leader determines that all parts of the environment must be explored no matter the cost, then they must decide which teams would investigate and whether the scouts would use a different strategy in dangerous areas.

The perception of risk could also vary depending on the kind of risk (Arabie & Maschmeyer, 1988). Perhaps the most high-impact risk is something like an explosion or sniper, but the higher likelihood comes from getting lost (Lawton et al., 1996). People tend to weigh losses more than gains (Tversky & Kahneman, 1981), so the leader is more likely to be cautious in the face of a large danger. People also tend to have large errors in estimating the likelihood of events and the impacts of those events (Dawson & Arkes, 1987; Hastie & Dawes, 2001), so it is vital to understand how the scouts and leader will respond to the specific risks in this scenario.

The effects of stress

The stress of the situation could lead to numerous distortions of spatial information, including the distance traveled, the reliability of memory, and the chance that a wrong turn would be taken. Previous research in navigation situations has largely focused on time stress, with mixed results. Some researchers have found limited effects of time pressure on spatial knowledge (Credé et al., 2019), while others have found that time pressure leads to increased reliance on well-known routes (Brunyé et al., 2017). However, the effects of other types of stress on navigation have only begun to be studied in depth (Boone, 2019).

On the other hand, the effects of stress on memory, perception, and decision-making are more well-known. For example, increases in cortisol—one of several stress-related hormones, although it has other functions as well—can lead to increased short-term memory but impairments in attention (Vedhara et al., 2000). Stress and other emotional factors can impair memory consolidation, which could prevent the encoding and retrieval of key spatial information (de Quervain et al., 1998; Lupien & Lepage, 2001; Roozendaal et al., 2009). Interestingly, research has also found that stress can impact memory differently depending on time phases: stress occurring prior or during the memory encoding stage generally impaired memory and stress occurring after memory encoding could improve memory (Shields et al., 2017). Therefore, it is important to measure stress-related responses at different stages of the task to have a thorough assessment of its impact. Stress can affect decision-making by biasing people toward habit-based decisions (Dias-Ferreira et al., 2009). The platoon leader needs to understand the effects of stress on decision-making and memory and take those effects into account when processing the incoming information.

The effects of the environment

The complexity of the environment could have a major effect on the formation of a common cognitive map (O'Neill, 1991a, 1991b). Indeed, an entire detailed analysis of central coordination could be made on just this topic alone. Complexity could come in the form of the number of intersections between roads (topological complexity), roads that are curved rather than straight, intersections that are not at right angles, environments with variation in elevation, a lack of clear local or distal landmarks, or a lack of visibility between different areas of the environment. A simple environment might be easier for the scouts to understand and for the leader to recreate than one with a lot of twists and turns. The complexity of the environment could also interact with the scouts' and the leader's spatial abilities such that those with higher spatial abilities would be more likely to learn the environment and create an accurate common map (Ishikawa & Montello, 2006; Weisberg et al., 2014; Wen et al., 2011). In addition, a city heavily damaged by war could be treated like a city built in a very unsystematic way, which may force people to rely more on local landmarks (Lawton, 2001).

Researchers have found that the complexity of the paths in the environment, such as the number of turns or the topological connections between locations, can greatly impact how well people are able to navigate (Carlson et al., 2010; O'Neill, 1991b; Slone et al., 2015; Weisman, 1981). To address these ideas, attempts have been made to quantify the complexity of environments (e.g., space syntax; Hillier et al., 1976). To fully appreciate the difficulty of the navigation challenge addressed here, a complete assessment of the complexity of the environment must be conducted. Alternatively, researchers could investigate how leaders and scouts coordinate and integrate spatial information in each specific environment.

In summary, a number of potential factors could reduce the accuracy of the collective map derived by the leader. Interestingly, we mention most of these factors because they likely reduce the quality of the leader's map, whereas very few factors serve to facilitate this process.

Approaches to studying these factors

Because our scenario delves into relatively uncharted territory, each one of these considerations could be studied on their own with a full line of research. Here, we propose three broad approaches to studying these considerations: real-world exploration of a new environment with remote communications systems, experimental scenarios of the leader integrating the information, and training scenarios to improve integration. These three approaches are all experimental to some degree, ranging from observing behaviors under varying conditions to a full training intervention.

The three approaches connect to the six factors identified earlier (spatial abilities, communication, conflicting information, risk, stress, and the environment) in various ways. We probe spatial abilities by observing how people with different levels of training act in this situation, as well as by explicitly manipulating the abilities of the group. Spatial communication runs throughout and is used primarily as an outcome measure to see what style of communication is used, for example, what kinds of questions the leader asks of the squad. However, communication is indirectly manipulated through spatial abilities and conflicting information. Conflicting information is more directly manipulated here as well, along with the presentation of irrelevant or incorrect information. Perception of risk is tested in conjunction with stress by manipulating danger levels and observing how stress response changes. Finally, the role of the environment is tested both through observing behavior in a particular terrain and by manipulating the type of environmental information the leader has access to. Through both observation and direct testing, we aim to examine these six factors directly. Our third approach—training—allows us to determine whether performance related to these factors can improve.

Real-world exploration of a new environment with remote communications systems

The first approach is broadly exploratory and qualitative but will lend insight into some of the challenges involved in our scenario and will establish some of the main issues in its problem space. This approach involves creating a smaller and safer version of the challenge. Pairs of scouts would be deployed to explore a new neighborhood (or in a virtual or mixed reality setting; Khooshabeh et al., 2017; Khooshabeh & Lucas, 2018), communicating information back to a central command. This approach makes a major contribution in understanding what types of information is communicated, how that information is interpreted, and what gaps remain in our understanding. The focus of these studies could be on the team leader, the scouts, or both. The three primary components of the analysis would be a detailed examination of the language used during the communication, the proportion of the environment covered, and the fidelity of the final map.

The spatial language used during this task would be analyzed for the types of spatial terms the scouts use. These terms include cardinal directions, distances in terms of a metric measure or time, and directions in terms of angles or related to an egocentric bearing. Landmark information would be analyzed for location, either in terms of latitude and longitude or relative location compared with other landmarks, and distinguishing features of the landmark. In addition, the language would be analyzed for the relative amounts of allocentric and egocentric terms. For example, the scouts could describe the space largely in terms of the route they took. For the team leader, analysis would focus on the kinds of clarifications and requests they make, such as directions, distances, or sizes, which can reveal where the communication is lacking.

Another outcome measure is the proportion of the total area that was covered during the experiment. This measure looks at the tradeoff between accuracy and coverage. For example, if the team covers only 25% of the area but in great detail, the outcomes could be very different than if the team covers the entire space but in less detail. In some situations, more detail might be more important, whereas in others, having a complete map is more valued. These observational studies would provide insight into the circumstances that lead to more or less exploration and coverage of the navigational area.

The third outcome measure is the comparison between the final composed map and an accurate map of the environment. Bidimensional regression (Tobler, 1994) between locations on the composed map and the actual location would measure metric placement errors. A map-drawing analyzer (e.g., Gardony et al., 2016) would be used for this process. In addition, the map would be analyzed for accurate topological connections between the roads, known as graph information (Chrastil & Warren, 2014). The number of identified landmarks would also be counted.

This outcome measure can be compared with the language analysis to assess how the spatial language during communication contributes to the final map. It is worth noting that the three outcome measurements stated here are just representative ones. Researchers should be encouraged to apply additional measures to get concrete answers to their specific questions. For example, each soldier could draw a sketch map at the end of the task to compare against the integrated final map to see what information has been changed and why.

In these real-world observational studies, pilot testing would be needed to determine the ideal amount of terrain that should be tested in a reasonable time frame. Presumably, it would be of interest to have a terrain that could not quite be covered in the allotted time, forcing the scouts to make decisions about which areas might be most important to explore and to how much detail is needed for any given section of the environment. Scouts would simply be instructed to provide as much information back to the leader as they think is necessary, with no guidelines as to how often they need to report back. However, the leader is free to request additional information as much as they need as well. Thus, the instructions are quite open-ended without much constraint. These open instructions allow for the experimenters to observe more naturalistic behavior, in whatever manner the scouts and leaders intuitively fall into during the course of testing. In real situations, soldiers may take more specific orders or instructions on whether breadth or accuracy is the objective in scouting an area. Therefore, researchers could also give different instructions to different groups of participants, such as "cover as wide a range as possible" or "make the recorded spatial information as accurate as possible," to study how specific instructions impact the scouts' collective spatial processing of information.

Three different experimental manipulations can be conducted from this paradigm.

The effects of training

This experimental manipulation examines the role of training. The scouts come from two broad categories, either untrained college students or people trained in spatial information (e.g., soldiers, first responders, and orienteers). In this experiment, comparisons would be made between scout groups. The leader could be kept constant in order to focus on the differences between scout groups or the leader could come from the same background as the troops in order to focus on the interactions between people with similar training. For example, one possible experimental design would fully cross the possibilities: leader-trained, scouts-trained; leader trained, scouts-untrained; leader-untrained, scouts-trained; and leader-untrained, scouts-untrained. The comparisons between groups focus on the three outcome measures outlined earlier, seeing how the spatial language, proportion explored, and final map fidelity vary under these differing conditions.

The involvement of the team leader

This experimental manipulation hones in on the team leader, such as how to ask the best questions to obtain the necessary information and what kind of communication is most effective. In this situation, the scouts would be confederates with the experimenter, trained to provide only a basic set of information to start. The scouts would only provide additional information when requested by the team leader. However, they have access to the full information of the environment—indeed, they do not actually need to be in the field at all. This experiment would test multiple team leaders to determine differences in the information requested and to determine what kinds of queries for information lead to the most accurate maps. For example, the team leader could be a trained first responder who might know what kind of information is needed or they could be relatively untrained but could hit upon crucial information by intuition or chance. For this experiment, the spatial abilities of the team leader would be extremely important to assess and control for. However, it is likely that those with higher spatial abilities would better be able to identify the needed spatial information than those with lower abilities.

Comparisons across different environments

The final experimental manipulation tests the contribution of the environment. In this experiment, the same scout team and leader would complete the task in several different environments (in counterbalanced order across groups). Thus, any differences in results can largely be attributed to the environment, rather than the composition or abilities of the scout team and leader. Comparisons would be made across differing terrain and environmental complexity, such as the topological complexity of the roads, how much of a grid system the streets are in, the amount of elevation changes, the visibility between different parts of the environment, how similar or distinct the landmarks and other buildings are, and the visibility of large external landmarks. Each of these factors could have substantial variation, and thus, the comparison of environmental conditions could explode into a huge experimental design. With that consideration, it is recommended to examine the variation within only one of these levels of complexity at a time, such as only examining differing degrees of elevation changes in the environment. This within-subjects design provides insight into what kind of information is acquired in these different environments, what types of spatial language interactions are best under certain environmental conditions, and how exploration behavior changes in different environments.

Experimental scenarios of the leader integrating the information

Our second approach examines how the team leader integrates the information, using controlled experimental methods. In both real and virtual environment scenarios, this approach manipulates the incoming information, what the leader knows about the abilities of the scouts, and the degree of stress involved. The leader would be tasked with either creating a new map from scratch or updating the previous (now fairly obsolete) map, and the leader may ask for additional information from the teams. In this paradigm, the information is controlled by the experimenter such that the information is either fixed for all leaders or the scout teams are confederates with the experimenter. Thus, these more experimental approaches test how leaders handle different levels of information or different environmental scenarios. While the observational approach keeps things fairly open ended and notices what happens in broad situations, the experimental approach focuses on the team leader and more directly manipulates what information they receive.

Outcome measures for this approach focus on the spatial language used, the proportion of the environment that has been updated, and the fidelity of the final map, the same as in the first experimental paradigm. In addition, physiological responses of the leader would be recorded to gauge stress levels. The physiological measures include heart rate, respiration rate, salivary cortisol levels, and galvanic skin conductance. Salivary cortisol would be collected prior to the study, multiple times during the experiment with the number and frequency determined by the specific question to be answered, and at the completion of the experiment. The other physiological measures would be acquired continuously through biomonitors attached to the leader. Stress-related responses measured across the whole task would help us better understand the effect of stress on the soldiers' performance.

Under this paradigm, five experimental manipulations can be conducted. These experiments mirror the factors identified in the first part of this chapter.

Some scout groups have better spatial abilities than others

This experiment would manipulate the spatial abilities of the different scout teams. Members of the scout teams would either be confederates or be pre-screened to achieve certain levels of spatial abilities. In some scenarios of this experiment, the leader would be aware of the spatial abilities of the teams, whereas in others, the leader must ascertain the spatial abilities of the team during the course of the information gathering stage. As a within-subjects design, the leader would have multiple experiences with different types of spatial abilities, so their reaction to differing levels of spatial abilities would be assessed, with a particular focus on spatial language and type of information requested, although the area covered and fidelity of the final map would also be vital factors to consider.

Conflicting accounts from different scout teams

This experiment tests how the leader deals with conflicting accounts from different scout teams. One scout team provides the information that conflicts with another team for several of the regions in the environment. The leader must figure out how to resolve the conflict, which could include determining which information is more reliable, reaching a compromise between the two, or asking for additional information that could resolve the issue. The type of additional information requested and the fidelity of the final map would be the primary outcome measures. The physiological response to this cognitive stressor would also be very important to understanding how people deal with conflicting spatial information.

Dealing with irrelevant, incorrect, or extraneous information

The scout teams would provide information that is not useful, either by being irrelevant or extraneous or by being completely incorrect. The leader must first be able to determine what information is incorrect or irrelevant, and then, they must obtain the correct information they need from the scout units. This experiment would measure the number of detections of incorrect or irrelevant information and the number and type of follow-up questions that the leader asks to gain clarification. It would also be of interest to determine whether the leader would begin to trust certain sources less when that source provides inconsistent information.

Changing risk levels and increased stress

In this experiment, the risk and stress levels are manipulated while keeping the leader constant. Risk levels would be modified by increasing the dangers of the virtual environment, including increased sniper fire, explosions, or environmental changes such as road closures, building collapses, or bad weather. These changing risk levels are expected to lead to increased stress for the leader, which would be monitored using the physiological measures. The larger aim of this experiment is to determine how the integration of information changes under stress. Thus, the primary outcome measures would be changed in the spatial language used to communicate with the scout teams, which are expected to be accompanied by reductions in the accuracy of the final common map.

Differing environmental conditions

Although the effects of the environment were examined in the first experimental phase by holding the scouts and leader constant, in this case, the focus is on the interaction between the environment and the information available to the leader. In this experiment, the leader would be given different sets of information for three different environments (counterbalanced). Unbeknownst to the leader, the environment is actually the same for all three, but they have access to differing degrees of information. In one case, the leader would be given street connections but not elevation or other terrain information. In another case, the leader would be given terrain information, but not street information. In the final case, the leader would be given both street and terrain information. Other variations on this idea can also be explored, using the variabilities in the environment described earlier.

Training scenarios to improve integration

Our third experimental approach is the most mechanistic because it involves conducting a training intervention to improve the outcomes of the scenario. Once the potential key features in this communication have been identified and tested in the first two experimental paradigms, we can develop training paradigms based on those hypothesized features. For example, suppose that the previous studies find that increased stress levels make a large impact on the leader's ability to form a common map, but the complexity of the environment does not particularly change the outcomes. In that case, the training would focus on how to mitigate the effects of stress, either through exercises to reduce the physiological responses to stress or by developing a task protocol that removes some of the burden from the individual leader. Thus, the particulars of the training depend somewhat on the outcomes of the previous two approaches and could take a variety of paths. For example, training efforts can be conducted to focus on the scout groups, on the team leader, or their interactions.

Indeed, even identifying a single characteristic to focus on to improve collective spatial cognition could lead to any number of training paradigms for improvement. Developing and testing one aspect of training is another area that could take years of dedicated research to perfect. In the example above of training to reduce stress levels, the development of exercises to reduce stress response would have many iterations of testing over several candidate theories for the biggest reductions in stress response. Testing would also have to be done on whether these reductions in stress response also improve the leader's ability to generate an accurate map.

Broadly speaking, training approaches would focus on either changing behaviors or developing protocols. Changing behaviors means that the members of the platoon would be trained in how to optimize this challenge by altering the spatial language that they use, learning best practices for balancing the incoming information, introducing breathing techniques to reduce stress, or providing training in different terrain types. Protocol development attempts to remove the cognitive load from individuals so that they follow an established set of procedures, which maximizes the best practices of communication. For example, this experimental approach could establish a script for the communication of spatial information or checklists that must be followed to ensure that all the information has been collected.

The primary outcome measure of the training approach would be improvements in the final common map made by the leader and the amount of area covered, as described earlier. Unlike the previous two approaches that also examined the language of the communication, here the primary concern is the final outcome. Language is certainly a factor when developing the training and protocols, but it is a means to the desired end of this approach of an accurate map of the environment. However, the extent to which the scouts and leaders stick to the established protocols and best practices of spatial language would also be assessed.

Training studies in this line would follow the same general approach. Baseline measures of map production would be assessed for all leaders, under the same experimental conditions (i.e., having the same terrain and scouts for all leaders). The leaders would then be divided into two groups: a training group or an active control group. The training group would undergo the training procedures that are to be tested. The active control group would not receive the training but would undergo some other type of activity that takes similar time and attention. For example, the active control group could undergo the current (unimproved) standards of training or could watch instructional videos about the importance of leadership. This control ensures that any effects observed are the result of the training specifically and not just general participation in a training study. After the training period, both groups would be tested on a new terrain and with a different scout group (although all participants would be tested using the same terrain and scouts). Prior to the study, pilot testing must be done to confirm that the pre- and post-test conditions are equally difficult. Performance would be compared between baseline and after training. The training group would be expected to show significant improvement compared to the control group.

In summary, these three experimental approaches provide a broad outline for examining the factors and constraints on the leader in this scenario. The first approach gathers information from the situation and hones in on the problem, observing how the interactions and common maps change under differing conditions. The second approach manipulates those factors experimentally, with an emphasis on how the leader responds to these various conditions. The third approach develops and tests training interventions to optimize the final outcome of a common map of the environment. The intervention approach can provide stringent tests of hypotheses about the cognitive mechanisms underlying the formation of a common cognitive map. Together, these experimental approaches follow the trajectory of studying a relatively new area of scientific inquiry from observation to mechanism.

Synthesis and conclusions

This chapter has largely focused on specifying the problem space and how to study it, which is a critical first step before prescriptive guidelines can be established. We proposed that spatial abilities, communication of spatial information, coordination of multiple sources of potentially conflicting information, the perception of risk, the effects of stress, and the complexity of the environment are the primary factors that would influence—usually in a negative way—the outcomes in a scenario like this. These factors must be examined experimentally to determine whether and how they contribute to the difficulties in forming a collective common map by a platoon exploring a destroyed urban area. We proposed three levels of experimental approaches—observational, controlled laboratory manipulations, and training interventions—that can be used to study these factors and develop optimal solutions to this challenge.

The study of collective spatial cognition is in its infancy. Thus, we have suggested both baby steps and larger pathways for growth and development. The richness of this problem space leaves open a wide opportunity for research for years to come.

References

Altman, D. G., & Bland, J. M. (1994). Diagnostic tests. 1: Sensitivity and specificity. *BMJ: British Medical Journal, 308*(6943), 1552.

Arabie, P., & Maschmeyer, C. (1988). Some current models for the perception and judgment of risk. *Organizational Behavior and Human Decision Processes, 41*(3), 300–329.

Aylett, M., & Turk, A. (2004). The smooth signal redundancy hypothesis: A functional explanation for relationships between redundancy, prosodic prominence, and duration in spontaneous speech. *Language and Speech, 47*(1), 31–56.

Bell, B. S., & Kozlowski, S. W. J. (2002). Adaptive guidance: Enhancing self-regulation, knowledge, and performance in technology-based training. *Personnel Psychology, 55*(2), 267–306.

Boone, A. (2019). *The influence of the human stress response on navigation strategy and efficiency*. Santa Barbara: The University of California.

Brodbeck, F. C., Kerschreiter, R., Mojzisch, A., & Schulz-Hardt, S. (2007). Group decision making under conditions of distributed knowledge: The information asymmetries model. *Academy of Management Review, 32*(2), 459–479.

Brown, D. E., & Duren, B. G. (1986). Conflicting information integration for decision support. *Decision Support Systems, 2*(4), 321–329.

Brunyé, T. T., Rapp, D. N., & Taylor, H. A. (2008). Representational flexibility and specificity following spatial descriptions of real-world environments. *Cognition, 108*(2), 418–443.

Brunyé, T. T., Wood, M. D., Houck, L. A., & Taylor, H. A. (2017). The path more travelled: Time pressure increases reliance on familiar route-based strategies during navigation. *Quarterly Journal of Experimental Psychology, 70*(8), 1439–1452.

Burger, J. M., Messian, N., Patel, S., Del Prado, A., & Anderson, C. (2004). What a coincidence! The effects of incidental similarity on compliance. *Personality and Social Psychology Bulletin, 30*(1), 35–43.

Burger, J. M., Soroka, S., Gonzago, K., Murphy, E., & Somervell, E. (2001). The effect of fleeting attraction on compliance to requests. *Personality and Social Psychology Bulletin, 27*(12), 1578–1586.

Butler, A. C., Karpicke, J. D., & Roediger III, H. L. (2008). Correcting a metacognitive error: Feedback increases retention of low-confidence correct responses. *Journal of Experimental Psychology: Learning, Memory, and Cognition, 34*(4), 918.

Butler, A. C., & Roediger, H. L. (2008). Feedback enhances the positive effects and reduces the negative effects of multiple-choice testing. *Memory & Cognition, 36*(3), 604–616.

Carlson, L. A., Holscher, C., Shipley, T. F., & Dalton, R. C. (2010). Getting lost in buildings. *Current Directions in Psychological Science*, *19*(5), 284–289.

Chrastil, E. R., & Warren, W. H. (2014). From cognitive maps to cognitive graphs. *PLoS One*, *9*(11).

Cialdini, R. B., & Trost, M. R. (1998). Social influence: Social norms, conformity and compliance. In D. T. Gilbert, S. T. Fiske, & G. Lindzey (Eds.), *The handbook of social psychology* (pp. 151–192). New York: McGraw-Hill.

Condon, D. M., Wilt, J., Cohen, C. A., Revelle, W., Hegarty, M., & Uttal, D. H. (2015). Sense of direction: General factor saturation and associations with the Big-Five traits. *Personality and Individual Differences*, *86*, 38–43.

Credé, S., Thrash, T., Hölscher, C., & Fabrikant, S. I. (2019). The acquisition of survey knowledge for local and global landmark configurations under time pressure. *Spatial Cognition & Computation*, 1–30.

Davidson, J. E. (1994). The role of metacognition in problem solving. In J. Metcalfe & A. P. Shimamura (Eds.), *Metacognition: Knowing about knowing* (pp. 208–226). Cambridge, MA: The MIT Press.

Dawson, N. V., & Arkes, H. R. (1987). Systematic errors in medical decision making. *Journal of General Internal Medicine*, *2*(3).

de Quervain, D. J.-F., Roozendaal, B., & McGaugh, J. L. (1998). Stress and glucocorticoids impair retrieval of long-term spatial memory. *Nature*, *394*(6695), 787–790.

Denis, M., Pazzaglia, F., Cornoldi, C., & Bertolo, L. (1999). Spatial discourse and navigation: An analysis of route directions in the city of Venice. *Applied Cognitive Psychology*, *13*(2), 145–174.

Dias-Ferreira, E., Sousa, J. C., Melo, I., Morgado, P., Mesquita, A. R., Cerqueira, J. J., Costa, R. M., & Sousa, N. (2009). Chronic stress causes frontostriatal reorganization and affects decision-making. *Science (New York, N.Y.)*, *325*(5940), 621–625.

Elgar, M. A. (1989). Predator vigilance and group size in mammals and birds: A critical review of the empirical evidence. *Biological Reviews*, *64*(1), 13–33.

Ellsberg, D. (1961). Risk, ambiguity, and the Savage axioms. *The Quarterly Journal of Economics*, 643–669.

Ellsberg, D. (2015). *Risk, ambiguity and decision*. Abdingdon: Routledge.

Erb, H.-P., Bohner, G., Rank, S., & Einwiller, S. (2002). Processing minority and majority communications: The role of conflict with prior attitudes. *Personality and Social Psychology Bulletin*, *28*(9), 1172–1182.

Forgas, J. P. (1999). Feeling and speaking: Mood effects on verbal communication strategies. *Personality and Social Psychology Bulletin*, *25*(7), 850–863.

Gaertner, S. L., Mann, J. A., Dovidio, J. F., Murrell, A. J., & Pomare, M. (1990). How does cooperation reduce intergroup bias? *Journal of Personality and Social Psychology*, *59*(4), 692.

Gagnon, K. T., Thomas, B. J., Munion, A., Creem-Regehr, S. H., Cashdan, E. A., & Stefanucci, J. K. (2018, June). Not all those who wander are lost: Spatial exploration patterns and their relationship to gender and spatial memory. *Cognition*, *180*.

Gardony, A. L., Taylor, H. A., & Brunyé, T. T. (2016). Gardony map drawing analyzer: Software for quantitative analysis of sketch maps. *Behavior Research Methods*, *48*(1), 151–177.

Gilbert, D. T. (1991). How mental systems believe. *American Psychologist*, *46*(2), 107.

Gwet, K. L. (2014). *Handbook of inter-rater reliability: The definitive guide to measuring the extent of agreement among raters*. Gaithersburg, MD: Advanced Analytics, LLC.

Hastie, R., & Dawes, R. (2001). *Rational choice in an uncertain world*. Thouand Oaks, CA: Sage Publications.
Hegarty, M., Richardson, A. E., Montello, D. R., Lovelace, K., & Subbiah, I. (2002). Development of a self-report measure of environmental spatial ability. *Intelligence*, *30*(5), 425–447.
Hillier, B., Leaman, A., Stansall, P., & Bedford, M. (1976). Space syntax. *Environment and Planning B: Planning and Design*, *3*(2).
Ishikawa, T., & Montello, D. R. (2006). Spatial knowledge acquisition from direct experience in the environment: Individual differences in the development of metric knowledge and the integration of separately learned places. *Cognitive Psychology*, *52*(2), 93–129.
Jaeger, T. F., & Levy, R. P. (2007). Speakers optimize information density through syntactic reduction. *Advances in Neural Information Processing Systems*, 849–856.
Khooshabeh, P., Choromanski, I., Neubauer, C., Krum, D. M., Spicer, R., & Campbell, J. (2017). Mixed reality training for tank platoon leader communication skills. *2017 IEEE Virtual Reality (VR)*, 333–334.
Khooshabeh, P., & Lucas, G. (2018). Virtual human role players for studying social factors in organizational decision making. *Frontiers in Psychology*, *9*, 194.
Lawton, C. A. (2001). Gender and regional differences in spatial referents used in direction giving. *Sex Roles*, *44*(5/6), 321–337.
Lawton, C. A., Charleston, S. I., & Zieles, A. S. (1996). Individual- and gender-related differences in indoor wayfinding. *Environment and Behavior*, *28*(2), 204–219.
Little, K. B. (1961). Confidence and reliability. *Educational and Psychological Measurement*, *21*(1), 95–101.
Lovelace, K. L., Hegarty, M., & Montello, D. R. (1999). *Elements of good route directions in familiar and unfamiliar environments* (pp. 65–82). Berlin and Heidelberg: Springer.
Lupien, S. J., & Lepage, M. (2001). Stress, memory, and the hippocampus: Can't live with it, can't live without it. *Behavioural Brain Research*, *127*(1–2), 137–158.
Mackie, D. M. (1987). Systematic and nonsystematic processing of majority and minority persuasive communications. *Journal of Personality and Social Psychology*, *53*(1), 41.
Mani, K., & Johnson-Laird, P. N. (1982). The mental representation of spatial descriptions. *Memory & Cognition*, *10*(2), 181–187.
Marchette, S. A., Bakker, A., & Shelton, A. L. (2011). Cognitive mappers to creatures of habit : Differential engagement of place and response learning mechanisms predicts human navigational behavior. *Journal of Neuroscience*, *31*(43), 15264–15268.
Markessini, J. (1996). Executive leadership in a changing world order: Requisite cognitive skills. *The First Literature Review*. Cae-Link Corp Alexandria Va Link Training Services Div.
Marshall-Mies, J. C., Fleishman, E. A., Martin, J. A., Zaccaro, S. J., Baughman, W. A., & McGee, M. L. (2000). Development and evaluation of cognitive and metacognitive measures for predicting leadership potential. *The Leadership Quarterly*, *11*(1), 135–153.
Mayseless, O. (2010). Attachment and the leader—follower relationship. *Journal of Social and Personal Relationships*, *27*(2), 271–280.
Mehlhorn, K., Newell, B. R., Todd, P. M., Lee, M. D., Morgan, K., Braithwaite, V. A., Hausmann, D., Fiedler, K., & Gonzalez, C. (2015). Unpacking the exploration–exploitation tradeoff: A synthesis of human and animal literatures. *Decision*, *2*(3), 191.

Meneghetti, C., Pazzaglia, F., & De Beni, R. (2011). Spatial mental representations derived from survey and route descriptions: When individuals prefer extrinsic frame of reference. *Learning and Individual Differences*, *21*(2), 150–157.

Ngo, C. T., Weisberg, S. M., Newcombe, N. S., & Olson, I. R. (2016). The relation between navigation strategy and associative memory: Individual differences approach. *Journal of Experimental Psychology: Learning, Memory, and Cognition*, *42*(4), 663–670.

O'Neill, M. J. (1991a). Effects of signage and floor plan configuration on wayfinding accuracy. *Environment and Behavior*, *23*(5), 553–574.

O'Neill, M. J. (1991b). Evaluation of a conceptual model of architectural legibility. *Environment and Behavior*, *23*(3), 259–284.

Plous, D. (1993). *The Psychology of Judgment and Decision Making*. Philadelphia, PA: Temple University Press.

Popper, M., Amit, K., Gal, R., Mishkal-Sinai, M., & Lisak, A. (2004). The capacity to lead: Major psychological differences between leaders and nonleaders. *Military Psychology*, *16*(4), 245–263.

Pulford, J., Siba, P. M., Mueller, I., & Hetzel, M. W. (2014). The exit interview as a proxy measure of malaria case management practice: Sensitivity and specificity relative to direct observation. *BMC Health Services Research*, *14*(1), 628.

Pulliam, R. H. (1973). On the advantages of flocking. *Journal of Theoretical Biology*, *38*.

Roozendaal, B., McEwen, B. S., & Chattarji, S. (2009). Stress, memory and the amygdala. *Nature Reviews Neuroscience*, *10*(6), 423–433.

Shields, G. S., Sazma, M. A., McCullough, A. M., & Yonelinas, A. P. (2017). The effects of acute stress on episodic memory: A meta-analysis and integrative review. *Psychological Bulletin*, *143*(6), 636.

Slone, E., Burles, F., Robinson, K., Levy, R. M., & Iaria, G. (2015). Floor plan connectivity influences wayfinding performance in virtual environments. *Environment and Behavior*, *47*(9), 1024–1053.

Stadtler, M., & Bromme, R. (2014). The content–source integration model: A taxonomic description of how readers comprehend conflicting scientific information. *Processing Inaccurate Information: Theoretical and Applied Perspectives from Cognitive Science and the Educational Sciences*, 379–402.

Taylor, H. A., & Tversky, B. (1992). Spatial mental models derived from survey and route descriptions. *Journal of Memory and Language*, *31*(2), 261–292.

Tobler, W. R. (1994). Bidimensional regression. *Geographical Analysis*, *26*(3), 187–212.

Tversky, A., & Kahneman, D. (1981). The framing of decisions and the psychology of choice. *Science*, *211*(4481), 453–458.

Vedhara, K., Hyde, J., Gilchrist, I., Tytherleigh, M., & Plummer, S. (2000). Acute stress, memory, attention and cortisol. *Psychoneuroendocrinology*, *25*(6), 535–549.

Vogel, A., & Jurafsky, D. (2010). *Learning to follow navigational directions*. Proceedings of the 48th Annual Meeting of the Association for Computational Linguistics, pp. 806–814.

Weisberg, S. M., Schinazi, V. R., Newcombe, N. S., Shipley, T. F., & Epstein, R. A. (2014). Variations in cognitive maps: Understanding individual differences in navigation. *Journal of Experimental Psychology: Learning Memory and Cognition*, *40*(3), 669–682.

Weisman, J. (1981). Evaluating architectural legibility. *Environment and Behavior*, *13*(2), 189–204.

Wen, W., Ishikawa, T., & Sato, T. (2011). Working memory in spatial knowledge acquisition: Differences in encoding processes and sense of direction. *Applied Cognitive Psychology*, *25*(4), 654–662.

Wolbers, T., & Hegarty, M. (2010). What determines our navigational abilities? *Trends in Cognitive Sciences*, *14*(3), 138–146.
Yukl, G., & Mahsud, R. (2010). Why flexible and adaptive leadership is essential. *Consulting Psychology Journal: Practice and Research*, *62*(2), 81.
Zhao, B., Rubinstein, B. I. P., Gemmell, J., & Han, J. (2012). A Bayesian approach to discovering truth from conflicting sources for data integration. *Proceedings of the VLDB Endowment*, *5*(6), 550–561.

8 Improving Wayfinding Through Transactive Memory Systems

Cynthia K. Maupin, Neil G. MacLaren, Gerald F. Goodwin, and Dorothy R. Carter

Introduction

Over the last five decades and more, researchers in geography and psychology have developed a rich understanding of how people perceive, use, think about, and navigate in space (cf. Gibson, 1954; Hart & Moore, 1973; Montello, 2001). This work has dwelled and developed primarily at the intersection of geography and cognitive psychology, and thus, it has largely focused on the spatial performance and capabilities of individuals. Indeed, great strides have been made in understanding both the impact of environmental features and human cognitive processing during wayfinding in real-world environments (e.g., Burgess, 2008; Golledge, 1999; Hart & Moore, 1973; Montello, & Raubal, 2013; Yesiltepe et al., 2021).

However, the prevalence of team-based organizational structures (e.g., Devine et al., 1999; Wuchty et al., 2007) requires researchers to shift how they consider spatial performance, to include not only individual capabilities but also collective capabilities for understanding and navigating through space. In comparison to the notable strides that have been made to understand individual spatial capabilities (e.g., Allen, 1999; Downs & Stea, 1973; Montello, 1993; Thorndyke & Hayes-Roth, 1982), a more limited body of research has begun to address how these concepts might exist in collectives (e.g., Dalton et al., 2019; Forlizzi et al., 2010; He et al., 2015; Hutchins, 1995); yet additional scholarly attention is needed in order to better understand the collective spatial cognition of teams operating in the natural world.

In parallel to these advancements regarding individual spatial capabilities, the concept of transactive memory systems (TMSs) has developed at the intersection of cognitive psychology and organizational science (Lewis & Herndon, 2011; Peltokorpi, 2008; Wegner, 1986, 1995), which may be useful in conceptualizing spatial capabilities at the team level. Transactive memory systems are cognitive structures that enable members of teams to understand how knowledge and capabilities, both shared and unique, are distributed among their fellow team members (Hollingshead, 2001; Wegner, 1986). Indeed, a rich body of research has developed to investigate the role of team transactive memory systems in understanding team coordination processes (e.g., DeChurch & Mesmer-Magnus, 2010; Lewis & Herndon, 2011;

DOI: 10.4324/9781003202738-11

Peltokorpi, 2008; Ren & Argote, 2011); thus, team collective cognition is ripe for integration with the collective spatial navigation literature.

Here, we leverage the transactive memory system concept as a launching point to elaborate how teams may perform spatial tasks—in particular, how they might distribute spatial information and processing within the team to perform an array of tasks relevant to effectively understanding and navigating a spatial environment. In the following sections, we (1) identify types of collectives for which spatial capabilities are critical, (2) describe the complex contexts these collectives operate within that require the ability to rely on collective spatial cognition, and (3) compare traditional team-based cognitive structures (i.e., transactive memory systems) with spatial cognition processes. Through this effort, we combine insights across multiple domains—including geography, cognitive psychology, and organizational science—to propose a comprehensive perspective for examining, understanding, and developing the collective spatial cognition abilities of action teams. We conclude with a series of propositions for future research endeavors at the intersection of collectives, cognition, and spatial navigation.

Spatial Cognition and Wayfinding

Understanding the intersection between human spatial cognition and wayfinding is a topic that has captured the attention of researchers across a variety of disciplines for a number of decades (Gibson, 1954; Golledge, 1999; Hart & Moore, 1973; Mark, 1993; Montello, 2001; Montello, & Raubal, 2013; O'Neill, 1991). From understanding the development of spatial awareness in children via a developmental psychology lens (e.g., Newcombe et al., 2013), to examining how tourists navigate a new city (e.g., Chang, 2013), to incorporating spatial capabilities to enhance wayfinding of robots (e.g., Epstein et al., 2015), this area has greatly advanced the way researchers understand individual spatial abilities. Yet understanding individuals' abilities to leverage their spatial cognition for wayfinding is only the first step toward understanding human spatial capabilities.

As people are inherently social creatures, much of our existence is experienced not as lone individuals, but rather in dyads (i.e., two people) or in groups (i.e., more than two people) (e.g., Bakeman & Beck, 1974; Burgess, 1984; James, 1951; Moreland, 2010; Wuchty et al., 2007). Thus, although individual spatial cognition and its impact on wayfinding provide a necessary foundation for understanding human navigation, the social dynamics that occur in groups can have a significant impact on this process as well (Dalton et al., 2019), which merits further investigation. Indeed, insights originating from spatial cognition and wayfinding at the individual level cannot be directly tied to actionable recommendations to enhance wayfinding for collectives without also considering the significant role social processes have on collectives' abilities to traverse through space.

Fortunately, a burgeoning area of research has begun to address the challenges associated with expanding spatial cognition from the individual level to the dyad, team, and larger group levels. Dalton and colleagues (2019) reviewed the literature in this area, concluding that although the number of studies was small, the studies that have examined collective navigation consistently noted that the social context of wayfinding was instrumental in influencing performance behaviors and outcomes. For example, He and colleagues (2015) studied pedestrian dyads navigating in unfamiliar urban areas. They found that successful versus unsuccessful dyads interacted in substantially different ways such that successful dyads communicated less and more succinctly and directly and unsuccessful dyads over-communicated, giving less useful directions. Additionally, Haghani and Sarvi (2017) examined how the physical space and social context impacted how groups navigated to escape during emergencies. Notably, they find that both physical environment and social cues interact together to impact the way individuals choose an evacuation route during emergency situations. Finally, although the context of the studies differed in some respects, studies by Forlizzi and colleagues (2010) and Bae and Montello (2019) had largely similar conclusions: relationships between specific individuals, as well as differences in task ability, led to notable differences in patterns and content of communication as well as measurable differences in objective performance outcomes.

A subtle feature common to the work reviewed here is that, in each case, the authors study a situation in which there is one navigation task to be accomplished, potentially divisible into several subordinate components. In fact, for much of the literature to date, differences in the relative position of study participants have not been of major focus. For instance, in the case of Forlizzi and colleagues (2010), the participants had no meaningful separation in position because the vehicle in which the participants were navigating was the relevant object moving through space. Likewise, in the Bae and Montello (2019) study, participants moved essentially side-by-side through space, were assessed on the basis of their ability to arrive at a certain point together, and faced no substantial individual differences in navigation challenges.

Our aim is to advance the developing research area on collective spatial cognition by clarifying the role that teams—and their multilevel processes—can have in spatial cognition and wayfinding. More specifically, we do so by exploring a specific type of team (i.e., the action team) in a dynamic context (i.e., complex, broken terrain) to demonstrate how insights from the team's literature (i.e., transactive memory systems) can be borrowed and integrated into the broader research paradigm for understanding collective spatial cognition. We demonstrate how integrating the teams, cognition, and spatial literature can promote exciting future research directions for geography, cognitive psychology, and organizational science.

Toward Collective Spatial Cognition

The Collective: Action Teams

Organizations that must perform effectively in spatially complex environments—such as the military, disaster relief agencies, and emergency response coalitions—rely on smaller subunits in the form of teams to effectively understand and navigate through space to accomplish goals. Salas and colleagues (1992) define teams as "a distinguishable set of two or more people who interact dynamically, interdependently, and adaptively toward a common and valued goal/objective/mission" (p. 4). Teams can function in a variety of contexts, with some requiring higher levels of environmental awareness and prowess than others.

A particular class of teams—action teams (Sundstrom et al., 1990)—have a significant spatial element to their performance. These teams are defined by their necessity to adapt to and coordinate within environments that are frequently dynamic, intense, and unpredictable, and thus require streamlined approaches for navigating these environments. Some examples of these teams might include military teams navigating through a war-torn urban landscape, first responders attempting to navigate through a rapidly burning building to locate and rescue inhabitants, or a disaster response team performing a wide array of tasks after a massive earthquake or wildfire. For discussion purposes, and keeping in mind parallel constraints and functional considerations in other types of non-military action teams, we will consider the spatial capabilities of teams that are "squad size." Infantry squads in the U.S. Army are typically composed of eight soldiers with a squad leader, while in the U.S. Marine Corps, they are composed of 12 marines with a similarly situated squad leader (e.g., Ancker & Scully, 2013). This is a prototypical size (9–13 personnel) for action teams in many military contexts around the world, as this structure is large enough for a squad to take on substantial military tasks and/or to allow for the squad to break into two or three smaller subunits (i.e., "fire teams") to facilitate accomplishing those tasks. This team size is also fairly common in other action team settings (e.g., firefighter companies and police squads).

Additionally, the moderate size of action teams—and their ability to break into smaller subunits—introduces multilevel complexities to their collective spatial navigation processes (see Figure 8.1). As an illustration, we consider a specific type of military task often assigned to an infantry squad: a patrol. The patrol, which can also refer to the unit doing the task, is a useful prototype for considering the multilevel nature of spatial navigation tasks because a patrol is typically operating separately from its parent unit, and a large component of the patrol's task is to move through space as a means to accomplish an assigned tactical mission (e.g., scouting the surrounding landscape). Patrol leaders must fix the position of their unit in space (i.e., a meso-level process); however, due to the patrol being composed of individuals with unique geographic positions (e.g., Ancker & Scully, 2013), the task is now a multilevel one: the relationship in space between the various patrol members may be small with respect to the overall distance traveled but is tactically meaningful. Furthermore, each individual faces their own wayfinding challenges (i.e., micro-level processes): each must fix their own position in space with respect to other members of the patrol—who may not always be visible—as well as dealing with route obstacles not faced by other patrol members, all while attending to their own tactical tasks (which are frequently distinct from the navigation task). In other words, there is a unit-level navigation task, typically addressed by a small number of patrol members, as well as many individual navigation tasks (i.e., as many as one per member of the patrol). Moreover, at a higher level, the patrol's

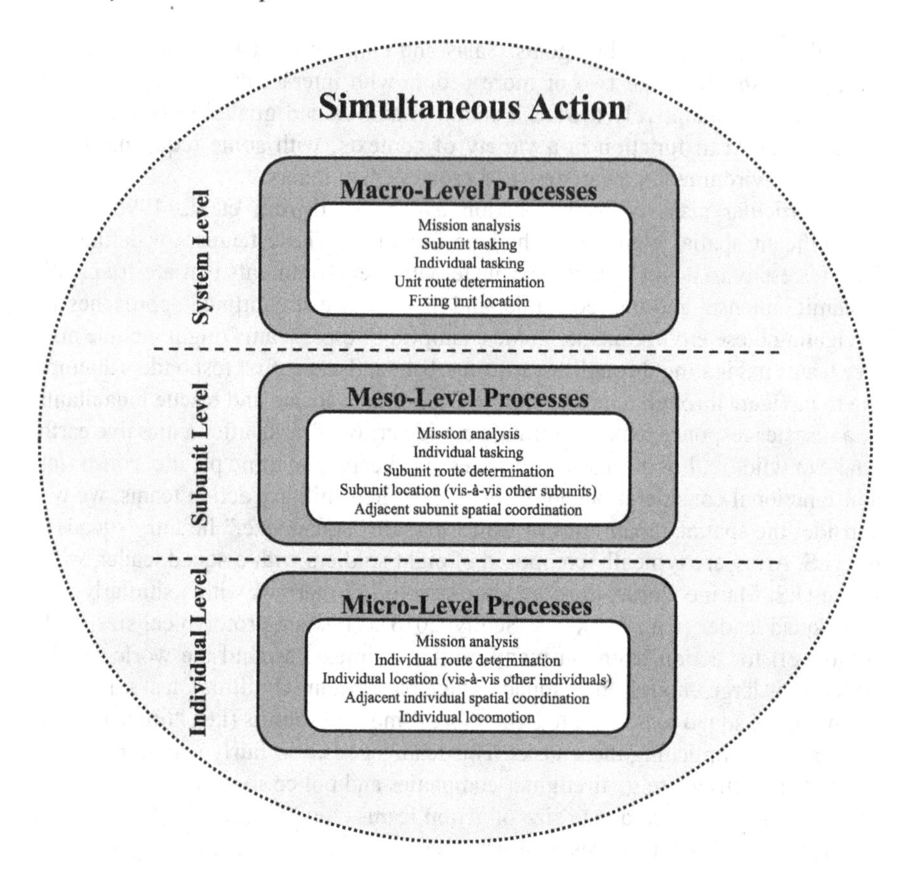

Figure 8.1 Multilevel simultaneous action processes for collective navigation of action teams.

parent unit may have several patrolling units, each of which must be aware of their location relative to each other, the parent unit, and other units in the same general area (i.e., macro-level processes).

Challenges frequently faced by these types of teams include coordinating movements with others when dispersed over varied terrain, sharing location or perspective-related information, and executing spatially relevant decisions, even when subgroups of the team are geographically dispersed (e.g., Flynn et al., 2008). The process of deciding how to separate into smaller subgroups while maintaining necessary spatial capabilities, which geographic areas those subgroups will be deployed to perform within, and how those subgroups will coordinate with the rest of their teammates in other locations is central to ensuring action team success. Although the leadership and skill development aspects of action teams have been researched extensively in the team's literature (e.g., Edmondson, 2003; Klein et al., 2006), decisions regarding geographic dispersion and the execution of navigation remain largely unexplored. Spatial capabilities represent unique skills that determine success in highly dynamic action team environments (Carroll, 1992); thus, collective spatial abilities—namely awareness of who has them, who does not, and an understanding of how to best utilize them—can determine success or failure in real-world action team missions.

The Space: Complex, Broken Terrain

Increasingly over the last 20 years and continuing into the future, action teams (e.g., military squads) are required to operate and remain functional and effective in complex, ambiguous contexts (Goodwin et al., 2018). As such, spatial skills are of premium importance, and to the extent to which larger units (e.g., platoons) are coordinating and wayfinding via subunits (e.g., squads), the spatial awareness component may become equally or even more important than other types of group knowledge. For instance, military squads are routinely asked to perform patrol tasks in limited visibility conditions brought on by night, bad weather, and close terrain (i.e., terrain where long-range visibility is not possible, such as terrain with forests, mountains, or urban canyons created by buildings; cf. Tsakiri et al., 1999)—even without adversary action, the wayfinding tasks presented to a military squad can be formidable, and the inherently adversarial nature of military squad deployment only compounds the difficulties they may face (Storr, 2009). For action teams, and particularly for military squads, having knowledge about which team members to trust with wayfinding can have life-and-death consequences (e.g., Lazzara et al., 2009). Meanwhile, knowing who has technical skills or facilitation skills, while traditionally important in team environments, may be less immediately pertinent for some tasks (e.g., an emergency evacuation from a dangerous area).

Future concepts for military operations incorporate features from many societal trends. For example, rapid, extensive urbanization worldwide has prompted a focus on training, tactics, and equipment appropriate to the particular challenges of urban maneuver (Ancker & Scully, 2013). With the trend toward increased urbanization, there is an increasingly mixed topographic environment: blends of dense districts with multi-storied buildings, urban parkland, suburban residential and semi-residential areas, industrial use areas, and subterranean features such as subway systems, utility systems for sewer and water, underground walkways and roads, and underground shopping and commercial facilities. When some or all of these environments have been subject to extended periods of active combat, the damage to the myriad structures across these environments can be substantial and cause significant obstacles to the basic movement and geo-location of military teams. More importantly, as any form of active combat continues in these complex environments, the changes to them as a result of intentional or unintentional damage can be substantial. In non-military contexts, there are a variety of terrain changing events that can lead to similar challenges—including earthquakes, wildfires, tsunamis, and other natural phenomena. These environments pose similar challenges for disaster response teams.

Returning to our hypothetical military squad performing a patrol task, a common navigational aid for such a unit in an urban environment is a grid reference graphic (GRG): geographic imagery with overlaid rectangle grids to provide positional information (e.g., Johantges et al., 2021). As damage to urban infrastructure causes the current landscape to deviate from available maps and imagery in substantive ways, available GRGs will become progressively poorer models of the environment—and this change can happen quickly in kinetic environments. Even without substantive changes to landmarks, disorientation is an ever-present risk for the patrol unit—especially if the subunits become separated—as even momentary loss of focus on position with respect to landmarks displayed on the GRG can quickly render the GRG nonsensical in a chaotic situation. Further adding to the potential for disorientation, GPS systems, which commonly aid military navigators in a variety of contexts, are susceptible to failure due to loss of signal, especially in urban or other compartmentalized terrain. Indeed, signal disruption by the adversary (or friendly) action and/or simple equipment malfunction (e.g., GPS denial; Grant et al., 2009) are all too common in the challenging environments in which infantry personnel operate.

For our purposes here, we envision a relatively simple mission for a military squad conducting a patrol through broken, highly urbanized terrain with multistory buildings, underground walkways and shopping areas, and streets blocked or partially blocked with broken-down vehicles or similar obstacles. The military squad will need the capability both to navigate through these spaces as a united group as well as when broken into two or three fire teams, depending on different aspects of their mission and the specific terrain encountered. While navigating as fire teams, they will need to remain spatially coordinated across the teams within the squad, so they do not get separated or move away from tightly coordinated formations due to the extreme risk encountered by having one fire team isolated from the others, even by as little as a block or two. Thus, military squads require the ability to collectively coordinate their actions, integrate their knowledge, and rely upon spatial capabilities as they traverse volatile terrain. To achieve this goal, we assert that leveraging collective cognition in the form of transactive memory systems may be especially beneficial for enabling the collective spatial navigation of action teams.

The Cognition: Transactive Memory Systems

Researchers have addressed the knowledge sharing capabilities of collectives with the concept of transactive memory systems (i.e., TMSs; Lewis & Herndon, 2011; Peltokorpi, 2008; Wegner, 1986, 1995), which are frequently described as a form of collective cognition that enable members of a team to know "who knows what" within their group. Here, we also expand this definition to also include knowing "who has what capabilities," which is a critical component of collective knowledge that has often been examined in TMS studies, but understated in previous TMS definitions. Transactive memory systems (also sometimes referred to as transactive knowledge systems; Brauner & Becker, 2006) are used to describe awareness of both shared and differentiated knowledge and skills, whereby members of a dyad or group specialize in certain types of knowledge or skills and have an awareness of their unique capabilities. Notably, this is different than cognitive systems that focus primarily on shared knowledge—such as team mental models (e.g., Klimoski & Mohammed, 1994; Mathieu et al., 2000; Mohammed et al., 2010)—where the importance is placed upon shared awareness and/or agreement about knowledge or tasks, as opposed to being aware of how knowledge is stored among team members. Both understanding who knows what and whether the team is in agreement on task processes are important for facilitating team success across different types of tasks and contexts, but our focus will primarily be on knowledge encoding and retrieval as opposed to shared agreement.

Transactive memory systems (TMSs) were originally developed to describe how couples use information about how knowledge is distributed across the pair, as well as how they leverage knowledge to enhance performance (Wegner, 1986). In fact, early work uncovered individuals in romantic partnerships often leveraged their partner as an external memory source such that each individual could specialize their knowledge to specific domains instead of necessitating that all knowledge be fully shared. For example, one partner may specialize in knowledge about the location of various household items, while the other partner might specialize in knowledge about the couple's social schedule. Each of these tasks could be cognitively taxing, but by splitting up the cognitive processes and relying upon one another to specialize in one task instead of both, the couple enhances their collective memory system.

Broadly, as the transactive memory concept has evolved, it has been focused on cognitive tasks performed by groups and teams. In the team's literature, transactive memory systems are defined as "a form of cognitive architecture that encompasses both the knowledge uniquely held by particular group members with a collective awareness of who knows what" (DeChurch & Mesmer-Magnus, 2010, p. 33). Accordingly, transactive memory systems have been used to describe how team members distribute the cognitive load of encoding, storing, retrieving, and communicating information relevant to performing a task (Austin, 2003; Lewis, 2003; Wegner, 1986). Indeed, the presence of a TMS within a work team can be evidenced by group dynamics, including the specialization of tasks, task coordination activities, and task credibility actions that demonstrate team members trust each other's specialized expertise (Liang et al., 1995; Moreland, 1999; Moreland et al., 1996).

The team's literature also provides compelling evidence to suggest that distributed cognition influences group performance (Lewis & Herndon, 2011). TMSs represent the division of cognitive labor by members of a team in order to promote shared understanding of both common and differentiated knowledge within the team, which may be leveraged to address changing task demands. In this way, TMSs have provided the cognitive explanation for the mediating link between team member knowledge and team performance such that teams who train as a collective perform better than those trained as individuals (e.g., Liang et al., 1995; Zhang et al., 2007). Accordingly, the concept of transactive memory systems has received considerable research attention due to its impact on team effectiveness across a myriad of domains (e.g., DeChurch & Mesmer-Magnus, 2010; Peltokorpi, 2008; Ren & Argote, 2011; Zhang et al., 2007).

Leveraging Transactive Memory Systems for Collective Spatial Cognition

Transactive memory systems provide a promising path to teams' success in dynamic environments, especially as teams leverage their transactive spatial capabilities in the form of coordinated wayfinding, which can be thought of as the combination of communication of knowledge and executing collective route navigation. Collective cognition in combination with coordinated wayfinding presents modern action teams with the potential to capitalize on spatial effectiveness in dynamic environments, promoting successful collective spatial navigation.

Spatial cognition researchers separate the navigation process into both the knowledge and skills required to successfully move through space (Klippel et al., 2010). Indeed, sources of spatial knowledge provide a foundation on which spatial capabilities are executed (Allen, 1999; Siegel & White, 1975). TMSs could provide an efficient mechanism for teams to maintain coordinated action, even when spatial challenges arise, through the integration of both spatial knowledge and spatial capability in complex environments. The mutual emphasis on knowledge and capability enables action teams to remain agile—knowledge alone can become outdated as changes in the environment introduce unpredictability into the equation, yet skills and capabilities can promote adaptability to enhance effective decision-making in uncertain environments.

Spatial knowledge can come in a variety of forms. In action teams operating in complex and broken terrain, knowledge regarding the location of landmarks, established routes to and from locations of interest, and surveys about the terrain are critical pieces of spatial knowledge for team member safety and mission success. Knowledge of landmarks has been shown to improve the navigation of both children and adults when wayfinding in unfamiliar terrain (Jansen-Osmann & Fuchs, 2006). Similarly, a large body of research demonstrates that knowledge of routes via direct experience or through second-hand information (e.g., videos and directions from someone else) enhances navigation in new environments (e.g., Gale et al., 1990; Goldin & Thorndyke, 1982). Finally, learning about an environment from a survey or map can be an important source of spatial information (Goldin & Thorndyke, 1982; Thorndyke & Hayes-Roth, 1982).

While spatial knowledge has been demonstrated to enhance navigation in static environments, dynamic environments introduce additional complexities that could impede effective navigation (Yamauchi & Beer, 1996). For instance, knowledge of a landmark can be useful as long as that landmark is still visible and recognizable. However, in the dynamic environments in which action teams operate, spatial knowledge may be unstable. In combat environments, the spaces through which teams are navigating may be drastically altered by explosions, building collapses, or other kinetic events, rendering once reliable landmarks and routes unrecognizable.

In order to promote better navigation performance in dynamic environments, action teams need to be able to leverage spatial capabilities in combination with spatial knowledge to be agile and adaptable when navigating changing terrain. Spatial capabilities are individual differences in how people mentally represent and manipulate spatial information to perform cognitive tasks (Hegarty & Waller, 2005). When navigating in a changing environment, capabilities, such as spatial perception, spatial working memory, and sense of direction, can enable action team members to adapt their navigation processes to fit their changing environment. Spatial perception has been defined as the manner in which people understand the distance, size, and shape of their environment (Proffitt, 2008). Spatial working memory is the encoding, maintenance, and retrieval of relevant information to guide goal-oriented navigation processes (Baddeley, 1992; Della Sala et al., 1999). Finally, a sense of direction is humans' ability to have a general understanding of their location in space and perform wayfinding based on that understanding (Cornell et al., 2003; Hegarty et al., 2002).

Each of these spatial capabilities can still function effectively in changing contexts to promote an action team's ability to successfully navigate through a dynamic environment. For instance, spatial perception can still be useful for creating detours (e.g., using different existing routes or creating new routes, such as going through a wall) in the absence of a usable route (e.g., Golledge, 1999). Additionally, spatial working memory can help team members to re-learn and update their mental maps of the current environment to make better navigation decisions (e.g., Blacker et al., 2017). Finally, a strong sense of direction can provide useful when operating in new territories, so instead of using landmarks (e.g., Zhao & Warren, 2015), action teams might rely upon cardinal directions to implement updated navigation strategies.

In an ideal world, all members of an action team would have high levels of both spatial knowledge and spatial capabilities, but this is rarely the case in real-world action teams. Instead, it is far more common for members of a team to specialize and for the team to collectively benefit from the advanced expertise that comes from combining individuals with different areas of expertise (Hackman & Katz, 2010; Kozlowski & Bell, 2003). For action teams, particularly those functioning in combat environments, a variety of important skills and knowledge are required for mission success (Klein et al., 2006). Thus, it is likely that spatial knowledge and skills would not be concentrated equally among all team members, but rather, these spatial abilities would be distributed among several team members, making it extremely important for their knowledge and capabilities to be leveraged effectively to benefit the team as a whole. Additionally, decisions about task assignment and which team members' expertise are used to lead others through a dynamic environment may shift as a result of the distribution of spatial knowledge and capabilities.

A transactive memory system is ideally suited to this task because it would enable the members of the action team to be aware of not only "who knows what" about the environment (i.e., spatial knowledge) but also "who can do what" to navigate effectively (i.e., spatial capabilities) and "how to do what" (i.e., executing the navigation plan) in order to promote the team's adaptability. Armed with this knowledge, action teams can benefit from coordination advantages. For instance, when an action team is divided into subunits, the team is aware of who has what types of spatial knowledge and capabilities, and the subunits can be divided such that there is a balance across them so that each subunit can achieve successful navigation performance.

When action teams have chosen to divide into smaller subunits, those subunits must also be able to maintain communication and coordinate their efforts among one another. Intentional dispersion of the action team based on a strong understanding of each member's spatial knowledge and capabilities can ensure that the scouting subunit is composed of individuals with both the knowledge of routes and the spatial perception to judge whether or not a route remains reliable for use by the rest of the action team. Importantly, these dispersal/aggregation cycles can happen at multiple scales. Lessons from wayfinding research with individuals suggest that similar cognitive mechanisms may support moving through a building or hundreds of meters of natural environment (Montello, 1993). Fire teams may be separated by only a few meters inside a building or by many more in open terrain, but the same subunit and collective wayfinding capabilities are required.

Additionally, decisions about team task priority and assignment may be influenced by the space through which an action team is navigating. Decision-making responsibilities may shift from individuals with formal authority (such as the squad leader) to those who have knowledge or skills that are better aligned with the task at hand (e.g., Friedrich et al., 2009). Having a well-formed transactive memory system would enable the action team to have the awareness necessary to make better decisions about shifting leadership responsibility in response to changing task demands, such as an unexpected loss of a landmark, when navigating dynamic terrain.

Each of these examples demonstrates how a robust transactive memory system can promote more successful navigational performance for action teams operating in complex, broken terrain. However, one of the challenges for effectively developing a transactive memory system is the ability to recognize when it is robust and when it requires further improvement. Indeed, the measurement of cognitive systems has been a continual challenge in the organizational science literature (Lewis & Herndon, 2011), and thus, we propose additional recommendations for overcoming these challenges for action teams in dynamic contexts.

Measuring and Developing TMSs for Effective Collective Spatial Navigation

Transactive memory systems are complex cognitive structures, and as such, they are difficult to capture empirically. Thus, team cognition researchers have come up with a variety of methods for measuring the presence of TMSs—including indirect measures, direct measures, and a combination of both (e.g., Garner, 2006; Hollingshead, 1998; Lewis & Herndon, 2011). We review these traditional measurement approaches and then expand upon them by demonstrating how novel forms of data that capture the members' spatial capabilities and collective coordination can be integrated into existing measurement approaches to determine the effectiveness of an action team's TMS during wayfinding.

The first method team cognition researchers have used to capture TMSs is via an indirect approach. Indirect measures of TMSs use either behavioral coding (Liang et al., 1995) or rating scales (Lewis, 2003) via a latent variable approach to capture (1) the differentiated structure of members' knowledge (i.e., specialization), (2) members' reliance on other members' knowledge (i.e., credibility), and (3) effective knowledge processing (i.e., coordination; Lewis & Herndon, 2011). Thus, when a team demonstrates behaviors that indicate their use of specialization, credibility, and coordination, researchers infer that they have created an effective TMS, as each person is aware of and can leverage each other's knowledge and skills effectively.

Transactive memory systems have also been measured via a more direct approach. The direct measurement approach begins by creating highly controlled environments requiring members of a group to specialize in terms of knowledge and skills in order to effectively coordinate with one another, and then, researchers measure the presence of a TMS by directly asking participants to report on any strategies they employed to solve knowledge and memory-based tasks (e.g., Hollingshead, 1998). The degree to which the content of these open-ended qualitative responses demonstrates specialization, credibility, and coordination activities indicates the degree to which a TMS has been developed in those groups (e.g., Lewis et al., 2007).

Finally, researchers have combined inferences from indirect approaches with direct objective measures to create social network representations of transactive memory systems in groups (e.g., Garner, 2006; Palazzolo, 2005). Network-based approaches for examining TMSs consider patterns of interactions among members of a team that demonstrate information sharing and knowledge retrieval, which can then be used to represent the cognitive structure of a team's TMS (Lewis & Herndon, 2011). Through this method, the actual patterns of information exchange can be compared to the team's performance to determine the degree to which their TMS structure is effective.

To capture transactive memory systems for spatial navigation, we propose expanding on techniques found in the TMS literature to incorporate more objective and dynamic methods for measuring TMSs. Specifically, we recommend leveraging continuously collected digital trace communications data combined with geographic information system (GIS) data to achieve three main goals: (1) create networked communication patterns that reveal the extent to which spatial skills knowledge is exchanged and integrated between members and subgroups of the collective, (2) examine from the communication context whether decisions regarding the formation of subgroups incorporate patterns of distributed spatial capabilities among the subgroups, and (3) determine spatial performance based on GIS markers, which can be combined with the previous metrics to create a holistic picture of coordinated wayfinding.

For action teams in dynamic environments, digital trace communication data could be gathered from radio transmissions, phone calls, or text messages. These data contain information about who was in communication with whom, the temporal order of communication, and also the communication content, which is a powerful combination of information that can be leveraged for both qualitative analysis (e.g., organizational discourse analysis; Fairhurst, 2008) or quantitative analysis (e.g., relational event modeling, Butts, 2008; network analysis, Wasserman & Faust, 1994). In fact, digital trace communication data has been used extensively in the organizational sciences to construct social network structures, infer patterns of leadership influence, and predict future social interactions based on relational event histories (e.g., Lazer et al., 2009; Maupin et al., 2020; Schecter & Contractor, 2019; Yoo et al., 2012).

Moreover, geographic information system (GIS) data is a critical piece of information for evaluating navigation performance of action teams. GIS data can include the coordinates of an action team's start point, end point, and a large number of points along their route (e.g., Zhou et al., 2019), or routes could be mathematically created as a series of time–space vectors consisting of their start points, the angle and distance of travel, and the time taken to travel (e.g., Fang et al., 2012). GIS data can also be represented as a complex system or network, which can be analyzed to infer valuable geospatial information (Curtin, 2007). Regardless of its form, timing and location-based GIS can uncover inefficiencies in navigation as well. For instance, an external monitor could examine the real-time location of an action team as they navigate through difficult terrain and determine whether their navigation patterns are more accurate and/or quicker when certain team members are in charge of various navigation tasks.

We provide an illustrative example (see Figure 8.2) to demonstrate the unique information that can be gained from leveraging both digital trace communication data and GIS data to diagnose effective versus ineffective TMSs for promoting navigation performance. In this example, two units (Units A and B) are collectively navigating through unfamiliar terrain with the goal of reuniting at a checkpoint (signified by the star) while keeping a safe distance from one another in case of an ambush. Based on this scenario, we provide an example of an effective TMS that is promoting navigation performance for the two units (Table 8.1) and an under-developed TMS that is hindering navigation performance for the two units (Table 8.2).

In Table 8.1, the units communicate effectively with one another, demonstrating that they are aware of both their positions within the grid and also their positions relative to one another, and they help each other to efficiently navigate around an obstacle that was not included in their grid reference graphic (GRG). Based on their GIS data, the pathway that each unit took to get to the rendezvous point was the most efficient and effective for both units. On the contrary, in Table 8.2, the units are communicating information to one another that is incorrect based on a poor sense of direction when reading the GRG, indicating that someone without the proper knowledge to translate the GRG into an effective navigational tool was entrusted with this responsibility. Accordingly, executing upon the incorrect navigation instructions results in one unit going the wrong direction, having to backtrack to their starting point, and then taking more than twice as long as the units in

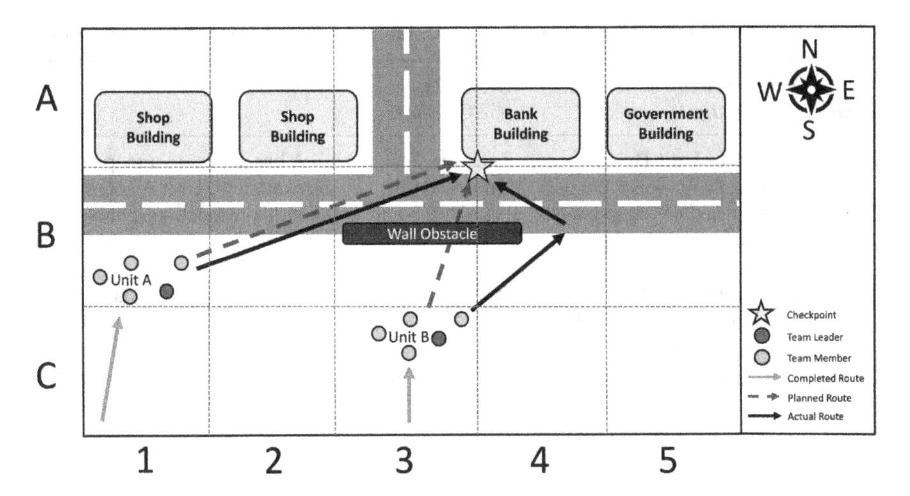

Figure 8.2 Illustration of two units navigating to a checkpoint.

Table 8.1 to meet at the checkpoint. When examining the GIS data in Table 8.2, it is apparent that, although one unit's path was efficient for getting to the checkpoint, the other unit's path was very inefficient and resulted in both units being unnecessarily exposed, potentially derailing their mission, which would not have been the case if they had a better-developed TMS.

By integrating the insights from digital trace communication data and GIS data, action teams can determine the degree to which their understanding of each team members' spatial knowledge and capabilities is accurate. In particular, action team members and/or leaders can evaluate how previous team decisions—such as dividing into subunits, assigning spatial tasks, and delegating the leadership responsibilities—have impacted their team processes and performance. For an action team with an under-developed transactive memory system, the communication data and the GIS data will reveal that they are not effectively utilizing the spatial strengths of different members of their team. On the contrary, for an action team with a highly developed transactive memory system, the communication and GIS data will reveal accurate and efficient communication and spatial navigation performance (cf. He et al., 2015).

Through leveraging insights from digital trace and GIS data, transactive memory systems can be diagnosed and evaluated, which can lead to opportunities for improving TMSs and ultimate gains in multilevel navigational performance. In order to better develop a robust TMS within an action team, steps can be taken during team training and onboarding to focus on uncovering each team members' unique spatial knowledge and capabilities along with their other areas of expertise. Furthermore, the action team leaders can intentionally assign training tasks to individuals who need to learn more about each other's capabilities in order to promote better navigational complementarities in the field. Preventative actions, such as spatial cognition-focused team training, can pay dividends in ultimate navigational performance and team success in dynamic environments.

Table 8.1 Fictional Example of Data From an Action Team With a Well-Developed TMS

Digital Trace Communication Data (Radio Transmissions)	*GIS Data (GRG Square)*	*Interpretation*
9:02 am Unit A to Unit B: "We are 40yds to your west. We will continue northeast 45yds to checkpoint. Map indicates checkpoint is 25yds due north of you."	**9:02 am** Unit A: B1 Unit B: C3	Unit A has access to a map of their environment. Unit A is correct about their location relative to Unit B.
9:03 am Unit B to Unit A: "Roger. Heading to checkpoint, but we have no visibility to the north: there is an obstacle. Do you have visibility?"	**9:03 am** Unit A: B1 Unit B: C3	Unit B has real-time updated information about a new obstacle not included on Unit A's map. Unit B understands Unit A's relative location to them and the obstacle. Unit B defers to Unit A for advice because Unit A has the better vantage point.
9:04 am Unit A to Unit B: "We have visibility. Go northeast around the obstacle, then go north across the road. Checkpoint is in front of the bank. Come to the west side."	**9:04 am** Unit A: B1 Unit B: B3	Unit A confirms their vantage point and uses the map to advise Unit B on an alternative route to the checkpoint.
9:05am Unit B to Unit A: "Roger. Passing obstacle now." Unit A to Unit B: "Roger. Signal when you reach the checkpoint and we'll move to you."	**9:05 am** Unit A: B1 Unit B: B4	Unit B is successfully able to follow Unit A's directions.
	9:10 am Unit A: B3 Unit B: B3	Units A and B are successfully reunited at the checkpoint.

Propositions and Future Directions

To advance scholarly understanding of collective spatial cognition through the lens of transactive memory systems, we introduce a series of research propositions that develop testable ideas based on the logic of our collective spatial navigation framework. First, the literature on team effectiveness is firm on the value of shared cognition in teams: team members who understand the technical aspects of their own roles and the ways in which their roles and their fellow team members' roles interact are able to make better decisions individually and contribute to a better team-level solution (Cannon-Bowers et al., 1993; Hutchins, 1995; Weick & Roberts, 1993). The development and implementation—and eventual near-universal acceptance—of crew resource management (CRM) training protocols provides a particularly important example (Helmreich & Foushee, 1993). CRM is a set of training and feedback mechanisms developed to improve decision-making and safe operation. This protocol was established first in multi-pilot aircraft and later in action teams

Table 8.2 Fictional Example of Data From an Action Team With an Under-Developed TMS

Digital Trace Communication Data (Radio Transmissions)	*GIS Data (GRG Square)*	*Interpretation*
9:02 am Unit A to Unit B: "We are 40yds to your north. We will continue southeast 45yds to checkpoint. Map indicates checkpoint is 25yds due east of you."	**9:02 am** Unit A: B1 Unit B: C3	Unit A has access to a map but is not reading it correctly based on their current location and their verbal description of their location. Unit A does not correctly understand their location relative to Unit B.
9:03am Unit B to Unit A: "Copy. Heading east 25yds."	**9:03 am** Unit A: B1 Unit B: C4	Unit B is relying on Unit A's instructions, which are incorrect, but they have to rely on Unit A because Unit A has the map and Unit B does not.
9:04 am Unit B to Unit A: "We have completed 25yds east, but do not see the checkpoint. Please advise."	**9:04 am** Unit A: B1 Unit B: C5	Unit B followed the directions by Unit A, but found those directions did not properly bring them to the checkpoint.
9:05 am Unit A to Unit B: "You should see the checkpoint. We have arrived. Where are you? What do you see?"	**9:05 am** Unit A: B3 Unit B: C5	Unit A does not know that they have misinformed Unit B because their directional markers are incorrect. Unit A asks for details about Unit B's location.
9:06 am Unit B to Unit A: "We see a road to our north. Maybe a building beyond that. What do you see?"	**9:06 am** Unit A: B3 Unit B: C5	Unit B describes their location but doesn't have a good vantage point for landmarks. Unit B attempts to determine Unit A's location based on their landmarks.
9:07 am Unit A to Unit B: "We are at the checkpoint but do not see you. Head back to your starting position 25yds to the west.	**9:07 am** Unit A: B3 Unit B: C5	Unit A instructs Unit B to backtrack their steps to try navigating to the checkpoint again.
9:08 am Unit B to Unit A: "We have returned to the starting point. You said 25yds east for the checkpoint? We see an obstacle to our north, and nothing to the east or west."	**9:08 am** Unit A: B3 Unit B: C3	Unit B goes back to their starting point and attempts to share landmark information.
9:09 am Unit A to Unit B: "You should see the road to your east." **Unit B to Unit A:** "We do not, we see an obstacle. Isn't the road to the north?"	**9:09am** Unit A: B3 Unit B: C3	Unit A tries to confirm Unit B's surroundings but is still reading the map incorrectly. Unit B attempts to correct their directional mistake by showing Unit A the road is north and not east.

(*Continued*)

Table 8.2 (Continued)

Digital Trace Communication Data (Radio Transmissions)	*GIS Data (GRG Square)*	*Interpretation*
9:10 am Unit A to Unit B: "Our mistake. You're right, the road IS to the north. The checkpoint is 25yds to the north of you."	**9:10am** Unit A: B3 Unit B: C3	Unit A realizes their mistake and reorients their relative position to Unit B while also instructing Unit B to the checkpoint.
9:10 am Unit B to Unit A: "We see an obstacle due north. Please advise?"	**9:10 am** Unit A: B3 Unit B: C3	Unit B identifies the obstacle that is not on Unit A's map.
9:11 am Unit A to Unit B: "We see the obstacle to our south. Change course to northeast, continue 15yds. Continue course northwest 30yds in front of Bank building to checkpoint."	**9:11 am** Unit A: B3 Unit B: C3	Unit A correctly identifies an alternative path for Unit B around the obstacle.
9:12 am Unit B to Unit A: "Copy. Following updated directions."	**9:12 am** Unit A: B3 Unit B: B4	Unit B follows the path around the obstacle.
	9:13 am Unit A: B3 Unit B: B3	After a detour, Units A and B are successfully reunited at the checkpoint.

more generally (Flin, 2010; Helmreich & Foushee, 1993). Importantly with respect to the team macrocognition literature (Cooke et al., 2013; Hutchins, 1995), CRM emphasizes open communication and information sharing among all team members, regardless of organizational status, as a pathway to improved performance outcomes.

Research on transactive memory systems provides a framework for understanding why training together as a team may provide the benefits it does. As an illustration, Liang et al. (1995) conducted an experiment where participants had to assemble a transistor radio, but some participants were allowed to train together, while others trained separately. Those who received training on the task as a group performed better than those who received training as individuals, due to the development of their transactive memory system. We anticipate similar benefits for action teams who train together during navigational training in order to enhance the development of their transactive memory system for performing spatial tasks. Training protocols suggested by Cannon-Bowers and colleagues (1993)—open discussion, cross training, and debriefing—are likely to be particularly important in team wayfinding due to the challenging, multilevel, interdependent nature of wayfinding inherent to the performance environment of action teams. Thus, we propose the following.

Proposition #1: Teams/squads who practice collective wayfinding together in training can enhance their transactive memory systems (TMSs) and promote better navigation performance in combat or other stressful situations.

However, action teams do not always remain intact when performing navigational tasks in the field. Despite the recognized potential of team-level training (e.g., Cannon-Bowers et al., 1993), practical constraints on training time and funding, as well as variation in individual aptitudes, suggest that action teams are unlikely to develop navigational capabilities of all team members to a high level. As an action team divides into subunits, potentially uneven distribution of individual wayfinding abilities may reduce the ability of the team to meet navigational performance expectations. Furthermore, it is possible that the loss of the unified team environment presents a risk to team performance outcomes (e.g., Liang et al., 1995).

We propose that rather than a team-level TMS restricting the action team to team-level employment, a multilevel view of the team wayfinding task suggests that the benefits of the TMS can be retained by "seeding" each group with the appropriate level of skill. Put differently, ensuring that each subunit has one (or several) member(s) with effective spatial capabilities enhances the likelihood of that subunit's wayfinding success. Additionally, the habits of communication and information sharing developed through team training are expected to carry over into a variety of situations (Cannon-Bowers et al., 1993; Flin, 2010). As the team breaks into subunits, a well-developed TMS would support the sharing of whatever skills were present within each subunit, as well as supporting the coordination of efforts between the subunits. Thus, careful distribution of individuals with critical levels of wayfinding skill may support the maintained effects of team training and TMS as the unit divides into subunits. Accordingly, we assert the following.

Proposition #2: Teams/squads that divide into smaller subunits for navigational tasks will experience better navigational performance when they limit their number of subunits to match the number of proficient wayfinders that can be assigned to each subunit.

One of the biggest challenges in a dynamic context is that the environment may be frequently changing (e.g., Stedmon et al., 2013). In order to effectively adapt to this dynamism, teams need to be able to train in a way that allows them to experience and overcome disorientation and misdirection. However, this is also a critical aspect of collective wayfinding that needs to be practiced as a group in order to be most effective. Indeed, similar to the development of cognitive systems, it is likely that the development of solutions to disorientation as a group increases the chances that someone on the team is able to perform effective navigation, even in contexts where disorientation is a possibility.

To prepare for and counteract the deficits in navigational performance that arise from disorientation in the field, action teams can benefit from effectively developing their metacognition to problem-solve effectively (Thompson & Cohen, 2012). Indeed, groups with a highly developed transactive memory system, which is built through training as a collective, will be aware of who they can rely upon to have a strong sense of direction and understanding of landmarks, even when there are visual or equipment impairments preventing the unit from being able to gather spatial information effectively from their environment. For instance, by training together in environments where vision is impaired (e.g., navigating through environments with lots of smoke, blizzards, or dust storms), teams will be much better equipped to learn effective strategies for overcoming those challenges instead of resorting to panic. Accordingly, should a team be navigating through terrain that is unexpectedly compromised in a real-world mission, the team members will be aware of what processes they should employ, and which people on their team they can rely on, to navigate them out of that disorienting situation. so they can continue their navigational tasks. Accordingly, we suggest the following.

Proposition #3: Teams/squads that practice recovering from disorientation as a group during training will experience better navigational performance in dynamic contexts.

One mechanism for realizing collective benefits to such differentiated information and ability is communication (e.g., Cooke et al., 2013). Indeed, it has been argued that communication is the observable manifestation of team cognition efforts (e.g., Cooke et al., 2007, 2013; Fiore & Salas, 2004; Hung, 2013). We have suggested that action teams must navigate in dynamic and often disorienting contexts. However, we have also noted that training together as a team has potential emergent benefits available through the differentiation of team member knowledge and capabilities. Additionally, research on individual-level wayfinding has emphasized the use of landmarks to effectively navigate through space (e.g., Chan et al., 2012).

As action teams adjust to changing environments and attempt to overcome disorientation, communication must be all the more succinct and effective to keep confusion to a minimum. The concept of transactive memory systems suggests that for communication to effectively support wayfinding, it should serve to transmit the most important information that will support improved wayfinding by those most able to use it; communication should therefore focus on affordances in the environment most useful for wayfinding. Although grid positions are often used for communication with outside entities, we suggest that for internal communications within the action teams, team members should focus on discussing details about identifiable spatial information (e.g., known routes and visible landmarks) as they change positions to assist in their team's collective orientation.

Proposition #4: Updating spatial knowledge with specific forms of spatial information will result in better navigational performance than grid references in team/squad-level navigation communication.

In summary, although there are a number of exciting future research directions at the intersection of collectives, cognition, and spatial navigation, we believe that these initial propositions will create an important foundation for action teams generally, and military squads specifically, that operate in dynamic contexts.

Discussion and Conclusions

With the growing emphasis on action teams in the military and other organizations (e.g., public safety and emergency response), there is a need to develop new understanding of the underlying skills and capabilities required to accomplish the tasks these teams perform in the real world. To date, team-focused research has largely ignored spatial information or skills (cf. Prince & Salas, 2000 for an exception), placing a primary emphasis on collective cognition and information processing to support decision-making teams and knowledge processing and synthesis teams (e.g., teams of intelligence analysts). Meanwhile, research on spatial abilities has been largely focused on the study of individuals (e.g., Allen, 1999), with some more recent work expanding into the study of collective navigation (e.g., Dalton et al., 2019; He et al., 2015), resulting in a ripe opportunity to continue to advance scholarly understanding of collective spatial cognition.

To address this need, we have adapted one of the well-developed concepts from the rich history of team cognition—transactive memory systems—to reflect this growing emphasis on skills and capabilities for teams to perform in the real world. While the integration of transactive memory systems with collective spatial navigation is relatively straightforward in principle, the deeper challenge lies in the operationalization and measurement of key elements of transactive spatial skills as the concept is empirically examined and modeled. To address this challenge, we introduced several avenues for capitalizing on robust methods and types of data from spatial cognition as well as team effectiveness research to ensure that the assessment approach reflects the developed wisdom of both domains, with particular attention given to the integration of digital trace communication data and GIS data for drawing inferences about effective TMS development and resulting navigation performance.

Future directions for this line of research would benefit from exploration of mathematical techniques to accurately capture coordinated movement of multiple individuals across complex terrain. Furthermore, integrating these elements in real time with methods of assessing the related collective cognition components would provide a particularly powerful tool for enabling collective spatial cognition in action teams. Through the introduction of transactive memory systems, we provide a foundation for incorporating spatial environments into team processes, promoting teams' abilities to benefit from collective spatial cognition and to improve their collective spatial navigation performance. We anticipate that this effort will be the first of many which explore the skills and capabilities required by teams performing in real-world contexts.

References

Allen, G. L. (1999). Spatial abilities, cognitive maps, and wayfinding. In R. G. Golledge (Ed.), *Wayfinding behavior: Cognitive mapping and other spatial processes* (pp. 46–80). Baltimore, MD: Johns Hopkins University Press.

Ancker III, C. J., & Scully, M. A. (2013). *Army doctrine publication 3–0: An opportunity to meet the challenges of the future*. Fort Leavenworth, KS: Army Combined Arms Center, Military Review.

Austin, J. R. (2003). Transactive memory in organizational groups: The effects of content, consensus, specialization, and accuracy on group performance. *Journal of Applied Psychology*, *88*, 866–878.

Baddeley, A. (1992). Working memory. *Science*, *255*(5044), 556–559.

Bae, C. J., & Montello, D. R. (2019). Dyadic route planning and navigation in collaborative wayfinding. In S. Timpf, C. Schlieder, M. Kattenbeck, B. Ludwig, & K. Stewart (Eds.), In *Proceedings of the 14th international conference on spatial information theory, COSIT 2019, September 9–13, 2019, Regensburg, Germany* (Vol. 142, pp. 1–20).

Bakeman, R., & Beck, S. (1974). The size of informal groups in public. *Environment and Behavior*, *6*(3), 378.

Blacker, K. J., Weisberg, S. M., Newcombe, N. S., & Courtney, S. M. (2017). Keeping track of where we are: Spatial working memory in navigation. *Visual Cognition*, *25*(7–8), 691–702.

Brauner, E., & Becker, A. (2006). Beyond knowledge sharing: The management of transactive knowledge systems. *Knowledge and Process Management*, *13*(1), 62–71.

Burgess, J. W. (1984). Do humans show a "species-typical" group size? Age, sex, and environmental differences in the size and composition of naturally-occurring causal groups. *Ethology and Sociobiology*, *5*, 1–57.

Burgess, N. (2008). Spatial cognition and the brain. *Annual Review of the New York Academy of Sciences*, *1124*, 77–97.

Butts, C. T. (2008). 4. A relational event framework for social action. *Sociological Methodology*, *38*(1), 155–200.

Cannon-Bowers, J. A., Converse, S., & Salas, E. (1993). Shared mental models in expert team decision making. *Individual and Group Decision Making: Current Issues*, *221*, 221–246.

Carroll, L. (1992). *Desperately seeking SA (No. AFRL-RH-AZ-JA-1992–0009)*. Mesa, AZ: Air Force Research Lab. Warfighter Readiness Research Division.

Chan, E., Baumann, O., Bellgrove, M. A., & Mattingley, J. B. (2012). From objects to landmarks: The function of visual location information in spatial navigation. *Frontiers in Psychology*, *3*, 304.

Chang, H. H. (2013). Wayfinding strategies and tourist anxiety in unfamiliar destinations. *Tourism Geographies*, *15*(3), 529–550.

Cooke, N. J., Gorman, J. C., Myers, C. W., & Duran, J. L. (2013). Interactive team cognition. *Cognitive Science*, *37*(2), 255–285.

Cooke, N. J., Gorman, J. C., & Winner, L. J. (2007). Team cognition. In F. Durso, R. Nickerson, S. Dumais, S. Lewandowsky, & T. Perfect (Eds.), *Handbook of applied cognition* (2nd ed., pp. 239–268). New York: Wiley.

Cornell, E. H., Sorenson, A., & Mio, T. (2003). Human sense of direction and wayfinding. *Annals of the Association of American Geographers*, *93*(2), 399–425.

Curtin, K. M. (2007). Network analysis in geographic information science: Review, assessment, and projections. *Cartography and Geographic Information Science*, *34*(2), 103–111.

Dalton, R. C., Hölscher, C., & Montello, D. R. (2019). Wayfinding as a social activity. *Frontiers in Psychology*, *10*, 142.

DeChurch, L. A., & Mesmer-Magnus, J. R. (2010). The cognitive underpinnings of effective teamwork: A meta-analysis. *Journal of Applied Psychology*, *95*(1), 32–53.

Della Sala, S., Gray, C., Baddeley, A., Allamano, N., & Wilson, L. (1999). Pattern span: A tool for unwelding visuo–spatial memory. *Neuropsychologia*, *37*(10), 1189–1199.

Devine, D. J., Clayton, L. D., Philips, J. L., Dunford, B. B., & Melner, S. B. (1999). Teams in organizations: Prevalence, characteristics, and effectiveness. *Small Group Research*, *30*(6), 678–711.

Downs, R. M., & Stea, D. (1973). Cognitive maps and spatial behavior: Process and products. In R. M. Downs & D. Stea (Eds.), *Image and environment* (pp. 8–26). Chicago: Aldine.

Edmondson, A. C. (2003). Speaking up in the operating room: How team leaders promote learning in interdisciplinary action teams. *Journal of Management Studies*, *40*(6), 1419–1452.

Epstein, S. L., Aroor, A., Evanusa, M., Sklar, E. I., & Parsons, S. (2015). Spatial abstraction for autonomous robot navigation. *Cognitive Processing*, *16*(1), 215–219.

Fang, Z., Li, Q., Zhang, X., & Shaw, S. L. (2012). A GIS data model for landmark-based pedestrian navigation. *International Journal of Geographical Information Science*, *26*(5), 817–838.

Fairhurst, G. T. (2008). Discursive leadership: A communication alternative to leadership psychology. *Management Communication Quarterly*, *21*(4), 510–521.

Fiore, S. M., & Salas, E. (2004). Why we need team cognition. In E. Salas & S. M. Fiore (Eds.), *Team cognition: Understanding the factors that drive process and performance* (pp. 235–248). Washington, DC: American Psychological Association.

Flin, R. (2010). CRM (non-technical) skills—applications for and beyond the flight deck. In *Crew resource management* (pp. 181–202). San Diego, CA: Academic Press.

Flynn, M. T., Juergens, R., & Cantrell, T. L. (2008). *Employing ISR SOF best practices. National defense university*. Washington, DC: Institute for National Strategic Studies.

Forlizzi, J., Barley, W. C., & Seder, T. (2010). Where should I turn: Moving from individual to collaborative navigation strategies to inform the interaction design of future navigation systems. In *Proceedings of the SIGCHI conference on human factors in computing systems* (pp. 1261–1270). New York: ACM.

Friedrich, T. L., Vessey, W. B., Schuelke, M. J., Ruark, G. A., & Mumford, M. D. (2009). A framework for understanding collective leadership: The selective utilization of leader and team expertise within networks. *The Leadership Quarterly*, *20*(6), 933–958.

Gale, N., Golledge, R. G., Pellegrino, J. W., & Doherty, S. (1990). The acquisition and integration of route knowledge in an unfamiliar neighborhood. *Journal of Environmental Psychology*, *10*(1), 3–25.

Garner, J. T. (2006). It's not what you know: A transactive memory analysis of knowledge networks at NASA. *Journal of Technical Writing and Communication*, *36*(4), 329–351.

Gibson, J. J. (1954). The visual perception of objective motion and subjective movement. *Psychological Review*, *61*, 304–314.

Goldin, S. E., & Thorndyke, P. W. (1982). Simulating navigation for spatial knowledge acquisition. *Human Factors*, *24*(4), 457–471.

Golledge, R. G. (Ed.). (1999). *Wayfinding behavior: Cognitive mapping and other spatial processes*. Baltimore, MD: JHU Press.

Goodwin, G. F., Blacksmith, N., & Coats, M. R. (2018). The science of teams in the military: Contributions from over 60 years of research. *American Psychologist*, *73*(4), 322.

Grant, A., Williams, P., Ward, N., & Basker, S. (2009). GPS jamming and the impact on maritime navigation. *The Journal of Navigation*, *62*(2), 173–187.

Hackman, J. R., & Katz, N. (2010). Group behavior and performance. In S. T. Fiske, D. T. Gilbert, & G. Lindzey (Eds.), *Handbook of social psychology* (Vol. 2, pp. 1208–1251). Hoboken, NJ: John Wiley & Sons.

Haghani, M., & Sarvi, M. (2017). Following the crowd or avoiding it? Empirical investigation of imitative behaviour in emergency escape of human crowds. *Animal Behaviour*, *124*, 47–56.

Hart, R. A., & Moore, G. T. (1973). The development of spatial cognition: A review. In R. M. Downs & D. Stea (Eds.), *Image and environment* (pp. 246–288). Chicago: Aldine.

He, G., Ishikawa, T., & Takemiya, M. (2015). Collaborative navigation in an unfamiliar environment with people having different spatial aptitudes. *Spatial Cognition and Computation, 15*, 285–307.

Hegarty, M., Richardson, A. E., Montello, D. R., Lovelace, K., & Subbiah, I. (2002). Development of a self-report measure of environmental spatial ability. *Intelligence, 30*(5), 425–447.

Hegarty, M., & Waller, D. A. (2005). Individual differences in spatial abilities. In P. Shah & A. Miyake (Eds.), *The Cambridge handbook of visuospatial thinking* (pp. 121–169). New York: Cambridge University Press.

Helmreich, R. L., & Foushee, H. C. (1993). Why crew resource management? Empirical and theoretical bases of human factors training in aviation. In E. Wiener, B. Kanki, & R. Helmreich (Eds.), *Cockpit resource management* (pp. 3–45). San Diego, CA: Academic Press.

Hollingshead, A. B. (1998). Communication, learning, and retrieval in transactive memory systems. *Journal of Experimental Social Psychology, 34*(5), 423–442.

Hollingshead, A. B. (2001). Cognitive interdependence and convergent expectations in transactive memory. *Journal of Personality and Social Psychology, 81*(6), 1080.

Hung, W. (2013). Team-based complex problem solving: A collective cognition perspective. *Educational Technology Research and Development, 61*(3), 365–384.

Hutchins, E. (1995). *Cognition in the wild*. Cambridge, MA: The MIT Press.

James, G. S. (1951). The comparison of several groups of observations when the ratios of the population variances are unknown. *Biometrika, 38*(3/4), 324–329.

Jansen-Osmann, P., & Fuchs, P. (2006). Wayfinding behavior and spatial knowledge of adults and children in a virtual environment: The role of landmarks. *Experimental Psychology, 53*(3), 171–181.

Johantges, A. D., Jonas, B. P., Oxendine, C., & O'Banion, M. S. (2021). Development of gridded reference graphics using machine learning and a customized geoprocessing workflow. In *2021 IEEE international geoscience and remote sensing symposium IGARSS, Brussels, Belgium* (pp. 3916–3919).

Klein, K. J., Ziegert, J. C., Knight, A. P., & Xiao, Y. (2006). Dynamic delegation: Shared, hierarchical, and deindividualized leadership in extreme action teams. *Administrative Science Quarterly, 51*, 590–621.

Klimoski, R., & Mohammed, S. (1994). Team mental model: Construct or metaphor? *Journal of Management, 20*(2), 403–437.

Klippel, A., Hirtle, S., & Davies, C. (2010). You-are-here maps: Creating spatial awareness through map-like representations. *Spatial Cognition & Computation, 10*(2–3), 83–93.

Kozlowski, S. W. J., & Bell, B. S. (2003). Work groups and teams in organizations. In W. C. Borman, D. R. Ilgen, & R. J. Klimoski (Eds.), *Handbook of psychology: Vol. 12. Industrial and organizational psychology* (pp. 333–375). London: Wiley.

Lazer, D., Pentland, A. S., Adamic, L., Aral, S., Barabasi, A. L., Brewer, D., . . . Van Alstyne, M. (2009). Life in the network: The coming age of computational social science. *Science (New York, NY), 323*(5915), 721.

Lazzara, E. H., Fiore, S., Wildman, J., Shuffler, M., & Salas, E. (2009). Managing trust in swiftly starting action teams. In *Proceedings of the human factors and ergonomics society annual meeting* (Vol. 53, No. 26, pp. 1922–1923). Sage, CA and Los Angeles, CA: Sage Publications.

Lewis, K. (2003). Measuring transactive memory systems in the field: Scale development and validation. *Journal of Applied Psychology, 88*(4), 587.

Lewis, K., Belliveau, M., Herndon, B., & Keller, J. (2007). Group cognition, membership change, and performance: Investigating the benefits and detriments of collective knowledge. *Organizational Behavior and Human Decision Processes*, *103*(2), 159–178.

Lewis, K., & Herndon, B. (2011). Transactive memory systems: Current issues and future research directions. *Organization Science*, *22*(5), 1254–1265.

Liang, D. W., Moreland, R., & Argote, L. (1995). Group versus individual training and group performance: The mediating role of transactive memory. *Personality and Social Psychology Bulletin*, *21*(4), 384–393.

Mark, D. M. (1993). Human spatial cognition. In D. Medyckyj-Scott & H. M. Hearnshaw (Eds.), *Human factors in geographical information systems* (pp. 51–60). London: Belhaven Press.

Mathieu, J. E., Heffner, T. S., Goodwin, G. F., Salas, E., & Cannon-Bowers, J. A. (2000). The influence of shared mental models on team process and performance. *Journal of Applied Psychology*, *85*(2), 273.

Maupin, C. K., McCusker, M. E., Slaughter, A. J., & Ruark, G. A. (2020). A tale of three approaches: Leveraging organizational discourse analysis, relational event modeling, and dynamic network analysis for collective leadership. *Human Relations*, *73*(4), 572–597.

Mohammed, S., Ferzandi, L., & Hamilton, K. (2010). Metaphor no more: A 15-year review of the team mental model construct. *Journal of Management*, *36*(4), 876–910.

Montello, D. R. (1993). Scale and multiple psychologies of space. In A. U. Frank & I. Campari (Eds.), *Spatial information theory* (pp. 312–321). Berlin: Springer.

Montello, D. R. (2001). Spatial cognition. In *International encyclopedia of the social & behavioral sciences* (pp. 14771–14775). Oxford: Pergamon Press.

Montello, D., & M. Raubal (2013). Functions and applications of spatial cognition. In D. Waller & L. Nadel (Eds.), *APA handbook of spatial cognition* (pp. 249–264). Washington, DC: APA.

Moreland, R. L. (1999). Transactive memory: Learning who knows what in work groups and organizations. In L. L. Thompson, J. M. Levine, & D. M. Messick (Eds.), *Shared cognition in organizations: The management of knowledge* (pp. 3–31). Mahwah, NJ: Erlbaum.

Moreland, R. L. (2010). Are dyads really groups? *Small Group Research*, *41*(2), 251–267.

Moreland, R. L., Argote, L., & Krishnan, R. (1996). Socially shared cognition at work: Transactive memory and group performance. In J. L. Nye & A. M. Brower (Eds.), *What's so social about social cognition? Social cognition research in small groups* (pp. 57–84). Thousand Oaks, CA: Sage.

Newcombe, N. S., Uttal, D. H., & Sauter, M. (2013). Spatial development. In P. D. Zelazo (Ed.), *Oxford handbook of developmental psychology* (pp. 564–590). New York, NY: Oxford University Press.

O'Neill, M. (1991). A biologically based model of spatial cognition and wayfinding. *Journal of Environmental Psychology*, *11*(4), 299–320.

Palazzolo, E. T. (2005). Organizing for information retrieval in transactive memory systems. *Communication Research*, *32*(6), 726–761.

Peltokorpi, V. (2008). Transactive memory systems. *Review of General Psychology*, *12*(4), 378–394.

Prince, C., & Salas, E. (2000). Team situational awareness, errors, and crew resource management: Research integration for training guidance. In M. R. Endsley & D. J. Garland (Eds.), *Situation awareness analysis and measurement* (pp. 290–309). Mahwah, NJ: LEA.

Proffitt, D. R. (2008). An action-specific approach to spatial perception. In R. L. Klatzky, B. MacWhinney, & M. Behrmann (Eds.), *Embodiment, ego-space, and action* (pp. 179–202). New York: Psychology Press.

Ren, Y., & Argote, L. (2011). Transactive memory systems 1985–2010: An integrative framework of key dimensions, antecedents, and consequences. *Academy of Management Annals*, *5*(1), 189–229.

Salas, E., Dickinson, T. L., Converse, S. A., & Tannenbaum, S. I. (1992). Toward an understanding of team performance and training. In R. W. Swezey & E. Salas (Eds.), *Teams: Their training and performance* (pp. 3–29). Norwood, NJ: Ablex.

Schecter, A., & Contractor, N. (2019). *Uncovering latent archetypes from digital trace sequences: An analytical method and empirical example*. ICIS 2019 Proceedings 7. https://aisel.aisnet.org/icis2019/data_science/data_science/7

Siegel, A. W., & White, S. H. (1975). The development of spatial representations of large-scale environments. In H. W. Reese (Ed.), *Advances in child development and behavior* (Vol. 10, pp. 9–55). New York: Academic.

Stedmon, A., Ryan, B., Fryer, P., McMillan, A., Sutherland, N., & Langley, A. (2013). Human factors and the human domain: Exploring aspects of human geography and human terrain in a military context. In *International conference on engineering psychology and cognitive ergonomics* (pp. 302–311). Berlin and Heidelberg: Springer.

Storr, J. (2009). *The human face of war*. London: Continuum.

Sundstrom, E., De Meuse, K. P., & Futrell, D. (1990). Work teams: Applications and effectiveness. *American Psychologist*, *45*(2), 120–133.

Thorndyke, P. W., & Hayes-Roth, B. (1982). Differences in spatial knowledge acquired from maps and navigation. *Cognitive Psychology*, *14*(4), 560–589.

Thompson, L., & Cohen, T. R. (2012). Metacognition in teams and organizations. *Social Metacognition*, 283–302.

Tsakiri, M., Kealy, A., & Stewart, M. (1999). Urban canyon vehicle navigation with integrated GPS/GLONASS/DR systems. *NAVIGATION, Journal of the Institute of Navigation*, *46*(3), 161–174.

Wasserman, S., & Faust, K. (1994). *Social network analysis: Methods and applications*. Cambridge: Cambridge University Press.

Wegner, D. M. (1986). Transactive memory: A contemporary analysis of the group mind. In B. Mullen & G. R. Goethals (Eds.), *Theories of group behavior* (pp. 185–205). New York: Springer-Verlag.

Wegner, D. M. (1995). A computer network model of human transactive memory. *Social Cognition*, *13*(3), 319–339.

Weick, K. E., & Roberts, K. H. (1993). Collective mind in organizations: Heedful interrelating on flight decks. *Administrative Science Quarterly*, *38*(3), 357–381.

Wuchty, S., Jones, B. F., & Uzzi, B. (2007). The increasing dominance of teams in production of knowledge. *Science*, *316*(5827), 1036–1039.

Yamauchi, B., & Beer, R. (1996). Spatial learning for navigation in dynamic environments. *IEEE Transactions on Systems, Man, and Cybernetics, Part B (Cybernetics)*, *26*(3), 496–505.

Yesiltepe, D., Dalton, R. C., & Torun, A. O. (2021). Landmarks in wayfinding: A review of the existing literature. *Cognitive Processing*, *22*, 369–410.

Yoo, Y., Boland Jr, R. J., Lyytinen, K., & Majchrzak, A. (2012). Organizing for innovation in the digitized world. *Organization Science*, *23*(5), 1398–1408.

Zhang, Z. X., Hempel, P. S., Han, Y. L., & Tjosvold, D. (2007). Transactive memory system links work team characteristics and performance. *Journal of Applied Psychology*, *92*(6), 1722.

Zhao, M., & Warren, W. H. (2015). How you get there from here: Interaction of visual landmarks and path integration in human navigation. *Psychological Science*, *26*(6), 915–924.

Zhou, S., Wang, R., Ding, J., Pan, X., Zhou, S., Fang, F., & Zhen, W. (2019). An approach for computing routes without complicated decision points in landmark-based pedestrian navigation. *International Journal of Geographical Information Science*, *33*(9), 1829–1846.

Teams

9 The Dynamics and Performance of Groups as Spatial Information Processors

Ernest S. Park and Verlin B. Hinsz

Introduction

In this chapter, we present our conceptualization of groups as spatial information processors. This perspective is important to consider because group members often have to collaboratively solve spatial problems in order to achieve their objectives (Dalton et al., 2019). Although the existing literature on spatial cognition is quite extensive, the research often focuses on individual-level phenomena and tends to overlook the potential for social influence (Dorfman et al., 2021; He et al., 2015). Therefore, one purpose of this chapter is to help fill this void by offering descriptions of what spatial cognition might look like in human collectives. We will focus on teams of unmanned aerial vehicle (UAV) operators as a particular type of workgroup and set of tasks to illustrate how members of this collective might process spatial information to produce group actions. As we explore groups as spatial information processors, we also draw attention to the specific ways in which we expect groups to influence spatial cognition. Given that our theory-based conjectures suggest that differences may emerge when tasks are performed in social compared to solitary contexts, we take this timely opportunity to encourage scholars to consider group influences to advance theories and research of spatial cognition.

The Collective Under Study

We chose teams of UAV operators as the collective under study because these teams solve spatial problems and remotely navigate their way through environments to accomplish group objectives. As such, this type of workgroup will serve as a basis of reference to help us convey many important features of our conceptualization of groups as spatial information processors.

Three-person teams are central to the operation of UAVs for many military surveillance assignments. In such cases, one member of the team is designated as the pilot, another occupies the role of sensor operator, and a third is the mission commander. Among other tasks, pilots are responsible for operating the aircraft controls and guiding the aircraft, so the sensor operator can focus on target locations and target engagement. Although pilots control the flight of the UAVs, sensor operators control cameras and sensors that are used to complete intelligence, surveillance, and reconnaissance tasks. Therefore, sensor operators search for targets

DOI: 10.4324/9781003202738-13

within designated locations are responsible for discriminating between valid and invalid cues and perform various analyses of targets that are detected. The mission commander is concerned with planning and preparing for the mission and focuses on gathering and exploiting the data that are the objective of the mission, all while communicating with agents external to the UAV team.

The Spatial Information, Task Demands, and Setting for the Collective Under Study

There are many components to the tasks that UAV operators must complete. The UAV team must develop an initial flight plan and be able to work together to revise such routes if needed (Hinsz et al., 2009). To make planning decisions and to direct the aircraft to specified locations, UAV operators must also acquire and communicate spatial knowledge that is relevant to their assignment. UAV teams need to be able to correctly recognize their targets and effectively position their aircraft so that the desired images can be obtained, delivered, and analyzed.

To complete their assignments, the team of UAV operators must coordinate their attention and actions to acquire data about ground targets (Gorman & Cooke, 2011). This task can be complicated because teams are working from remote locations and are only provided with limited sets of information. The nature of the visual information and spatial information that are supplied from a UAV tends to also be atypical because the images sent to the operators are often taken from an aerial perspective (Fincannon et al., 2011). Consequently, past work suggests that it can be both difficult and effortful to form accurate perceptions of the physical environment and to navigate effectively and efficiently (Casper & Murphy, 2003).

Visual information from remote locations is presented to UAV operators on multiple screens, and operators must communicate and make specific requests about the spatial information that they need from one another. For example, as the pilot navigates the aircraft, the sensor operator often provides input and updates about the chosen flight path. As a pilot controls the UAV to approach a target, the sensor operator also supplies the pilot with visual information that relates to timing, distance, and the potential for obstructions that could obscure the camera's view. Therefore, team members are presented with a range of information and are responsible for attending to, selecting, and exchanging spatial information that facilitates both object localization and aircraft navigation (Fincannon et al., 2011).

To meet the demands of these types of tasks, spatial perspective taking is required. When group members coordinate in collaborative settings, they have to retrieve spatial information from memory, or the environment, and convey it to others to solve spatial problems. But the spatial viewpoint that a person occupies often differs from the vantage point of others, so alternative spatial perspectives sometimes have to be considered for efficient communication (Duran et al., 2016). Although it seems like this might require deliberation and a great deal of effort, research shows that some people engage in spatial perspective taking rather spontaneously. For example, in two studies, Tversky and Martin Hard (2009) presented participants with a photographed scene and asked them to describe the spatial relations between two objects (e.g., book and bottle) on a table (e.g., "In relation to the bottle, where is the book?"). In a condition in which the image included the mere presence of a person who was in a position to act on the objects, roughly 25 and 50% of the participants from this condition (Studies 1 and 2, respectively) spontaneously described the spatial relations from the photographed person's point of view (e.g., "to his left" and "to the left according to the way he is facing").

The environments that people traverse or describe are often interpreted and organized in memory in terms of a reference system that encodes spatial relations from a preferred direction (Galati & Avraamides, 2013). There are many factors that can influence the preferred direction that is adopted, including representational information such as the environment's shape or the intrinsic structure of the spatial configuration (e.g., Mou & McNamara, 2002). In collaborative tasks, it has been noted that social information (e.g., teammate's viewpoint, cognitive abilities, and status) can also play a role in determining one's organizing direction when people are sufficiently motivated and able to represent their partner's viewpoint (Galati & Avraamides, 2013). Interestingly, research shows that on collaborative tasks, people actually weigh and integrate these different forms of information to determine the appropriate perspective for organizing spatial knowledge, so it can be expressed and easily understood by collaborators (Galati & Avraamides, 2015). In the process of adopting a spatial perspective, it is believed that social information is considered to help foster a shared understanding of the situation, presumably so collective effort can be minimized on collaborative tasks (Galati & Avraamides, 2013). Spatial perspective taking promotes effective communication and coordination, and it is also required to complete other tasks that UAV teams perform, like collaborative navigation.

Collaborative navigation occurs when group members have to know where they are and what direction they are headed in and inform others in the group where to go (e.g., person in a car's passenger seat providing navigational directions to the driver; He et al., 2015). In UAV missions, we believe that collaborative navigation places a particular emphasis on wayfinding, where orientation needs to be maintained relative to a distal environment (vs. locomotion, which depends on sensory motor systems interacting with immediate surroundings; Dalton et al., 2019). Collaborative navigation is likely to take place frequently during UAV missions. For example, it occurs when the sensor operator needs an alternative view of a target and has to provide navigational instructions to the pilot as they revise their approach. Similarly, collaborative navigation might be needed when the mission commander receives new orders, and the crew has to work together to determine how to best re-route their aircraft.

Collaborative navigation can be challenging for people as they traverse their immediate physical environment, and it may be difficult for members of UAV teams as well. When operating a vehicle remotely, team members lack direct access to the proximal and contextual information that is typically acquired to determine a person or group's location. Navigational information might also be distributed unevenly across members of a UAV team, depending on the roles they occupy. So in this setting, collaborative navigation is likely to require spatial perspective taking and a system of communication that is mutually understood and easy to express with precision. Therefore, it is important to develop ways of enhancing the motivation and ability levels of team members, so they can function effectively as a unit (cf., Hackman, 1987). With a UAV team and mission in mind, in the following sections, we describe groups as spatial information processors and discuss factors that are likely to influence collaboration and group performance on spatial problems.

Group Member Contributions to Spatial Problems

There are many factors that can impact group actions and outcomes when members collaborate to perform spatial tasks. To appreciate this type of group experience, a range of aspects relating to the tasks, environment, and nature of the group itself should be considered. To shed light on some of these variables, we draw upon theory and research related to group performance on tasks that are informationally rich and complex. The theory of combinations of contributions (Hinsz & Ladbury, 2012) suggests that two key components are important for the prediction of group performance: contributions and combinations. The types of contributions that impact group performance are those that correspond to the critical demands of the group, task, or interaction. In conjunction, it is important to determine the relevant processes that are used to combine (e.g., aggregate, pool, transform, and integrate) the various contributions to produce group-level outcomes. For solving spatial knowledge-related problems, the contributions that are likely to correspond to the solutions are those that involve spatial cognition. In other words, to navigate and surveil targets effectively, relevant knowledge about spatial environments has to be appropriately acquired, organized, utilized, and updated.

Because members of a UAV team occupy different roles and have different duties, the critical contributions that are expected from each will differ. For example, UAV pilots are responsible for navigation and have to maneuver their vehicle through physical space to reach target destinations. Thus, pilots need to acquire spatial information that can be utilized to determine the orientation of their vehicle, and collect information about the UAV's environment, so they can see how it is situated in relation to other places and things. Pilots need to assess the aircraft's direction and speed, and attend to cues that affect these variables, so they can sense where the UAV will be located and headed at various points during missions. With the constant movement and changing spatial relations of vehicles in flight, pilots should also frequently update their knowledge, so they form a relatively accurate and ongoing understanding of their dynamic situations. UAV pilots use navigational aids such as GPS-enabled devices to acquire these types of information (Babel, 2014), but this is not to say that pilot competencies have no bearing on team performance for spatial problems.

Because UAVs are operated remotely, pilots rely heavily on maps to navigate their vehicles. Maps present spatial relations between locations using an aerial view. Therefore, the spatial knowledge acquired from these tools differs from the spatial information obtained from direct navigational experience in an immediate environment (Thorndyke & Hayes-Roth, 1982). Interestingly, this suggests that a pilot's map-reading skills might contribute more to successful UAV navigation than their actual sense of direction. Past work shows that sense of direction is more connected to spatial learning about large-scale areas from direct experience and relates less to learning from visual media or virtual environments (Hegarty et al., 2006).

Given the importance of map reading and learning, we expect contributions from pilots with strong spatial visualization skills to be particularly indicative of successful UAV navigation. Strong spatial visualizers rely on schematic diagrams when solving spatial problems, are good at mental rotation, and have an easier time learning and remembering the spatial relations between objects (Aggarwal & Woolley, 2013). Spatial visualizers generate and process images analytically rather than holistically and think about spatial problems in more granular, detail-oriented ways. Spatial visualization relates to a process focus and tendency to identify subtasks that need to be completed (Kozhevnikov, 2007), so spatial visualizers should excel at map reading and learning (Kozhevnikov et al., 2010).

Although pilots who engage in spatial perspective-taking and spatial visualization are likely to be effective teammates, object visualization is an ability that should contribute to the performance of a UAV's sensor operator. Object visualization refers to the ability to represent the literal appearance of objects in terms of their precise form, size, shape, color, and brightness (Kozhevnikov et al., 2010). With object processing, physical cues receive much attention and are bound together in memory. This is in contrast to spatial processing in which information about spatial locations, spatial relations and movement are encoded and maintained (Kozhevnikov et al., 2010).

Although object and spatial visualization were once characterized as orthogonal individual differences, this is no longer the case. Many now believe that high levels of one ability are accompanied by low levels of the other (e.g., Kozhevnikov et al., 2010). One reason an object-spatial visualization tradeoff might emerge is that both types of visualization involve visual attention, which is a limited-capacity shared resource. So, rather than being independent, if a person is predisposed to either object or spatial visualization during functional integration of the dorsal and ventral systems in early childhood, one type of visualization might develop at the expense of the other (Kozhevnikov et al., 2010).

As object visualizers developed their ability to maintain large amounts of pictorial details, it is suspected that their ability to encode spatial information is somewhat sacrificed, with the reverse presumably occurring for spatial visualizers (Blazhenkova & Kozhevnikov, 2009). So, unlike spatial visualizers who generate and process images analytically and part by part, object visualizers encode and process images holistically as single perceptual units (Kozhevnikov, 2007). In the process, large quantities of pictorial details are encoded and remembered, so it makes sense that object visualizers excel at recognizing objects, even when the presented images are degraded (Blazhenkova & Kozhevnikov, 2009). Thus, object visualization should contribute to the effectiveness of a sensor operator because this ability should enhance their capacity to imagine and remember the appearance of designated targets. It should also allow them to quickly locate and correctly identify targets upon presentation, even when the quality of visual information that is acquired is suboptimal.

We can see that group member contributions should correspond to performance when UAV pilots are spatial visualizers and sensor operators are object visualizers. Because pilots plan routes and navigate, if they are spatial visualizers, they will be comfortable relying on spatial maps of an environment and will have an easy time deciphering and remembering spatial information presented in this format. In contrast, object visualizers are more likely to plan and navigate routes by selecting and recalling a sequence of distinct, singular landmarks. This tendency to attend to, remember, and utilize landmarks during navigation further supports our notion that object visualization will enhance the effectiveness of sensor operators. There are times when sensor operators play a larger role in navigation. One instance would be when UAV teams are faced with long-term GPS outages and thus have to rely on landmark-based visual navigation (Babel, 2014).

To engage in route planning with this visual navigation method, a sequence of landmarks having fixed and known positions are determined, and geo-referenced images of these landmarks are stored in an onboard computer. These landmarks appear at specified intervals and serve as navigation updates. Because their positions are fixed and known, when landmarks are encountered, operators can localize the UAV (Babel, 2014). To do so, when the UAV is in flight, the sensor operator collects onboard images using the aircraft's camera. This spatial information that is gathered allows the team to estimate their position by comparing these acquired onboard images with the geo-referenced ones.

When navigating, reliance on too many landmarks is typically thought to be inefficient, distracting, and cumbersome (Ishikawa & Nakamura, 2012). However, when UAV teams use landmark-based navigation, it is critical that intervals between two navigation updates (i.e., the distance between two consecutive landmarks on the flight path) are not too large. If the distance between landmarks is too large, an upcoming landmark may be outside of the camera's field of view. When this happens, it may be difficult for the sensor operator to determine the location of the UAV, resulting in position updates that may be increasingly inaccurate. Then, navigation errors can cascade, resulting in an increased risk of the aircraft getting lost (Babel, 2014). Because UAV teams using this method of navigation rely on many landmarks, sensor operators need to be able to remember and identify them accurately. So, when navigational aids are unavailable to the pilot, it is critical for the sensor operator to contribute by acquiring and sharing the spatial information that is utilized for landmark-based navigation.

In most cases though, the primary duty of sensor operators is to control the UAV cameras to search for and image designated targets. When military missions call for the UAV team to locate and surveil hostile targets on the ground, the corresponding contributions from the sensor operator will largely center on their ability to detect threats. When mistakes are costly, it is especially important to have the ability to make accurate distinctions between targets that are legitimately threatening versus those that are not. This requires effortful vigilance, so monitoring an environment for threats can be quite depleting when missions are long. Because mistakes increase when cognitive resources are low, it is important to identify individual differences that correspond to the ability to remain vigilant (Ein-Dor & Perry, 2014). This ability would allow sensor operators to detect valid threats more effectively, so target information can be shared with and used by the group.

For example, attachment anxiety is one individual difference that explains variations in sensitivity and responses to threat (Mikulincer & Shaver, 2007). Attachment anxiety relates to the activation of the attachment behavioral system, which induces people to seek proximity to significant others when they feel protection is needed. Those high in attachment anxiety have a heightened sense of vulnerability, and therefore, they often seek out others for help and care. These individuals are described as frequently using a "sentinel schema," which makes them chronically alert and in a frequent state of monitoring of their environment (Feeney & Noller, 1990). Anxious individuals are hypersensitive to subtle cues that may denote threat, and if detected, their sentinel strategies activate and they become quick to alert others to this potential for harm (Ein-Dor et al., 2011).

People high in attachment anxiety are vigilant in monitoring for threats, and they are also more sensitive, quicker, and more accurate in detecting a range of such cues (Ein-Dor & Perry, 2014). Research using small groups found that the group member with the highest attachment anxiety was the most likely to detect the presence of smoke that seemingly came from a nearby computer. It was also this participant who was then motivated to alert fellow group members to this threat in their immediate environment (Ein-Dor et al., 2011). Because they are relatively accustomed to being vigilant, people high in attachment anxiety are likely to enter this attentional state readily and to sustain it with greater ease. So, vigilance will be less detrimental to performance for those high in attachment anxiety. Research supports this presumption.

In one study, participants were placed in realistic shooting simulations where they had to quickly distinguish between friend (person holding an object that was the size of a gun) and foe (a person holding a gun). The simulated environment was immersive and physiologically arousing, distinctions between valid and invalid threats were subtle, and participants were required to make quick decisions to shoot or not shoot targets. In this difficult task, people high in attachment anxiety were more able to quickly identify potential threats and they made more accurate shooting decisions (Ein-Dor et al., 2017). Therefore, we expect characteristics such as attachment anxiety to correspond with more effective contributions from group members who need to be vigilant and frequently monitor the environment.

While the pilot navigates the UAV and the sensor operator vigilantly surveils the environment for designated sites, the mission commander directs the group. This group member generates plans, monitors mission progress, makes updates and revisions as information is collected, and analyzes these data so appropriate decisions can be made and executed. The mission commander is also likely to play a role in promoting cohesion and coordination. In other words, one way-mission commanders contribute to group performance is by collecting the work of group members and combing their efforts to produce unified, collective actions.

Combinations of Group Member Contributions

The theory of combinations of contributions states that effective group performance on spatial problems will occur when group members offer relevant contributions and when these contributions are combined appropriately (Hinsz & Ladbury, 2012). In the previous section, we described some of the contributions that team members make as they navigate spatial problems. We also presented select examples of characteristics that are expected to correspond to higher quality contributions from team members. In the current section, we offer a few examples of how group member contributions might be combined in order to demonstrate some useful ways of conceptualizing groups as spatial information processors.

To understand how spatial knowledge is utilized by a group, it is important to consider how groups process the information. In the generic information processing model (Hinsz et al., 1997), this would include processes, such as acquisition (e.g., information distribution among members), attention (e.g., divided or shared), encoding (e.g., collaborative memory), storage (e.g., mental or situational models), retrieval (e.g., socially cued remembering), and processing workspace (e.g., information integration and inference), which are conditioned by the processing objective (e.g., member or group goals) and feedback (e.g., global or specific to member and/or subtask).

For a UAV team to process spatial information, they must first attend to it. When important pieces of information are neglected, team performance will certainly suffer. So, as team members acquire relevant information, they need to offer it to the group, so it can be attended to and utilized. When a team member withholds vital information, the mission commander or others need to be able to recognize this fact and then request the information, so it can be shared and processed. Because members of UAV teams occupy different roles and are delegated different duties, it is likely that this division-of-labor will make it salient to the team that each member and their contributions serve important functions. Also, when team members work on interdependent tasks, they have to attend to others' contributions because they need this information to perform their own duties. Feelings of indispensability and accountability should motivate team members to willingly offer their contributions to the group and should increase their willingness to attend to the contributions of others (Williams & Karau, 1991). Task motivation should be especially high if mission commanders foster cohesiveness and people find their membership and participation meaningful (Adarves-Yorno et al., 2013). Thus, in many instances, the contribution from team members should receive attention from others and be utilized, particularly when tasks require interdependence.

For example, to capture images of a target, the mission commander needs to be assigned orders that provide physical descriptions of the desired target, along with the target location. The mission commander then needs to share this information with the team. The pilot needs to gather and share navigational and spatial information, so the sensor operator can collect the requested images. The sensor operator needs to recognize the target visually, capture the physical information that was requested, and then share this knowledge with the team. As team members offer and combine their contributions, as a unified team they can execute actions that they could not complete alone (e.g., navigate their vehicle to a designated location and collect images of a target). After the desired images are obtained, the team would also evaluate their performance by comparing it to their stated goal, so they can decide whether their goal has been met or not (e.g., the desired image was captured). Because team members cannot perform their duties without attending to the contributions from others and because team performance requires contributions from all, the three-person UAV team should be quite adept at collectively processing spatial information and solving spatial problems such as navigation and surveillance.

That said, there certainly may be instances when groups are inadequate or ineffective as spatial information processors. This can occur when important spatial information is not attended to because it is not mentioned by a group member or it is not utilized because it is not remembered or valued by the group. To understand how this can happen during collaborative endeavors, it may be useful to first consider some of the forces that underlie interpersonal coordination.

To coordinate and perform collective actions, a variety of team member tendencies need to converge so that teammates form a shared understanding of their situation (Park et al., 2012). To reach consensus about goals, decisions, and behaviors, a degree of social sharedness and mutual understanding is needed (Kameda et al., 1997). The adoption of shared mental models (Hinsz, 1995; Hinsz et al., 1997; Klimoski & Mohammed, 1994) and a shared reality (Tindale & Kameda, 2000) are required for group identities and effective group work, so convergence and alignment should take place often and somewhat easily in group contexts. As previously discussed, spatial perspective taking has been argued to be functional and occur spontaneously for such reasons (Galati & Avraamides, 2013; Tversky & Martin Hard, 2009).

One way that group members can increase social sharedness is by attending to and valuing the same stimuli (Kameda et al., 1997). In past work on decision-making groups, participants were asked to select the best job candidate among three possible choices (Stasser & Titus, 1985). Some information about these candidates was supplied to every group member, and this knowledge was shared and known by all. However, other pieces of candidate information were only distributed to select group members, so this information was not shared nor known by all. As group members engaged in discussion to reach consensus about their preferred candidate, the information that was shared and known by all group members was more likely to be mentioned, repeated, and utilized. Furthermore, research shows that when people offer shared knowledge during discussion, they are perceived to be more competent and credible by their fellow group members (Wittenbaum et al., 1999).

As people make contributions and offer information to the group, there can be a tendency to focus on content that is mutually known and can be validated by others. This push to mention and utilize shared knowledge may be reinforced when group members reap social rewards for their contribution. On one hand, this pattern of communication might promote agreement, cohesion, and perceived efficacy, but at times, these benefits may come with some costs. As items of shared information get mentioned, they may get combined and used prematurely to form corresponding preferences and beliefs (e.g., the shared information that is mentioned and attended to are all consistent with a particular viewpoint, so a preference is formed). If this occurs, group members who possess any unshared knowledge that counters this emerging viewpoint may be reluctant to mention the unique information they hold. If the information is not mentioned, it cannot be attended to or utilized by the group. This can be detrimental to group decisions and performance when the unshared knowledge that is uniquely held by a member is critical and needs to be integrated for successful performance.

For example, imagine a UAV team is collecting images of a hostile weapons factory, so it can later be destroyed. The mission commander reiterates the team's goal for this target, and the pilot and sensor operator offer verbal confirmation. As the target is approached, the pilot acquires navigational information and announces it to the group. As the other members confirm this information, the sensor operator prepares for their task and begins to collect the visual images. As the visual information is captured, the sensor operator shares the results with the team for review. The team confirms that their goal has been met, so the mission commander reminds the pilot about the next target and the pilot confirms and begins to navigate accordingly. On the basis of the information that is discussed, the team determines they are on schedule and are eager to proceed as planned.

In this scenario, the information that was being contributed to the group consisted mostly of shared information. The updates that were provided were pieces of information that team members already knew or confirmed what they expected. There is nothing inherently inappropriate with this type of communication, and updates and validation are certainly important for team functioning. But, when the information that is mentioned and utilized is primarily shared or affirming in nature, it is possible that the team members will feel overly confident about their own and collective abilities. Furthermore, as the team focuses on their efficacy and progress, team members may become reluctant to contribute any information that undermines these desired outcomes. This may be particularly true when that information is unshared. And, if the withheld information is critical, team performance may suffer.

For example, imagine the previously described scenario with the following variation. As the pilot begins to route the UAV to the next target, the sensor operator catches a glimpse of children playing on the roof of a building adjacent to the weapons factory. This building was reported to be abandoned, but the sensor operator now suspects it may be occupied by some families. The sensor operator has acquired unshared visual information that other team members do not possess. Given plans to destroy the weapons factory, this unshared information would be useful to mention.

Yet there are reasons why the sensor operator may be reluctant to introduce this new unshared information. If they offer this information, they may be challenged by their fellow team members because the information is unshared rather than shared. This may undermine the cohesiveness and satisfaction that was felt.

Furthermore, because the unshared information contradicts the knowledge that the group previously held (e.g., the building was abandoned), the sensor operator may also doubt the validity of the new information they acquired. And, if the sensor operator were to mention the unshared information, the team would have to return to the prior target to investigate the situation more thoroughly. This would require re-routing the UAV and disrupting the progress that had been established.

For reasons like these, it may be difficult for team members to offer or integrate information that could undermine the existing position or progress of the group. But, if critical information is not shared, then updated information will not be combined with other contributions and group products may be suboptimal. And importantly, most group members may be unaware of this fact if the withheld information was unshared. Although coordination necessitates that some tendencies converge, so team members can adopt a shared identity and shared mental models, it is critical for people to feel encouraged and able to contribute novel information and opposing views. One way that this can be fostered is through the assignment of roles (Park et al., 2006). Recall that when team members are aware that their unique contributions are instrumental, they are more willing to provide those contributions. Furthermore, when team members are assigned specific roles, a transactive memory system (TMS) is also more likely to develop. This type of collective memory system can improve a team's collective ability to remember and process information (Liang et al., 1995).

Before a team can attend to information and utilize it, the information has to first be remembered, so it can be discussed. A transactive memory system (TMS) consists of the combination of individual memory systems and communications between individuals (Wegner, 1986). A TMS fosters a shared division of cognitive labor to encode, store, and retrieve knowledge among a collection of group members who are perceived to have different areas of expertise. As a group is presented with information, group members tacitly coordinate their efforts, so they can utilize an efficient and more expansive collective memory system (Wittenbaum et al., 1996).

In essence, when information is presented, group members presume that the person who possesses the relevant expertise will be the one responsible for remembering it. Rather than having all members encode this knowledge, the expert learns the information, while the other non-expert members simply encode where that information can be found ("who" possess it) and how to retrieve it (e.g., jargon needed to request and retrieve it). This allows individual group members to conserve energy by primarily encoding "who knows what" when the information is not related to their specialization. And, when information does concern their area of expertise, cognitive resources and capacity are more likely to be available and utilized.

Thus, UAV teams that operate in information-rich environments may have processing advantages if they can utilize a TMS. For example, if a team performs a long mission involving surveillance of many targets, the members of the team can distribute the spatial knowledge that they encode by primarily remembering the information that corresponds to their role. For example, the pilot may end up remembering navigational information and might encode information about spatial relations. Because it will be presumed that the pilot will acquire and remember this information, the sensor operator can attend to and remember visual information that relates to the physical characteristics of objects and scenes that are of potential interest. And, because team members will feel accountable for maintaining this knowledge, their sense of responsibility to the group will motivate them to remember it and offer it during group interactions.

Group Influences on Spatial Cognition

The different research traditions we reviewed illustrate how the information processing in groups conceptualization can help uncover a number of processes and phenomena that contribute to a broader and more interdisciplinary understanding of spatial thinking in groups. But, to appreciate collective spatial cognition, it is not sufficient to only focus on the ways in which groups generate collective outcomes. Instead, it is also critical to acknowledge the potential influences that groups may have on the spatial cognitions of individual members, particularly when individual differences are not strong enough to negate the social influences. Along these lines, we believe that it will be fruitful to discuss a growing body of work that demonstrates a range of ways in which group contexts systematically influence attention and information processing.

Motivational Systems Theory of Group Involvement (MST-GI; Hinsz et al., 2019; Park & Hinsz, 2006) predicts that participation in consensus-seeking groups will increase approach motivation tendencies. This is directly relevant to collective spatial cognition because this theory anticipates that members of UAV teams will experience increased efficacy and will be particularly attentive to visual cues that are representative of rewards and gains. If group members are more attentive and responsive to this type of information, it is expected that spatial information that is relevant to the acquisition of desirable targets will be particularly salient and likely to be communicated. These cues could include any feature of a target that allows perceivers to readily identify potential rewards (e.g., rows of plants that are indicative of crops if the team is searching for sources of food; evidence of mining if the team is searching for desirable minerals). Furthermore, the eagerness that accompanies approach motivation is theorized to direct attention toward spatial information that allows assignments to be completed in an expeditious way (e.g., efficient navigation routes). In essence, the MST-GI predicts that members of UAV teams will be relatively approach oriented and, thus, visually attentive to cues with a positive valence (e.g., enhanced detection and recognition of "friendly" or allied forces) and cues associated with rapid task completion (e.g., discovery of shortcuts that individuals might typically overlook because they involve risk).

Although this influence of groups might generally be construed as beneficial, it is argued that the desirability of this motivational tendency depends on the nature of the task (i.e., does the task involve the accumulation of gains or prevention of losses). If members of a group are inclined to quickly perceive an ambiguous stimulus (e.g., a group of people) as a reward cue (e.g., a group of allied forces the search and rescue team is trying to find), they will subsequently be more likely to pursue it. Whether that tendency is constructive or not would depend on the nature of the mission and the costs associated with making inaccurate judgments.

The MST-GI also predicts that participation in consensus-seeking groups will reduce inhibitory tendencies and decrease avoidance motivation in situations that involve potential threats. This is due to the relative feelings of certainty and psychological safety that tend to be induced in group settings (e.g., Chou & Nordgren, 2017). When a sense of certainty is established through social validation or affirmation, many ambiguous aspects of perception inherently decline (Shteynberg, 2018). As people feel more certain about the perceptions they are constructing, confidence and a sense of correctness increase in a corresponding manner. Therefore, when people work together in groups, they should feel less confusion about the spatial judgments they are making and will feel less inhibited and less reluctant to express and act on such information. This is consistent with past work that shows that people in collaborative groups tend to be overconfident in their judgmental abilities and tend to diminish the value of counter-attitudinal information when it comes from sources outside of their group (Minson & Mueller, 2012). Group contexts are predicted to not only reduce inhibitory tendencies, but through feelings of certainty and safety, groups are also expected to reduce vigilance and avoidance tendencies in members. Supportive of this claim, research from different labs has shown that people are less vigilant in groups (Beckes & Coan, 2011) and are more willing to take risks in the presence of others (Chou & Nordgren, 2017). Thus, theory and research suggest that compared to lone individuals, members of UAV teams will be less likely to visually perceive threat cues, or will take longer to notice them, and will be less likely to integrate them when making judgments or decisions.

Consistent with some of the tenets of the MST-GI, other lines of work have also documented systematic ways that group contexts influence visual perceptions. To provide some background, it may be informative to first mention that past work has shown that prior to the onset of a behavior, visual perceptions of the physical world can be distorted when people are depleted in terms of physical or psychosocial resources (Schnall et al., 2008). Perception has been argued to function within a behavioral "economy of action" (Proffitt, 2006). To promote energetic efficiency, perception relates spatial contexts (e.g., heights, distances, gradients, and perceived weight) to the physical demands of the situation and to the perceiver's physical and psychological states. So, to the extent people feel low in efficacy and believe that they will be unable to successfully traverse distances or climb hills, visual perceptions are distorted to make distances seem farther and gradients seem steeper.

This perceptual bias to exacerbate the intensity of situational demands is thought to occur in an anticipatory manner to convince perceivers to consider alternative courses of action when imagined capabilities are low. Other researchers have relied on a functional perspective to explain perceptual distortions as well. For example, researchers have found that sources of threat are visually perceived to be closer and physically larger than they actually are (Matthews & Mackintosh, 2004). The presumed purpose behind this visual illusion is to induce urgency and to motivate perceivers to respond earlier to threats, so they can be successfully avoided. So, not only do gradients appear steeper than they actually are when perceivers are burdened and at the bottom of a hill (Proffitt et al., 2003), but also spiders seem closer than they actually are when perceivers are spider phobics (Riskind et al., 1995). This type of work highlights the impact that psychological forces can have on spatial cognition and shows why it is important to identify and acknowledge psychological variables that are likely to shape how people acquire and apply knowledge about their environments.

Importantly, researchers have begun to examine these types of perceptual biases within the contexts of groups. For example, it has been found that people perceive hills to be less steep when they are in the presence of a friend, compared to when they are at the bottom of the hill alone (Schnall et al., 2008). Furthermore, it was found that the interpersonal closeness of these relationships was negatively correlated with participants' steepness perceptions. Others have also found that the presence of group members decreases the perceptions of physical formidability when judging threatening targets (Fessler & Holbrook, 2013). In this work, males were either alone or in the presence of group members and watched a brief video that featured scenes with a terrorist from an extremist group. Participants were then asked to describe the perceived stature of the portrayed terrorist by selecting from an array of body sizes and shapes. Participants who were alone selected images that were more physically formidable (taller, wider, and more muscular) than participants who were in groups.

Therefore, literature and past work show that group contexts can systematically influence the types of spatial information that group members respond to and see, as well as the types of spatial information they are likely to value, convey, and remember. Within settings that involve threats, group contexts have also been shown to moderate some of the visual distortions that emerge as observers construct perceptions of distance, size, and gradient. Consequently, although we previously conceptualized groups as spatial information processors, we also simultaneously see utility in appreciating group influences on spatial cognition, a position informed by a Motivational Systems Theory of Group Involvement.

Concluding Comments

Although this chapter focuses on a particular type of group and set of activities associated with teams of UAV operators, the applicability of the theories and principles that we discussed are not limited to this specific collective or set of tasks. We consider theories and evidence that describe how group contexts systematically influence attention and the visual perceptions that individuals form. As that content is presented along with our discussion of the group processes that are relevant to the combination of contributions, we highlighted how the nature of group work is inherently distinct from the experience of working alone. Therefore, through this chapter, we demonstrate that visual perceptions and spatial cognitions that form in groups are likely to differ from those that are formed alone. These differences often result from psychological influences that are specific to group settings, along with the particular processes that are used by groups to generate collaborative responses.

Therefore, we hoped to have drawn attention to relatively novel concepts for collective spatial cognition and encouraged scholars to consider how spatial cognition should be approached differently when such activity occurs within a collective and the information is shared among its members.

Authors' Notes

Preparation of aspects of this chapter was supported by a grant from the Air Force Research Laboratory, under agreement F49620–03–1-0353. The views and conclusions contained herein are those of the authors and should not be interpreted as necessarily representing the official policies or endorsements, either expressed or implied, of the Air Force Research Laboratory or the U.S. Government. Direct inquiries to Ernest S. Park of the Department of Psychology, Grand Valley State University, Allendale, Michigan 49401; e-mail: parker@gvsu.edu.

References

Adarves-Yorno, I., Jetten, J., Postemes, T., & Haslam, S. A. (2013). What are we fighting for? The effects of framing on ingroup identification and allegiance. *Journal of Social Psychology*, *153*, 25–37.

Aggarwal, I., & Woolley, A. W. (2013). Do you see what I see? The effect of members' cognitive style on team processes and errors in task execution. *Organizational Behavior and Human Decision Processes*, *122*, 92–99.

Babel, L. (2014). Flight path planning for unmanned aerial vehicles with landmark-based visual navigation. *Robotics and Autonomous Systems*, *62*, 142–150.

Beckes, L., & Coan, J. A. (2011). Social baseline theory: The role of social proximity in emotion and economy of action. *Social and Personality Psychology Compass*, *5*, 976–988.

Blazhenkova, O., & Kozhevnikov, M. (2009). The new object-spatial-verbal cognitive style model: Theory and measurement. *Applied Cognitive Psychology*, *23*, 638–663.

Casper, J., & Murphy, R. R., (2003). Human-robot interactions during the robot-assisted urban search and rescue response at the world trade center. *Transaction on Systems, Man, and Cybernetics*, *33*, 367–385.

Chou, E. Y., & Nordgren, L. F., (2017). Safety in numbers: Why the mere physical presence of others affects risk taking behaviors. *Journal of Behavioral Decision Making*, *30*, 671–682.

Dalton, R. C., Hölscher, C., & Montello, D. R. (2019). Wayfaring as a social activity. *Frontiers in Psychology*, *10*, 1–14.

Dorfman, A., Weiss, O., Hagbi, Z., Levi, A., & Eilam, D. (2021). Social spatial cognition. *Neuroscience and Biobehavioral Reviews*, *121*, 277–290.

Duran, N., Dale, R., & Galati, A. (2016). Toward integrative dynamic models for adaptive perspective taking. *Topics in Cognitive Science*, 1–19.

Ein-Dor, T., Mikulincer, M., & Shaver, P. R. (2011). Effective reaction to danger: Attachment insecurities predict behavioral reactions to an experimentally induced threat above and beyond general personality traits. *Social Psychological and Personality Science*, *2*, 467–474.

Ein-Dor, T., & Perry-Paldi, A. (2014). Full house of fears: Evidence that people high in attachment anxiety are more accurate in detecting deceit. *Journal of Personality*, *82*, 83–92.

Ein-Dor, T., Perry-Paldi, A., & Hirschberger, G. (2017). Friend or foe? Evidence that anxious people are better at distinguishing targets from non-targets. *European Journal of Social Psychology, 47*, 783–788.

Feeney, J. A., & Noller, P. (1990). Attachment style as a predictor of adult romantic relationships. *Journal of Personality and Social Psychology, 58*, 281–291.

Fessler, D. M. T., & Holbrook, C. (2013). Friends shrink foes: The presence of comrades decreases the envisioned physical formidability of an opponent. *Psychological Science, 24*, 797–802.

Fincannon, T., Keebler, J. R., Jentsch, F., Phillips, E., & Evans, A. W. III. (2011). Team size, team role, communication modality, and team coordination in the distributed operation of multiple heterogeneous unmanned vehicles. *Journal of Cognitive Engineering and Decision Making, 5*, 106–131.

Galati, A., & Avraamides, M. N. (2013). Flexible spatial perspective-taking: Conversational partners weigh multiple cues in collaborative tasks. *Frontiers in Human Neuroscience, 7*, 1–16.

Galati, A., & Avraamides, M. N. (2015). Social and representational cues jointly influence spatial perspective-taking. *Cognitive Science, 39*, 739–765.

Gorman, J. C., & Cooke, N. J. (2011). Changes in team cognition after a retention interval: The benefits of mixing it up. *Journal of Experimental Psychology: Applied, 17*, 303–319.

Hackman, J. R. (1987). The design of work teams. In J. Lorsch (Ed.), *Handbook of organizational behavior* (pp. 315–342). Englewood Cliffs, NJ: Prentice-Hall.

He, G., Ishikawa, T., & Takemiya, M. (2015). Collaborative navigation in an unfamiliar environment with people having different spatial aptitudes. *Spatial Cognition and Computation, 15*, 285–307.

Hegarty, M., Montello, D. R., Richardson, A. E., Ishikawa, T., & Lovelace, K. (2006). Spatial abilities at different scales: Individual differences in aptitude-test performance and spatial layout learning. *Intelligence, 34*, 151–176.

Hinsz, V. B. (1995). Mental models of groups as social systems: Considerations of specification and assessment. *Small Group Research, 26*, 200–233.

Hinsz, V. B., & Ladbury, J. L. (2012). Combinations of contributions for sharing cognitions in teams. In E. Salas, S. M. Fiore, & M. P. Letsky (Eds.), *Theories of team cognition: Cross-disciplinary perspectives* (pp. 245–270). New York: Routledge.

Hinsz, V. B., Park, E., Leung, A. K., & Ladbury, J. (2019). Cultural disposition influences in workgroups: A motivational systems theory of group involvement perspective. *Small Group Research, 50*, 81–137.

Hinsz, V. B., Tindale, R. S., & Vollrath, D. A. (1997). The emerging conceptualization of groups as information processors. *Psychological Bulletin, 121*, 43–64.

Hinsz, V. B., Wallace, D. M., & Ladbury, J. L. (2009). Team performance in dynamic task environments. In G. P. Hodgkinson & J. K. Ford (Eds.), *International review of industrial and organizational psychology* (Vol. 24, pp. 183–216). New York: Wiley.

Ishikawa, T., & Nakamura, U. (2012). Landmark selection in the environment: Relationship with object characteristics and sense of direction. *Spatial Cognition and Computation, 12*, 1–22.

Kameda, T., Ohtsubo, Y., & Takezawa, M. (1997). Centrality in sociocognitive networks and social influence: An illustration in a group-decision making context. *Journal of Personality and Social Psychology, 73*, 296–309.

Klimoski, R., & Mohammed, S. (1994). Team mental model: Construct or metaphor? *Journal of Management, 20*, 403–437.

Kozhevnikov, M. (2007). Cognitive styles in the context of modern psychology: Toward an integrative framework. *Psychological Bulletin, 133*, 464–481.
Kozhevnikov, M., Blazhenkova, O., & Becker, M. (2010). Trade-off in object versus spatial visualization abilities: Restriction in the development of visual-processing resources. *Psychonomic Bulletin and Review, 17*, 29–35.
Liang, D. W., Moreland, R., & Argote, L. (1995). Group versus individual training and group performance: The mediating factor of transactive memory. *Personality and Social Psychology Bulletin, 21*, 384–393.
Matthews, A., & Mackintosh, B. (2004). Take a closer look: Emotion modifies the boundary extension effect. *Emotion, 4*, 36–45.
Mikulincer, M., & Shaver, P. R. (2007). *Attachment in adulthood: Structure, dynamics, and change*. New York: NY. Guilford Press.
Minson, J. A., & Mueller, J. S. (2012). The cost of collaboration: Why joint decision making exacerbates rejection of outside information. *Psychological Science, 23*, 219–224.
Mou, W., & McNamara, T. P. (2002). Intrinsic frames of reference in spatial memory. *Journal of Experimental Psychology: Learning, Memory, and Cognition, 28* (*1*), 162–170.
Park, E. S., & Hinsz, V. B. (2006). "Strength and safety in numbers": A theoretical perspective on group influences on approach and avoidance motivation. *Motivation and Emotion, 30*, 135–142.
Park, E. S., Hinsz, V. B., & Ladbury, J. (2006). Enhancing coordination and collaboration in remotely operated vehicle (ROV) teams. In N. J. Cooke, H. Pringle, H. Pedersen, & O. Connor (Eds.), *Advances in human performance and cognitive engineering research: Human factors of remotely operated vehicles* (pp. 299–310). North Holland: Elsevier.
Park, E. S., Tindale, R. S., & Hinsz, V. B. (2012). Interpersonal cognitive consistency and the sharing of cognition in groups. In B. Gawronski & F. Strack (Eds.), *Cognitive consistency: A fundamental principle in social cognition* (pp. 445–466). New York, NY. Guilford Press.
Proffitt, D. R. (2006). Embodied perception and the economy of action. *Perspectives on Psychological Science, 1*, 110–122.
Proffitt, D. R., Stefanucci, J., Banton, T., & Epstein, W. (2003). The role of effort in perceived distance. *Psychological Science, 14*, 106–112.
Riskind, J. H., Moore, R., & Bowley, L. (1995). The looming of spiders: The fearful distortion of movement and menace. *Behaviour Research and Therapy, 33*, 171–178.
Schnall, S., Harber, K. D., Stefanucci, J. K., & Proffitt, D. R. (2008). Social support and the perception of geographical slant. *Journal of Experimental Social Psychology, 44*, 1246–1255.
Shteynberg, G. (2018). A collective perspective: Shared attention and the mind. *Current Opinion in Psychology, 23*, 93–97.
Stasser, G., & Titus, W. (1985). Pooling of unshared information in group decision making: Bias in information sampling during discussion. *Journal of Personality and Social Psychology, 48*, 1467–1478.
Tindale, R. S., & Kameda, T. (2000). 'Social sharedness' as a unifying theme for information processing in groups. *Group Processes and Intergroup Relations, 3*, 123–140.
Thorndyke, P. W., & Hayes-Roth, B. (1982). Differences in spatial knowledge acquired from maps and navigation. *Cognitive Psychology, 14*, 560–589.
Tversky, B., & Martin Hard, B. (2009). Embodied and disembodied cognition: Spatial perspective taking. *Cognition, 110*, 124–129.

Wegner, D. M. (1986). Transactive memory: A contemporary analysis of the group mind. In B. Mullen & G. R. Goethals (Eds.), *Theories of group behavior* (pp. 185–208). New York, NY: Springer Verlag.
Williams, K. D., & Karau, S. J. (1991). Social loafing and social compensation: The effects of expectations of co-worker performance. *Journal of Personality and Social Psychology*, *61*, 570–581.
Wittenbaum, G. M., Hubbell, A. P., & Zuckerman, C. (1999). Mutual enhancement: Toward an understanding of the collective preference for shared information. *Journal of Personality and Social Psychology*, *77*, 967–978.
Wittenbaum, G. M., Stasser, G., & Merry, C. (1996). Tacit coordination in anticipation of small group task completion. *Journal of Experimental Social Psychology*, *32*, 129–152.

10 A Review of Multiteam Systems With an Eye Toward Applications for Collective Spatial Reasoning

Michael R. Baumann, Donald R. Kretz, and Qiliang He

Introduction

Many important tasks are too big for any one individual to complete alone. For such tasks, society often turns to groups, teams, and other collectives. Such collectives are common in and have been studied in business, science, and other domains. Much of the research in this area is in the tradition of what social psychologists refer to as "socially shared cognition." In the most general sense, this includes collaborative efforts in which two or more people combine information with each other and share the effort of doing so (cf. Hinsz et al., 1997). These processes are often examined with respect to how performance is influenced by different forms of distributing effort (e.g., duplication of effort vs. division of effort) and by members developing or having a shared understanding of the task. As such, this literature has significant overlap with what some call team cognition (Fiore & Salas, 2004), interactive team cognition (Cooke et al., 2013), and collaborative aspects of distributed cognition (Hutchins, 1991). In spite of the range of tasks and types of collectives to which this perspective has been applied, spatial cognition in such collectives has been largely overlooked. This gap in the literature is of great practical importance as many tasks carried out in collectives involve a spatial component. The goal of the current effort is to take a step toward addressing that gap.

The motivation for this chapter was to address challenges in spatial cognition faced by large collectives, specifically in military and emergency services domains. In each of these domains, getting people to the right place at the right time is vital, but the environment is complex and continually changing. The continual change quickly renders mapping data out of date, reducing the usefulness of many technological conveniences. Visibility is often poor and spaces confined, limiting the practicality of updating mapping data via drone-mounted cameras or similar technology. For example, military operations often involve entry into unfamiliar areas occupied by hostile actors. The locations of those actors may be unknown and often change over time. Hostile actors may create obstacles to change what lines of travel are possible, and lines of sight to reference points may be blocked by a variety of factors, including terrain, structures, weather conditions (e.g., fog), smoke, or intentional obscuration. In a similar vein, when fighting fires in a large building, the primary foci (hotspots) of the fires are initially unknown and change

DOI: 10.4324/9781003202738-14

as the fire burns. The locations of trapped occupants are also often unknown and change as occupants attempt their own escapes. The fire is itself a moving obstacle and changes structural features and navigable routes through structural damage it causes to hallways, stairs, and so on. Lines of sight may be blocked by intact elements of the building, collapses, or smoke. Achieving optimal performance requires efficiently building and updating collective mental representations of the space in support of distributing and re-distributing personnel throughout that space as needed to address threats and protect or rescue assets. In short, in each scenario, reasoning about and combining spatial information in the context of wayfinding (i.e., the process of determining and following routes through a given area) is extremely important.

In both military (DeCostanza et al., 2014; DiRosa, 2013) and emergency services settings (Owen et al., 2013), individuals act together in small groups, and these small groups then coordinate with each other within a larger system. These conditions are typical of a kind of collective referred to as a *multiteam system* (MTS; Mathieu et al., 2001). In infantry operations, personnel are often divided into 3–4 person teams where members work together toward a particular objective. These teams combine their efforts as part of a larger collective with larger goals (a two-team squad), and this process continues to form larger structures (squads into platoons, platoons into companies, companies into brigades, and so on). Turning to emergency services, firefighters in larger cities are often divided into 4–6 person companies. Members coordinate within each company to achieve specific objectives, but also coordinate across companies to fight a fire. At a smaller fire, often one company (a "ladder" company) enters, conducts search and rescue, and scouts the fire, while another company (an "engine" company) sets up hose lines to attack the fire. The engine company chooses where to focus the attack based in part on information from the ladder company. Larger fires may require additional companies of each type, and more complex emergencies may require additional types of companies (e.g., hazmat) or additional emergency specializations (e.g., police, personnel with medical training beyond that of firefighters). In short, each of these contexts often involves coordinating inputs and actions of members within small groups *and* between those groups. Therefore, exploring collective spatial cognition in these domains requires both an understanding of how individuals combine information within groups and teams and of how efforts of groups and teams are combined in MTS.

Groups and Teams

How people collaboratively process information has been studied in one form or another since the 1920s (cf. Watson, 1928; Shaw, 1932). As stated by Cooke et al. (2008), it involves both "in the head" and "between the heads" aspects. That is, it involves both what individual team members do and think (in the head) and how those team members interact to combine their knowledge, reasoning, and products thereof (between the heads). Understanding combination processes is key to understanding performance. Steiner (1966) proposed several combination models based on the nature of the task. These included simple sums of member inputs for tasks such as a tug-of-war (an additive model), models in which only the best or worst member mattered, such as fastest in a foot race or slowest climber (disjunctive and conjunctive models, respectively), and models in which performance depended on the fit between members (complementary models). Complementary models are particularly relevant in the domains we focus on here. In complementary models, effort can be divided across individuals such that the collective ability exceeds that of any individual member or component. That is, no one member needs to have all the necessary information or abilities for the collective to succeed. Each piece of necessary information and each ability must be present *somewhere* in the group, but each can be held by a different member. Returning to our firefighter example, one member can have the right knowledge, skills, and equipment to determine how best to attack the fire; another for suppressing the fire; another for providing medical treatment to people rescued from the fire; and so on, rather than one member having it all. In a military unit, the expertise represented is different (e.g., an Army Special Operations unit has intelligence specialists, communication specialists, weapons specialists, medics, and engineers), but the principle remains. This makes it possible for a team to behave as a composite of the best aspects of each member.

Several decades of research exists on how and when members combine their information (e.g., knowledge and perspectives), what information gets combined, and the impact thereof on performance. Although most of this research has been done in the context of decision-making or problem solving, it is likely to be useful for understanding sharing of spatial information and combining spatial reasoning efforts in collectives. In group problem solving, each member often has some task-relevant information he or she knows that others do not ("unique" information) as well as some known to all members ("common" information). Pooling unique information benefits the group both directly (i.e., more information to use to solve the problem at hand) and indirectly, by making it possible to integrate unique information in ways that allow the group to infer information no members previously knew (Fraidin, 2004; van Ginkel & van Knippenberg, 2008). Members may also learn from each other how to better use the information they have (Schultze et al., 2012), thus raising each member's potential performance. Indeed, meta-analyses have shown that the extent to which members discuss and use unique information has a positive association with desirable group outcomes, ranging from group cohesion to performance across a range of task types (Mesmer-Magnus & DeChurch, 2009; Mesmer-Magnus et al., 2017). Unfortunately, meta-analyses have also shown that groups often do a poor job of combining members' unique information (e.g., Lu et al., 2012; Mesmer-Magnus & DeChurch, 2009). This is a common issue in conventional military units as well (retired Sergeant Major G. Patti, personal communication, August 19, 2021). To make matters worse, there is evidence that, when both types of information are discussed, less weight is given to the unique information discussed than to the common information discussed (e.g., Baumann & Bonner, 2013; Chernyshenko et al., 2003; Mojzisch et al., 2010). Together, these reduce groups' and teams' effectiveness on many information tasks.

As the research on unique versus common information shows, knowing *which* information to share is important. This can be facilitated by members having similar mental models of the situation at hand. Referred to as *shared mental models* by some authors (e.g., Cannon-Bowers et al., 1993; Mathieu et al., 2000) and *team mental models* by others (Klimoski & Mohammed, 1994; Mohammed et al., 2010), these include members' thoughts regarding the task, technology, expected interactions, and each other. When members have similar models, they are better able to anticipate what information others in their group need (Mohammed et al., 2010). In addition to facilitating information sharing, groups in which members have similar mental models often experience better coordination and performance (e.g., Mathieu et al., 2000).

It is important to note that people can have similar mental models about member *differences*. For example, in "transactive memory" (Wegner, 1986), members are responsible for different areas of information and each member's mental model of the group includes knowledge of that division of responsibility (i.e., who knows what). These models can be formed in several ways, including through long-term interaction (Wegner et al., 1991), reputation (Moreland & Myaskovsky, 2000), or inferences conveyed by formal titles (Brandon & Hollingshead, 2004). These models are updated through various communication processes between members (Brandon & Hollingshead, 2004). Each member focuses on learning information within their area of responsibility and the group turns to the appropriate member for that information at retrieval. The models of who knows what, processes for maintaining and updating those models, and processes for storing and retrieving information are collectively referred to as a transactive memory system.

Research in this vein has shown that knowledge of differences in responsibilities (or expertise) increases the discussion (Stasser et al., 1995), use (Baumann & Bonner, 2013), and integration (Fraidin, 2004) of information across group members. When task demands stretch or exceed individual member abilities, transactive memory is associated with better performance on tasks ranging from decision-making (Fraidin, 2004) to assembling radios (Liang et al., 1995), with performance benefits generally increasing with the accuracy of the model of who knows/is responsible for what (Austin, 2003). As noted in the previous paragraph, these models are often built from direct interaction between members but can also be based on indirect information. Using indirect information can help build transactive memory more quickly by providing initial information but can also limit model completeness and accuracy. Two people with the same title may vary significantly in knowledge and skills or in relevant extra-role knowledge of benefit to the group. Likewise, reputation and stereotypes can be inaccurate and lead to poor initial models. Poor initial models may be problematic. Once models are formed, they impact information search (Snyder & Swan, 1978) and interpretation (Lord & Taylor, 2009) in self-sustaining ways, making inaccuracies difficult to detect and correct (cf. Hollingshead & Fraidin, 2003). Together, these issues may make the resulting transactive memory system less beneficial.

In military and emergency services operations, members are often distributed in such a way that no one member sees the entire operation space. Rather, each member must rely on observations from other members to determine the locations of key entities (e.g., hostile actors or individuals in need of rescue), obstacles (e.g., natural or deliberately placed barriers, collapses, or fires), and routes. Even within a co-located group, different members may have different lines of sight, be looking in different directions while transiting a given area, or otherwise have different pieces of the map. In short, members sharing their unique information is important for forming an overall shared mental model of the space. For convenience, we call these *shared spatial mental models*. These, in turn, are likely to help members better understand what information they need to convey to others (e.g., things that deviate from the model). Members understanding who is responsible for what aspects of the space (as in transactive memory) would likely facilitate efficient sharing of information (i.e., pooling unique information) and allow member inputs to combine in a complementary fashion. Taken together, these group states and processes provide the potential to draw on a greater amount of spatial information (i.e., the sum of all unique spatial information held by members), increase the potential for correctly inferring information no member directly observed (i.e., by integrating observations), and create opportunities for members to learn from each other what to observe, what to communicate, and how best to communicate to create the most accurate overall spatial models.

Before moving on to multiteam systems (MTS), it is worth highlighting a few other important points about groups. First, in addition to shared mental models and transactive memory being important for combination of members' inputs, leadership can also play an important role. Indeed, laboratory groups whose leaders actively query members and repeat/connect information mentioned back to the group tend to show greater sharing and integration of unique information (e.g., Larson et al., 1998). These leader behaviors are important in the field as well (G. Patti, personal communication, August 19, 2021). Second, groups often demonstrate error-correction properties (Hinsz et al., 1997). Findings on a range of decision-making and problem solving tasks are consistent with the notion that errors in reasoning made by one member may be noticed and corrected by others (see Laughlin, 2011, for a review). Taken together with groups' potential for combining member expertise, this gives groups the potential for outperforming individual members on a variety of tasks. In the context of combining spatial information and collaborative spatial reasoning, this may manifest as correcting mis-stated information (e.g., a "west" that should be "east") or flawed reasoning (e.g., noticing a miscalculation of heading).

Multiteam Systems

Multiteam systems (MTS) are collectives made up of multiple distinct groups working together toward one or more superordinate goals. It has been argued that MTSs are common in military command hierarchies (DeCostanza et al., 2014; DiRosa, 2013). Indeed, military units may form a hierarchy of MTS (DeCostanza et al., 2014). A platoon is an MTS of subordinate squads but is also a component team within the larger MTS of its parent company, which in turn is a component team of its brigade and so on. MTSs are also common in emergency services contexts (Owen et al., 2013). Firefighters are typically organized into small groups known as companies, with different companies having different specializations and therefore different responsibilities. When fighting a fire, one company is typically responsible for entry, search and rescue, and reconnaissance (typically a ladder company), and another is responsible for fire suppression (usually an engine company). Individual members work together within the company on some aspect of the overall task (e.g., for the ladder company, search and rescue). However, the overall task requires coordination between companies as well. In this example, the ladder company identifies both stranded civilians and hot spots. This information influences where the engine company targets its efforts. Likewise, where the engine company targets its efforts influences where the ladder company can search. In larger emergencies, these companies may in turn be embedded in larger systems coordinating multiple ladder, engine, and other fire companies alongside police and specialized medical personnel.

Although MTS have some similarities to traditional groups and teams, hereafter called groups, there are also important differences (Shuffler & Carter, 2018). Many of these derive from the fact that groups within an MTS, referred to as *component teams* of the MTS, are more than simply subgroups or factions. Members of an MTS typically identify with their component team more strongly than with the overall system (DiRosa, 2013; Lanaj et al., 2018). Furthermore, although component teams in an MTS work together toward one or more superordinate goals, component teams frequently have their own proximal goals. The superordinate and proximal goals are not necessarily in complete alignment. It has been theorized that members prioritize the goals of their component team over those of the MTS (e.g., Kanfer & Kerry, 2012; Rico et al., 2017), analogous to individuals putting personal goals over team goals in the literature on traditional teams (DeShon et al., 2004). It has long been argued that this differential prioritization may create conflict between component teams, and recent case studies of military MTS (Wijnmaalen et al., 2018) obtained findings consistent with that notion.

In addition to MTS differing from traditional groups in terms of identification and motives, MTS differ from traditional groups in communication patterns. In traditional groups, it is common for each member to talk directly to any other member as the need arises. This is common within component teams of an MTS as well. However, in many MTS, members of one component team do not talk directly to *just any* member of another component team. Cross-team communication tends to be routed through specific individuals. For example, two members of the same platoon likely communicate directly with each other, but to get a message to another platoon, they relay it through the platoon leader or their radio operator/maintainer. These points of contact are called *boundary spanners* (Davison & Hollenbeck, 2012). As our example shows, these points of contact are often (but do not have to be) component leaders. Although this practice decreases the overall number of messages between members and can help performance by reducing information overload (Davison et al., 2012), it also creates an additional step at which information may be lost or mis-stated (i.e., the relay through the boundary spanner) compared to direct communication between members. See Figure 10.1 for a graphical representation of the differences between groups and MTS.

In our discussion of traditional groups and teams, we noted the importance of shared mental models and transactive memory. Each develops spontaneously and naturally in teams whose members are co-located and frequently interact (Mohammed et al., 2010; Brandon & Hollingshead, 2004, respectively). However, in many MTSs, different component teams are in different physical locations and unable to easily have spontaneous face-to-face communication with or direct observation of each other. This slows the development of shared mental models (Maynard & Gilson, 2014) and transactive memory systems (O'Leary & Mortensen, 2010) in groups and leads to even stronger information sharing biases in MTS than in group settings (Shuffler & Carter, 2018). Assuming that a model was already in place, separation would also inhibit model updating by limiting opportunities to observe others (in this case, other groups) and how they have (or have not) changed over time. We argue that the military and emergency services settings we are focusing on have features that are functionally analogous to geographic separation. In military operations, different squads may scout different portions of the area of operations and be sufficiently distant from or sufficiently obscured from each other (e.g., by terrain or buildings) to make face-to-face communication difficult, hand gestures impractical, and observation impractical. In firefighting, even in relatively small buildings, twisting hallways, separation across different floors, or the fire itself may have much the same effect. The respective features of each setting limit team/MTS members' ability to develop shared mental models and to use transactive memory for the *specific* operation space.

Sufficient interaction *prior to* operations could provide component teams a sense of which teams will generally have what sorts of knowledge, potentially

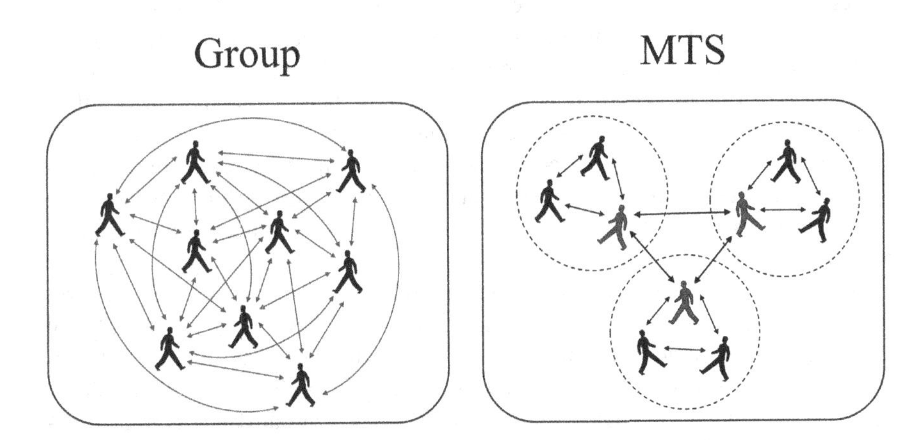

Figure 10.1 Communication in groups vs. MTS. In groups (left), communication is typically all-channel, that is, any two members can speak directly. In MTS (right), communication within teams is typically all-channel, but *between* teams is often through a designated relay known as a *boundary spanner* (depicted in gray).

facilitating development of shared mental models and transactive memory systems during the operation. However, such interaction is often lacking in our contexts. Military personnel associate more with other members of their immediate unit than of other components of their system (i.e., within platoon more than across platoons) and different components may be housed in different sections of the home installation. In larger MTS (e.g., brigades within a division), components may even be housed at different installations. Likewise for emergency services, a single fire station typically only houses two or three companies, and most interaction is within rather than across company or station. The gap is even more pronounced for MTS involving multiple services (e.g., fire, police, and specialized medical personnel).

In addition to impeding development of shared mental models and of transactive memory systems, and sharing and use of unique information, the lack of co-location common in MTS may exacerbate tendencies to identify more strongly with one's component team than with the MTS (Wijnmaalen et al., 2018). In the teams literature, it has been suggested that lack of co-location increases the risk of miscommunication and thereby increases the risk of misunderstandings and conflict (e.g., Hinds & Mortensen, 2005). Consistent with that notion, trust develops more slowly in geographically dispersed teams than co-located ones (e.g., Jarvenpaa et al., 1998; Jarvenpaa & Leidner, 1999). Over-identifying with the component team relative to the MTS (Mell et al., 2018) and conflict between component teams (Wijnmaalen et al., 2018) each potentially contribute to a reduction in information sharing. Each of these is a challenge to effective MTS performance in general and likely a challenge to wayfinding by MTS as well.

Analogous to in teams, some of the issues in MTS may be mitigated by the presence of a leadership team (e.g., in military settings, a headquarters unit or command group). Frequent communication of the leadership team with each component team allows for better integration of component teams' inputs (Davison et al., 2012). However, this integration would still need to be pushed out to other component teams to maintain coordination absent shared mental models. Furthermore, although the leadership team may have been co-located with some components prior to action, it is unlikely to have been co-located with all of them and will generally lack co-location during action. As such, the issues of miscommunication, trust, and conflict between component teams may potentially apply between regular component teams and a leadership team as well. The literature shows many benefits of having a leadership team, but it is not a panacea.

Although MTS face several obstacles in military and emergency services settings, MTS also have several desirable features for these contexts. Traditional teams make it possible to combine inputs from individuals who differ in knowledge or abilities in ways that allow teams to do things no individual could do alone. However, sometimes, one team is not enough. In each of our contexts, the operation consists of sufficiently large or complex subtasks that each subtask requires a team. In each, multiple tasks often need to be completed simultaneously across an area that is too large for any single team to cover while still maintaining sufficient contact and interaction to function as a team. An MTS addresses these issues.

Much as how individuals within a team may specialize in different parts of the task, each *component team* of an MTS may specialize in a different aspect of the overall task (Shuffler & Carter, 2018), including different types of spatial knowledge or spatial knowledge for different parts of the operation space.

In military and emergency services settings, an MTS structure is likely to be highly desirable and have a greater chance of success overall than individuals or teams alone would have. However, to achieve this will require carefully considering what is known about MTS and what is known about spatial reasoning to determine what can be done to best support these MTS in accomplishing the spatial aspects of their missions. Although a large body of research has been conducted on how traditional teams pool, integrate, and otherwise make use of members' information, and a smaller body has examined this for MTS, the vast majority of research on each has been done in settings in which the information exchanged is limited to simple facts (e.g., profitability of a product, a job candidate's alma mater, the alibi of a person of interest in a criminal investigation). To the best of our knowledge, no studies have looked at this issue in the context of spatial information, spatial reasoning, or wayfinding. This is important because the research on teams and MTS has shown that process and performance vary considerably with the nature of the task the team or MTS is performing (see Laughlin, 2011; Rico et al., 2018, respectively). Therefore, before we can address the question of wayfinding or any spatial task in MTS, we must first examine the nature of spatial reasoning in the selected contexts.

Special Considerations for Spatial Tasks and Our Selected Environments

Most research on groups and MTS has focused on informational tasks such as problem solving or decision-making. Laboratory studies of groups often give participants lists of information about each of several choices in a decision-making task and observe which information participants discuss (see Lu et al., 2012, for a recent review). Field studies of MTS often focus on what information is being passed between component teams, by whom, and via what medium (Shuffler et al., 2015). Although research has studied tasks that involved a spatial component (e.g., simulated command and control tasks, case studies of firefighters), the spatial component *as a spatial component* has been largely overlooked.

There are far too many aspects of spatial cognition to address in a single chapter. The issue of most interest to the authors is how people coordinate their efforts while moving through a dynamic area of operations to achieve mission objectives. Doing so includes coordinating efforts to determine and follow paths to objectives, a task generally referred to as *wayfinding* (Lynch, 1960). Rather than review all aspects of spatial cognition, we focus on aspects of spatial cognition that we believe are particularly likely to influence the ability of MTS to coordinate their efforts during wayfinding. Specifically, we focus on the ways that people conceive of, represent, communicate, and interpret communications about space, and how these influence the shared spatial mental models that MTS form.

Wayfinding

Wayfinding, a term introduced by Lynch (1960), is "the process of determining and following a path or route between an origin and a destination" (Golledge, 1999, p. 6). It's about determining and knowing "where you are, where you want to go, and how to get there" (Dalton et al., 2019, p. 1). Wayfinding requires that one form and manipulate internal representations of space by monitoring both external cues such as landmarks and terrain and internal cues such as memories and inferences. While sometimes used synonymously with navigation, wayfinding can be distinguished as the higher cognitive component of navigation (the lower being locomotion) (Lynch, 1960; Montello, 2001; Wiener et al., 2009). In military and emergency services settings, what may begin as aided wayfinding (i.e., assisted by maps, imagery, signs, and so on) can quickly turn into an unaided task as the environment changes.

Many attempts have been made to describe the cognitive processes involved in wayfinding. The processes can loosely be grouped into perception of spatial cues (both environmental and internal), computing relative distances and directions (including to self) accounting for one's own movement, and spatial representations within one's current focus of attention and in memory (Wolbers & Hegarty, 2010). For example, Allen (1999) discusses the role of relating current location to past movement and working memory in different types of wayfinding. Montello (1998) discusses processes for the development of metric and relational spatial knowledge and integration of spatial knowledge into more elaborate knowledge structures. The analogous processes for a collective would be members perceiving spatial cues, combining across members to make judgments of relational distance and direction, integrating the knowledge and judgments into spatial mental models at the component team level representing their portion of the area of operations, then integrating those models to the MTS level to form a shared mental model of the overall area of operations. The component team and MTS shared spatial mental models would then be represented in memory (likely different members' memory), and the relevant model(s) and member(s) consulted when determining how to move through the area of operations. If component teams are not successfully wayfinding, they have no information to contribute to the shared spatial mental models of the larger MTS. If they are not communicating their information, their potentially valuable information is not available to the MTS or other components.

The nature of collective wayfinding in the context of an MTS is very similar to the version of *social wayfinding* Dalton et al. (2019) referred to as *strong, synchronous* social wayfinding. In Dalton et al., strong synchronous social wayfinding is presented as wayfinding by a co-located group in which people are traveling together. In the context of an MTS, the component teams are in the same general space but not closely co-located. That is, collective wayfinding involves combining the information acquired by multiple teams that are working in different parts of the environment (see Figure 10.2). As such, in an MTS, component teams each learn and develop different pieces of the space and then integrate those pieces into

Figure 10.2 Group vs. MTS moving through area of operations. In groups (left), members stay together while exploring or traveling. In an MTS (right), component teams can separate to explore or travel through multiple areas simultaneously to cover more area/complete more objectives per unit time.

a shared spatial mental model. Assuming that the component teams communicate well and each component team is aware of the area of responsibility of each other team, the MTS can divide the labor of learning and recalling spatial representations much as a group might do through transactive memory systems. Because the component teams are not limited to being co-located, the MTS as a whole is able to cover a greater proportion of the area of operations at any given time compared to the same number of personnel operating as a traditional team. This provides an opportunity for the MTS to gain multiple perspectives on each portion of the area of operations. For example, component team A may see the northern side of a particular feature of the environment, and team B may see the southern side. If the teams communicate their observations to each other, they have an opportunity to develop a more detailed and accurate mental model of the space. Because both teams now have this model, it becomes a shared spatial mental model. In a similar vein, by integrating the two sets of information, they may infer information about an unexplored portion of the environment, resulting in an even more detailed model. The discussion and integration of information also creates a potential for error correction. If one component team provides information that is logically inconsistent with information provided by another, it may trigger discussion that reveals a miscalculation or miscommunication.

Spatial Representations

Cognitive representations of spatial information at the individual level can be stratified into three levels: discrete points in space (e.g., landmarks, street names, origin and destination points), a sequence of points in space (i.e., a path or "route"), or an area (i.e., a "survey") (He, McNamara, Bodenheimer et al., 2019; Siegel & White, 1975; Wiener et al., 2009). Humans can use these representations to store memories and build mental models of space often called *cognitive maps* (Thorndyke & Hayes-Roth, 1982; Tversky, 1993). In the terminology used in the groups literature, these would be spatial mental models. Once spatial mental models are formed, people can modify them, expand them, or integrate them with new information that becomes available. However, forming, using, and modifying mental models becomes increasingly difficult as model complexity increases (i.e., as the number of objects and associations it contains increases). Known as a *fan effect*, the difficulty is amplified as more models are integrated to form larger models (e.g., discrete routes are integrated through spatial reasoning; Han & Becker, 2014). Research has shown that mental models about space and spatial relations can be easily distorted even within individual (Montello, 2005; Tversky et al., 1999). As such, we would expect individuals to have difficulty maintaining, updating, and rapidly accessing full cognitive maps of large or complex areas of operations such as those common in military and emergency services contexts. However, an MTS could overcome this using a division of responsibility akin to transactive memory. Rather than every member keeping the entire map in their own memory, each component team can keep a detailed map of their area of responsibility and retrieve information on other areas when needed from the component responsible for that area.

Some people are better than others at forming spatial mental models (Ishikawa & Montello, 2006; Weisberg & Newcombe, 2018). On one end are London taxi drivers, who have to pass a test of their memory of the city's labyrinthine 25,000 streets. On the opposite end, some people have a very difficult time orienting themselves even in extremely familiar surroundings, despite the absence of any acquired brain damage or neurological disorder (Iaria & Burles, 2016). A number of factors, such as working memory capacity, personality, or gender, could influence one's ability to form spatial mental models (Weisberg & Newcombe, 2018). It is also theorized that good navigators can take advantage of the spatial cues flexibly, switching between cues and types of cues used on the basis of task demand or cue availability (Wolbers & Hegarty, 2010), whereas bad navigators might have memory limitations that prevent them from forming or retrieving useful representations (He & Brown, 2020).

Communicating Spatial Information

To achieve the full potential of an MTS for collective wayfinding, members must communicate spatial information and spatial knowledge effectively both within and between components in the MTS. Because communication in such settings is typically via a voice-only medium (e.g., radio), spoken language mediates the transfer of spatial information. While an examination of specific linguistic patterns and structures is not within the scope of this chapter, the cognitive implications are worth noting. To successfully convey spatial representations in this manner requires the sender to encode spatial ideas into words that a distant and unseen individual can decode back into a spatial representation. This requires the sender and receiver to have a common understanding of and adapt to each other's use of descriptive language and choice of geographical referents such as landmarks. Part of this involves choosing a frame of reference (Klatzky, 1998). To be effective, this frame must be understood by both the communicator and the receiver. However, multiple frames are possible. In an allocentric reference frame, objects are represented along dimensions that are extrinsic to the navigator (e.g., the entrance is northeast of the exit). In an egocentric reference frame, objects are represented along dimensions relative to the navigator (e.g., the entrance is to my right). Switching between allocentric and egocentric reference frames is shown to be error-prone (He & McNamara, 2018a; He et al., 2016). Because MTS rely on boundary spanners for inter-component communication, boundary spanners have the added burden of making themselves understood both by members of the local component team and by component teams in other parts of the area of operations.

Individual differences and preferences can influence the quality of interpersonal and intergroup communication and impact a receivers' ability to understand the communication. For example, dyads using voice-only communication are more successful when composed of members with similar levels of spatial reasoning ability and beliefs about what makes a good landmark (He et al., 2015) and when directions are more general rather than highly precise (Bloom et al., 1996). Furthermore, the semantics of different spatial frames of reference (e.g., allocentric versus egocentric) offer different perspectives for conveying the same spatial information.

Communication can be complicated by various factors. Allen (1999) emphasizes the importance of comprehending maps and directions since a group relies on each individual's ability to associate objects in the environment to objects on a map, associate one's own location and direction of movement to a point and heading on map, and apply verbal and reasoning abilities in addition to visual memory. Weisberg and Newcombe (2018) notes that individuals differ in their natural ability to encode accurate internal maps, integrate route information, and make crucial between-route spatial inferences. As such, some receivers will have more difficulty understanding a message than others. The presentation of information in the development of shared spatial mental models matters as well. For example, if points and connections are received discontinuously, individuals often have more difficulty processing them (i.e., a *continuity effect* occurs) (Nejasmic et al., 2015). Furthermore, the higher the relational complexity of the model (in terms of the number of entities and corresponding associations), the longer the retrieval time (i.e., the "fan effect") (Radvansky & Zacks, 1991). The bottom line is that without a proper and integrated internal map, there is a strong risk that communicated spatial information will be compromised.

Operational Considerations

Military and emergency services operations often involve many distracting stimuli and situational stressors. These can make it difficult for individuals to form complete representations of their surroundings (cf. Wright, 1974). Because both the environment and the location of the teams within it change over time, the spatial mental model of any part of the area of operation may become unimportant (we are no longer there) or inaccurate (that way is now shut) over time. Each of these factors increases the difficulty of developing accurate spatial representations and communicating those representations to others. The error checking properties of teams and of MTS may help mitigate these problems by allowing for the comparison of different teams' spatial mental models and integration across them. Where communication reveals inconsistencies, queries can be sent to the relevant component teams to resolve them.

As we noted earlier, communication is important to the development of shared mental models and transactive memory systems as well as to maintaining both their accuracy and sharedness among component teams of an MTS. Unfortunately, in many military and emergency services settings, communication channels can become degraded. Communication may be blocked deliberately by hostiles (e.g., jammed), the environment (e.g., structural features or collapse), or misadventure (e.g., damage to a radio). This creates gaps in communication, during which shared spatial mental models can become out of date, the similarity of each component teams' model to other teams' models may decrease, or both. It also creates a risk for relying too heavily on transactive memory systems—if the component responsible for an area cannot be contacted, their knowledge cannot be accessed. However, a well-developed shared mental model often allows component teams to function in a relatively coordinated fashion for some time without communication. Indeed, one of the benefits of shared mental models in general is that they allow the anticipation of others' actions and responses, thereby facilitating coordinated action absent direct communication (Cannon-Bowers et al., 1993). Likewise, partial overlaps in responsibility can be built into transactive memory systems to serve as backups for unavailability of the responsible component (Baumann & Bonner, 2017).

Finally, although we have been discussing wayfinding in MTS in a general sense, the exact processes will vary by environment. Wiener et al. (2009) point out that the cognitive resources required for operating outdoors differ from those when operating inside buildings and structures both in terms of spatial knowledge and problem solving strategies. The operational context dictates the sort of cognitive models and system of directions and referents that would be most effective.

Each of these elements stresses the importance of pooling and integrating information, particularly unique information, as thoroughly as possible between components. If done well, the shared mental model of the operational environment will be accurate and coherent. However, if the information is poorly pooled or poorly integrated, the overall shared spatial mental model will likely be inaccurate. If information is not well communicated to component teams, each team will have its own unique spatial mental model rather than one that is in common (shared) among teams. Each of these reduces the likelihood of coordination between teams and thus jeopardizes success.

Recommendations for Spatial Cognition in MTS and Avenues for Research

The literature provides several recommendations regarding how best to build and support MTS. Some of these would seem to apply to MTS engaged in wayfinding as well. For example, building group cohesion can be helpful for group performance (e.g., see Beal et al., 2003, for a meta-analysis). The co-location of members within the same component team of an MTS and the spontaneous, face-to-face interaction co-location permits tend to support the development of group cohesion (cf. Grossman, 2014). The limited interaction across component teams would tend to inhibit the development of an overall MTS cohesion. Studies of military MTS have emphasized the importance of both within-component team and cross-component team cohesion (e.g., DeCostanza et al., 2014; Wijnmaalen et al., 2018). Wijnmallen et al. recommend providing interventions to increase both within- and cross-component cohesion prior to deploying an MTS. However, it is important to note that being simultaneously very high in within-component cohesion and very high in cross-component cohesion may have detrimental effects on MTS performance (DiRosa, 2013). That is, high cross-component cohesion may at some point become what DeChurch and Zaccaro (2013) refer to as a *countervailing* factor. This describes a factor that may have positive effects at one level (in this case, within component) and yet have negative effects at another (in this case, cross component). As such, determining how much within- and cross-component cohesion is desirable for MTS engaged in spatial cognition tasks may be a fruitful area for future research.

Research Recommendation 1: Determine the optimal level of within- and between-component cohesion for MTS wayfinding.

Both in the broader MTS literature and in the current context specifically, boundary spanners play a key role. It is important that communication between component teams is well managed. Information from each component team that is relevant to other component teams needs to be provided to ("shared with") the latter by the former. Failures in information sharing have been cited as key factors in mission failure, including in high-profile disasters (e.g., the loss of the space shuttle *Columbia*; Columbia Accident Investigation Board, 2003). However, sharing all information with every component team, or having each member of each team share directly with each member of each other component rather than through boundary spanners, is likely to lead to information overload (Davison et al., 2012). This makes boundary spanners vital in the function of an MTS, both when communicating horizontally to other components of the same level and when communicating to a leadership group if present. It has been argued that it is important for boundary spanners to have a sufficiently general knowledge base to be able to communicate effectively both with the members of the component team they represent and with members of other component teams (Olabsi & Lewis, 2018). In a similar vein, MTS are typically more effective when boundary spanners have shared multiteam interaction mental models (Murase et al., 2014). That is, MTS perform better when the boundary spanner for each component team understands the other teams' needs and expectations regarding how the component teams will interact and coordinate.

The literature on shared mental models in teams (as reviewed in Mohammed et al., 2010) would suggest that having boundary spanners interact often with both their own team and with boundary spanners from the other component teams prior to deploying may help boundary spanners develop the within- and cross-component shared mental models they will need for effective communication. This is likely to impact their capacity to support the equivalent of a transactive memory system incorporating the component teams. The communication tools chosen will also affect shared mental model development. For example, Jiménez-Rodríguez (2012) found that shared mental models developed more quickly when the communication medium was more retrievable (e.g., a text one could look back at) than when the tool imposed a greater memory load (e.g., videoconferencing). Other scholars have suggested MTS charters specifying norms and expectations for communication between teams also contribute to effective communication between boundary spanners (Asencio et al., 2012). Yet others have emphasized the value of creating a separate leadership team (a component team tasked with coordination, such as a headquarters or command group) that can serve as a hub through which component teams coordinate and to whom to delegate the effort of integrating information (Davison et al., 2012), an idea echoed by a subject matter expert kind enough to comment on an earlier draft of our chapter.

Research Recommendation 2: Determine the best medium to convey spatial information between boundary spanners to support development of shared spatial mental models and transactive memory-like systems across boundary spanners.

Research Recommendation 3: Assess the benefits of a leadership team to coordinate spatial information between boundary spanners relative to direct coordination between boundary spanners.

In the context of wayfinding in MTS, the question of what exactly boundary spanners should be communicating is somewhat complex. The literature on spatial reasoning has shown wide individual differences and situational differences in how people think about space (Montello, 2005; Tversky et al., 1999). In a similar vein, the literature on giving and following spatial directions has shown that the match between the sender's and perceiver's conceptualizations of space is extremely important. As such, the first step in determining how boundary spanners should communicate is to determine what the most effective spatial representation is in the contexts in which the MTS will be wayfinding. Once that is determined, those who will be serving as boundary spanners should receive specialized training to ensure that they are well versed in communication strategies supporting said representations. Whether boundary spanners are communicating directly with their counterpart on another component team or via a leadership team, having the same understanding of how to view space and how to communicate spatial relations will be vital.

Research Recommendation 4: Determine the optimal communication strategies (i.e., strategies most aligned with best spatial representations) for boundary spanners to use.

Research Recommendation 5: Develop training to support the consistent use of the communication strategies identified in Recommendation 4.

One of the main benefits of either a team or MTS is the potential to integrate information from across different members and component teams, respectively. Although integration is made possible by communication between individuals, the integration itself still takes place within an individual's mind. In an MTS, that person would likely be a boundary spanner or, if a separate leadership team is part of the MTS, a member of the leadership team. Unfortunately, humans, in general, are not good at integrating spatial information into a precise and coherent representation (Foo et al., 2005; Han & Becker, 2014; Meilinger et al., 2011). In addition, the person doing the integrating is often relying on reports from others rather than direct observation. This may make the integration process even more challenging (Chrastil & Warren, 2012). Together, this makes it vital that the people tasked with performing the integration are particularly skilled at integrating information across sources. Weisberg et al. (2014) present a navigation task that can identify people who are good at integrating locations of buildings across remote regions. However, the spatial information given in this task was from a first-person perspective, and it is unknown whether the good integrators identified by this task are also good integrators when the spatial information is delivered by a proxy. Being able to integrate spatial information delivered via proxy is essential for boundary spanners in the spatial navigation context.

Research Recommendation 6: Identify techniques for assessing skill at integrating spatial information received via various media, particularly verbal reports.

Research Recommendation 7: Identify techniques for improving skill at integrating spatial information received via various media, particularly verbal reports.

Although a consistent reference frame for communicating between boundary spanners may be useful, it is also likely that MTS would benefit from boundary spanners who are able to switch between frames accurately and swiftly. First, it is unlikely that every member of every team will use the same reference frame in every communication with their boundary spanner. Second, the best method for communicating within team may differ from the best method for communicating between teams. For example, when the location of interest is near and immediate action is required, egocentric reference frames may be more useful (Crawford et al., 2011; e.g., turn left at the next crossing and the target is 100 feet on your right). In contrast, when the location of interest is far away and there are few salient landmarks along the route, an allocentric reference frame may be better (Gallistel, 1990; move two miles northeast). The former situation is likely to be more common within a team and the latter between teams, making it likely that boundary spanners will need to be skilled in each. In a similar vein, it is also likely to be useful for boundary spanners to be skilled at using spatial updating strategies that convert the information received from any reference frame into any other so that all information can be stored in a common frame (He & McNamara, 2018b). To our knowledge, there is no published test or questionnaire of individuals' speed and accuracy in switching between reference frames.

Research Recommendation 8: Develop techniques for assessing speed and accuracy of switching between and integrating across reference frames.

The settings on which we have focused are inherently dynamic (e.g., a steeple used for orientation yesterday may be gone today; the geometry of a room may change due to structural damage). As such, the visual cues that can be used for orientation or localization in these environments could be very different than in more stable settings. Therefore, personnel on the ground need to take advantage of the available cues in a flexible way. That is, when one kind of cue changes or becomes obscured (e.g., a landmark), personnel need to be able to switch to using other cues (e.g., the geometry). Cue-use flexibility is a relatively new concept in the spatial cognition literature, but a recent study found that self-reported sense of direction (SOD; Hegarty et al., 2002; Pazzaglia & De Beni, 2001) could predict such use (He, McNamara, Bodenheimer et al., 2019). Many more studies are needed to examine how flexible humans can be in using unfamiliar spatial cues, and whether SOD is still predictive of such ability in circumstances different from the one in He, McNamara, and Brown (2019). Moreover, it is unknown whether other psychological attributes such as working memory capacity can also predict cue-use flexibility and whether these attributes can be combined to better predict such flexibility. Another important but understudied topic is how cue-use flexibility interacts with pressure (Brunyé et al., 2016; Credé et al., 2019). As soldiers and first responders are often under various levels of time pressure to carry out the mission, it is important to investigate how stress level influences the cue-use flexibility and their reports of spatial information.

Research Recommendation 9: Investigate what other factors besides sense of direction can predict cue-use flexibility and how different levels of stress affect such flexibility.

Final Thoughts

Our review has focused on MTS and the opportunities and challenges they create for wayfinding. An MTS can be a truly powerful structure for combining members' inputs in ways that allow for the successful performance of highly complex tasks such as those found in military and emergency services environments. The combined processing power of not just teams, but a team of teams, could potentially overcome a number of difficulties present in these settings. For example, previously, we mentioned the fan effect. In this effect, the more complex the spatial mental model, the more difficult it is to store the model in memory and the longer the retrieval time when attempting to access aspects of the model. Any one individual attempting to store a detailed spatial mental model of an entire battlefield or large-scale emergency would clearly be hindered by this effect. Distributing the load among several individuals (a group or team), as is common in transactive memory systems, would reduce the effect. Distributing the load among several teams, each with responsibility for a different portion of the overall area of operation, makes it possible for the MTS, collectively, to store and quickly access detailed information about much more of that area of operations. Nor is this likely to simply be an additive effect. Rather, different members and different component teams can specialize in learning and storing different types of information and applying that information in a complementary fashion. To do this, of course, requires overcoming a number of challenges. Component teams must each develop good spatial mental models of their component's area, these models must be communicated to other teams, and each component team must be able to correctly interpret and apply them.

To overcome those challenges, each team will need one or more members skilled at integrating spatial representations from members of the team into a spatial mental model of their local area. Each team will also need one or more members who are skilled at communicating spatial information within their team, and one or more members able to serve as a boundary spanner with the skills to communicate to other component teams. At least one member of each component team, likely a boundary spanner, must be skilled at integrating information from the component team's spatial mental model with information relayed by boundary spanners from other component teams. There is reason to believe that at least one member, again likely the boundary spanner, will need to be skilled at switching between reference frames to support communication both within and between component teams. These represent significant potential challenges to overcome. We believe that the research recommendations we have proposed represent important steps in determining how best to overcome these challenges.

We chose to focus our review on informational aspects of collective wayfinding in MTS. The main reasons were that we see wayfinding as a key component of success in military and emergency services settings, see combining information from disparate sources as an important part of collective wayfinding, and that MTS are common structures in military and emergency services settings. However, it is important to keep in mind that teams and MTS are also methods for combining members' *skills* and *abilities*. As such, many of our recommendations for research focus on the skills of one or a few key members (e.g., boundary spanners) at integrating and communicating spatial information. We do not expect all members to need all the skills noted any more than one would expect every player on a hockey team to need the skills required of good goalie. In some contexts, one member with each skill on each team may be sufficient. However, in contexts in which a member may suddenly become unavailable (e.g., due to injury), having a second member with each skill on each team is desirable. Unfortunately, although the literature on how teams and MTS use information is large, the literature on combining members' skills and abilities is less developed. This does not make the question of how members combine skills and abilities any less important, but it does limit our ability to make sound inferences on how it is likely to apply to collective spatial tasks.

As we hope we have demonstrated, MTS are significantly different from both individuals and groups. Furthermore, they represent a significantly different kind of entity than has been studied thus far in collective spatial cognition. Spatial cognition in an MTS context presents significant challenges. However, MTS also provide an opportunity to overcome challenges to spatial cognition at the individual or group levels. We believe that an MTS perspective provides a powerful lens for understanding and optimizing spatial cognition in military and emergency services contexts.

References

Allen, G. L. (1999). Cognitive abilities in the service of wayfinding: A functional approach. *The Professional Geographer, 51*, 555–561. doi:10.1111/0033–0124.00192|

Asencio, R., Carter, D. R., DeChurch, L. A., Zaccaro, S. J., & Fiore, S. M. (2012). Charting a course for collaboration: A multiteam perspective. *Translational Behavioral Medicine, 2*, 487–494. doi:10.1007/s13142-012-0170-3

Austin, J. R. (2003). Transactive memory in organizational groups: The effects of content, consensus, specialization, and accuracy on group performance. *Journal of Applied Psychology, 88*, 866–878. doi:10.1037/0021–9010.88.5.866

Baumann, M. R., & Bonner, B. L (2013). Member awareness of expertise, information sharing, information weighting, & group decision making. *Small Group Research, 44*, 532–562. doi:10.1177/1046496413494415

Baumann, M. R., & Bonner, B. L. (2017). An expectancy theory approach to group coordination: Expertise, task features, and member behavior. *Journal of Behavioral Decision Making, 30*, 407–419. doi:10.1002/bdm.1954

Beal, D. J., Cohen, R. R., Burke, M. J., & McLendon, C. L. (2003). Cohesion and performance in groups: A meta-analytic clarification of construct relations. *Journal of Applied Psychology, 88*, 989–1004. doi:10.1037/0021–9010.88.6.989

Bloom, P., Peterson, M. A., Nadel, L., & Garrett, M. F. (Eds.). (1996). *Language and space*. Cambridge, MA: The MIT Press.

Brandon, D. P., & Hollingshead, A. B. (2004). Transactive memory systems in organizations: Matching tasks, expertise, and people. *Organization Science, 15*, 633–644. doi:10.128/orsc.1040.0069

Brunyé, T. T., Wood, M. D., Houck, L. A., & Taylor, H. A. (2016). The path more travelled: Time pressure increases reliance on familiar route-based strategies during navigation. *The Quarterly Journal of Experimental Psychology, 70*(8), 1439–1452. doi:10.1080/17470218.2016.1187637

Cannon-Bowers, J. A., Salas, E., & Converse, S. (1993). Shared mental models in expert team decision making. In N. J. Castellan, Jr. (Ed.), *Individual and group decision making: Current issues* (pp. 221–246). Hillsdale, NJ: Lawrence Erlbaum Associates, Inc.

Chernyshenko, O. S., Miner, A. G., Baumann, M. R., & Sniezek, J. A. (2003). The impact of information distribution, ownership and discussion on group member judgment: The differential due weighting model. *Organizational Behavior and Human Decision Processes, 91*, 12–25. doi:10.1016/S0749-5978(02)00533-2

Chrastil, E. R., & Warren, W. H. (2012). Active and passive contributions to spatial learning. *Psychonomic Bulletin & Review, 19*(1), 1–23. doi:10.3758/s13423-011-0182-x

Columbia Accident Investigation Board. (2003). *Columbia accident investigation board report* (Online report, Vol. 1). https://www.nasa.gov/columbia/home/CAIB_Vol1.html

Cooke, M. J., Gorman, J. C., & Kiekel, P. A. (2008). Communication as team-level cognitive processing. In M. P. Letsky, N. W. Warner, S. M. Fiore, & C. A. P. Smith (Eds.), *Macrognition in teams: Theories and methodologies* (pp. 51–64). London, UK: CRC Press. doi:10.1201/9781315593166-4

Cooke, M. J., Gorman, J. C., Myers, C. W., & Duran, J. L. (2013). Interactive team cognition. *Cognitive Science: A Multidisciplinary Journal, 37*, 255–285. doi:10.1111/cogs.12009

Crawford, J. D., Henriques, D. Y., & Medendorp, W. P. (2011). Three-dimensional transformations for goal-directed action. *Annual Review of Neuroscience, 34*, 309–331. doi:10.1146/annurev-neuro-061010–113749

Credé, S., Thrash, T., Hölscher, C., & Fabrikant, S. I. (2019). The acquisition of survey knowledge for local and global landmark configurations under time pressure. *Spatial Cognition & Computation*, 1–30. doi:10.1080/13875868.2019.1569016

Dalton, R. C., Hölscher, C., & Montello, D. R. (2019). Wayfinding as a social activity. *Frontiers in Psychology, 10*, 142. doi:10.3389/fpsyg.2019.00142

Davison, R. B., & Hollenbeck, J. R. (2012). Boundary spanning in the domain of multiteam systems. In S. J. Zacarro, M. A. Marks, & L. DeChurch (Eds.), *Multiteam systems: An organization form for dynamic and complex environments* (pp. 323–364). New York, NY: Routledge.

Davison, R. B., Hollenbeck, J. R., Barnes, C. M., Sleesman, D. J., & Ilgen, D. R. (2012). Coordinated action in multiteam systems. *Journal of Applied Psychology, 97*, 808–824. doi:10.1037/a0026682

DeChurch, L. A., & Zaccaro, S. J. (2013, July). Innovation in scientific multiteam systems: Confluent and countervailing forces. In *National Academy of Sciences workshop on science team dynamics and effectiveness*. Washington, DC: National Research Council. http://sites.nationalacademies.org/cs/groups/dbassesite/documents/webpage/dbasse_083773.pdf

DeCostanza, A., DiRosa, G., Jimenez-Rodriguez, M., & Cianciolo, A. (2014). No mission too difficult: Army units within exponentially complex multiteam systems. In M. L. Shuffler, E. Salas, & R. Rico (Eds.), *Pushing the boundaries: Multiteam systems in research & practice* (pp. 61–76). Bingley, UK: Emerald Group Publishing. doi:10.1108/S1534-085620140000016003

DeShon, R. P., Kozlowski, S. W. J., Schmidt, A. M., Milner, K. R., & Wiechmann, D. (2004). A multiple-goal, multilevel model of feedback effects on the regulation of individual and team performance. *Journal of Applied Psychology, 89*, 1035–1056. doi:10.1037/0021–9010.89.6.1035

DiRosa, G. (2013). *Emergent phenomena in multiteam systems: An examination of between-team cohesion* (Unpublished doctoral dissertation). George Mason University, Fairfax, VA.

Fiore, S. M., & Salas, E. (2004). Why we need team cognition. In E. Salas & S. M. Fiore (Eds.), *Team cognition: Understanding the factors that drive process and performance* (pp. 235–248). Washington, DC: American Psychological Association. doi:10.1037/10690–011

Foo, P., Warren, W. H., Duchon, A., & Tarr, M. J. (2005). Do humans integrate routes into a cognitive map? Map- versus landmark-based navigation of novel shortcuts. *Journal of Experimental Psychology. Learning, Memory, and Cognition, 31*(2), 195–215. doi:10.1037/0278–7393.31.2.195

Fraidin, S. N. (2004). When is one head better than two? Interdependent information in group decision making. *Organizational Behavior and Human Decision Processes, 93*, 102–113. doi:10.1016/j.obhdp.2003.12.003

Gallistel, C. R. (1990). *The organization of learning*. Cambridge, MA: The MIT Press.

Golledge, R. G. (1999). Human wayfinding and cognitive maps. In R. Golledge (Ed.), *Wayfinding behavior: Cognitive mapping and other spatial processes* (pp. 5–45). Baltimore, MD: Johns Hopkins University Press.

Grossman, R. (2014). *How do teams become cohesive? A meta-analysis of cohesion's antecedents* (Doctoral Dissertation), University of Central Florida. STARS Electronic Theses and Dissertations, p. 4609.

Han, X., & Becker, S. (2014). One spatial map or many? Spatial coding of connected environments. *Journal of Experimental Psychology: Learning, Memory, and Cognition, 40*(2), 511–531. doi:10.1037/a0035259

He, G., Ishikawa, T., & Takemiya, M. (2015). Collaborative navigation in an unfamiliar environment with people having different spatial aptitudes. *Spatial Cognition & Computation, 15*, 285–307. doi:10.1080/13875868.2015.1072537

He, Q., & Brown, T. I. (2020). Heterogeneous correlations between hippocampus volume and cognitive map accuracy among healthy young adults. *Cortex, 124*, 167–175. doi:10.1016/j.cortex.2019.11.011

He, Q., & McNamara, T. P. (2018a). Virtual orientation overrides physical orientation to define a reference frame in spatial updating. *Frontiers in Human Neuroscience*, 12. doi:10.3389/fnhum.2018.00269

He, Q., & McNamara, T. P. (2018b). Spatial updating strategy affects the reference frame in path integration. *Psychonomic Bulletin & Review, 25*(3), 1073–1079. doi:10.3758/s13423-017-1307-7

He, Q., McNamara, T. P., Bodenheimer, B., & Klippel, A. (2019). Acquisition and transfer of spatial knowledge during wayfinding. *Journal of Experimental Psychology: Learning, Memory, and Cognition, 45*(8), 1364–1386.

He, Q., McNamara, T. P., & Brown, T. I. (2019). Manipulating the visibility of barriers to improve spatial navigation efficiency and cognitive mapping. *Scientific Reports, 9*(1), 1–12. doi:10.1038/s41598-019-48098-0

He, Q., McNamara, T. P., & Kelly, J. W. (2016). Environmental and idiothetic cues to reference frame selection in path integration. In *Lecture notes in computer science. Spatial cognition X* (pp. 137–156). doi:10.1007/978–3-319–68189–4_9

Hegarty, M., Richardson, A. E., Montello, D. R., Lovelace, K., & Subbiah, I. (2002). Development of a self-report measure of environmental spatial ability. *Intelligence, 30*(5), 425–447. doi:10.1016/S0160–2896(02)00116–2

Hinds, P. J. & Mortensen, M. (2005) Understanding conflict in geographically distributed teams: The moderating effects of shared identity, shared context, and spontaneous communication. *Organization Science, 16*, 290–307. doi:10.1287/orsc.1050.0122

Hinsz, V. B., Tindale, R. S., & Vollrath, D. A. (1997). The emerging conceptualization of groups as information processors. *Psychological Bulletin, 121*, 43–64. doi:10.1037/0033–2909.121.1.43

Hollingshead, A. B., & Fraidin, S. N. (2003). Gender stereotypes and assumptions about expertise in transactive memory. *Journal of Experimental Social Psychology, 39*, 355–363. doi:10.1016/S0022–1031(02)00549–8

Hutchins, E. (1991). The social organization of distributed cognition. In L. B. Resnick, J. M. Levine, & S. D. Teasley (Eds.), *Perspectives on socially shared cognition* (pp. 283–307). Washington, DC: American Psychological Association. doi:10.1037/10096–012

Iaria, G., & Burles, F. (2016). Developmental topographical disorientation. *Trends in Cognitive Sciences, 20*(10), 720–722. doi:10.1016/j.tics.2016.07.004

Ishikawa, T., & Montello, D. R. (2006). Spatial knowledge acquisition from direct experience in the environment: Individual differences in the development of metric knowledge and the integration of separately learned places. *Cognitive Psychology*, 52, 93–129. doi:10.1016/j.cogpsych.2005.08.003

Jarvenpaa, S. L., Knoll, K., & Leidner, D. E. (1998). Is anybody out there? Antecedents of trust in global virtual teams. *Journal of Management Information Systems, 14*, 29–64. doi:10.1080/07421222.1998.11518185

Jarvenpaa, S. L., & Leidner, D. E. (1999). Communication and trust in global virtual teams. *Organization Science, 10*, 791–815. doi:10.1287/orsc.10.6.791

Jiménez-Rodríguez, M. (2012). *Two pathways to performance: Affective- and motivationally-driven development in virtual multiteam systems* (Unpublished doctoral dissertation). University of Central Florida, Orlando.
Kanfer, R., & Kerry, M. (2012). Motivation in multiteam systems. In S. J. Zacarro, M. A. Marks, & L. DeChurch (Eds.), *Multiteam systems: An organization form for dynamic and complex environments* (pp. 95–122). New York, NY: Routledge.
Klatzky, R. L. (1998). Allocentric and egocentric spatial representations: Definitions, distinctions, and interconnections. In C. Freksa, C. Habel, & K. F. Wender (Eds.), *Spatial cognition: An interdisciplinary approach to representing and processing spatial knowledge* (pp. 1–17). doi:10.1007/3-540-69342-4_1
Klimoski, R., & Mohammed, S. (1994). Team mental model: Construct or metaphor? *Journal of Management, 20*, 403–437. doi:10.1016/0149–2063(94)90021–3
Lanaj, K., Foulk, T., & Hollenbeck, J. (2018). The benefits of not seeing eye to eye with leadership: Divergence in risk preferences impacts multiteam system behavior and performance. *Academy of Management Journal, 61*, 1554–1582. doi:10.5465/amj.2015.0946
Larson, J. R. Jr., Christensen, C., Franz, T. M., & Abbott, A. S. (1998). Diagnosing groups: The pooling, management, and impact of shared and unshared case information in team-based medical decision making. *Journal of Personality and Social Psychology, 75*, 93–180. doi:10.1037/0022–3514.75.1.93
Laughlin, P. R. (2011). *Group problem solving*. Princeton, NJ: Princeton University Press.
Liang, D. W., Moreland, R., & Argote, L. (1995). Group versus individual training and group performance: The mediating role of transactive memory. *Personality and Social Psychology Bulletin, 2*, 384–393. doi:10.1177/0146167295214009
Lord, C. G., & Taylor, C. A. (2009). Biased assimilation: Effects of assumptions and expectation on the interpretation of new evidence. *Journal of Social and Personality Compass, 3*, 827–841. doi:10.1111/j.1751–9004.2009.00203.x
Lu, L., Yuan, Y. C., & McLeod, P. L (2012). Twenty-five years of hidden profiles in group decision making: A meta-analysis. *Personality and Social Psychology Review, 16*, 54–75. doi:10.1177/1088868311417243
Lynch, K. (1960). *The image of the city* (Vol. 11). Cambridge, MA: MIT Press.
Mathieu, J. E., Heffner, T. S., Goodwin, G. F., Salas, E., & Cannon-Bowers, J. A. (2000). The influence of shared mental models on team process and performance. *Journal of Applied Psychology, 85*, 273–283. doi:10.1037/0021–9010.85.2.273
Mathieu, J. E., Marks, M. A., & Zacarro, S. J. (2001). Multi-team systems. In N. Anderson, D. Ones, H. K. Sinangil, & C. Viswesvaran (Eds.), *International handbook of work and organizational psychology* (pp. 289–313). London: Sage.
Maynard, M. T., & Gilson, L. L. (2014). The role of shared mental model development in understanding virtual team effectiveness. *Group & Organization Management, 39*, 3–32. doi:10.1177/1059601113475361
Meilinger, T., Berthoz, A., & Wiener, J. M. (2011). The integration of spatial information across different viewpoints. *Memory & Cognition*, 39, 1042–1054. doi:10.3758/s13421–011–0088-x
Mell., J., DeChurch, L. A., & Leenders, R. (2018, August). *Information sharing and performance in multiteam systems: The role of identity asymmetries*. Paper presented at the 2018 meeting of the Academy of Management, Chicago, IL. doi:10.5465/AMBPP.2018.10481
Mesmer-Magnus, J. R., & DeChurch, L. A. (2009). Information sharing and team performance: A meta-analysis. *Journal of Applied Psychology, 94*, 535–546. doi:10.1037/a0013773

Mesmer-Magnus, J. R., Niler, A. A., Plummer, G., Larson, L. E., & DeChurch, L. A. (2017). The cognitive underpinnings of effective teamwork: A continuation. *Career Development International, 22,* 507–519. doi:10.1108/CDI-08-2017-0140

Mohammed, S., Ferzandi, L., & Hamilton, K. (2010). Metaphor no more: A 15-year review of the team mental model construct. *Journal of Management, 36,* 876–910. doi:10.1177/0149206309356804

Mojzisch, A., Grouneva, L., & Schulz-Hardt, S. (2010). Biased evaluation of information during discussion: Disentangling the effects of preference consistency, social validation, and ownership of information. *European Journal of Social Psychology, 40*, 946–956. doi:10.1002/ejsp.660

Montello, D. R. (1998). A new framework for understanding the acquisition of spatial knowledge in large-scale environments. In M. J. Egenhofer & R. G. Golledge (Eds.), *Spatial and temporal reasoning in geographic information systems* (pp. 143–154). New York: Oxford University Press.

Montello, D. R. (2001). Spatial cognition. In N. J. Smelser & P. B. Baltes (Eds.), *International encyclopedia of the social and behavioral sciences* (pp. 14771–14775). Oxford: Pergamon Press.

Montello, D. R. (2005). Spatial cognition. In J. D. Wright (Ed.), *International encyclopedia of the social & behavioral sciences* (2nd ed., pp. 111–115). Oxford: Elsevier. doi:10.1017/CBO9780511610448.008

Moreland, R. L., & Myaskovsky, L. (2000). Exploring the performance benefits of group training: Transactive memory or improved communication. *Organizational Behavior and Human Decision Processes, 82*, 117–133. doi:10.1006/obhd.2000.2891

Murase, T., Carter, D. R., DeChurch, L. A., & Marks, M. A. (2014). Mind the gap: The role of leadership in multiteam system collective cognition. *The Leadership Quarterly, 25*, 972–986. doi:10.1016/j.leaqua.2014.06.003

Nejasmic, J., Bucher, L., & Knauff, M. (2015). The construction of spatial mental models - A new view on the continuity effect. *The Quarterly Journal of Experimental Psychology, 68*, 1794–1812. doi:10.1080/17470218.2014.991335

O'Leary, M. B. & Mortensen, M. (2010). Go (con)figure: Subgroups, imbalance, and isolates in geographically dispersed teams. *Organization Science, 21*, 115–131. doi:10.1287/orsc.1090.0434

Olabsi, J., & Lewis, K. (2018). Within- and between-team coordination via transactive memory systems and boundary spanning. *Group & Organization Management, 43,* 691–717. doi:10.1177/1059601118793750

Owen, C., Bearman, C., Brooks, B., Chapman, J., Paton, D., & Hossain, L. (2013). Developing a research framework for complex multi–team coordination in emergency management. *International Journal of Emergency Management, 9*, 1–17. doi:10.1504/IJEM.2013.054098

Pazzaglia, F., & De Beni, R. (2001). Strategies of processing spatial information in survey and landmark-centred individuals. *European Journal of Cognitive Psychology, 13*, 493–508. doi:10.1080/09541440125778

Radvansky, G. A., & Zacks, R. T. (1991). Mental models and the fan effect. *Journal of Experimental Psychology: Learning, Memory, and Cognition, 17*, 940. doi:10.1037/0278–7393.17.5.940

Rico, R., Hinsz, V. B., Burke, S., & Salas, E. (2017). A multilevel model of multiteam motivation and performance. *Organizational Psychology Review, 73*, 197–226. doi:10.1177/2041386616665456

Rico, R., Hinsz, V. B., Daison, R. B., & Salas, E. (2018). Structural influences upon coordination and performance in multiteam systems. *Human Resource Management Review, 28,* 332–346. doi:10.1016/j.hrmr.2017.02.001

Schultze, T., Mojzisch, A., & Schulz-Hardt, S. (2012). Why groups perform better than individuals at quantitative judgment tasks: Group-to-individual transfer as an alternative to differential weighting. *Organizational Behavior and Human Decision Processes, 118,* 24–36. doi:10.1016/j.obhdp.2011.12.006

Shaw, M. E. (1932). Comparison of individual and small groups in the rational solution of complex problems. *American Journal of Psychology, 44*, 491–504. doi:10.2307/1415351

Shuffler, M. L., & Carter, D. R. (2018). Teamwork situation in multiteam systems: Key lessons learned and future opportunities. *American Psychologist, 73*, 390–406. doi:10.1037/amp0000322

Shuffler, M. L., Jiménez-Rodríguez, M., & Kramer, W. S. (2015). The science of multiteam systems: A review and future research agenda. *Small Group Research, 46*, 659–699. doi:10.1177/1046496415603455

Siegel, A. W., & White, S. H. (1975). The development of spatial representations of large-scale environments. In H. W. Reese (Ed.), *Advances in child development and behavior 10* (pp. 9–55). doi:10.1016/j.cities.2019.01.006

Snyder, M., & Swan, W. B. Jr. (1978). Hypothesis-testing processes in social interaction. *Journal of Personality and Social Psychology, 36,* 1202–1212. doi:10.1037/0022–3514.36.11.1202

Stasser, G., Stewart, D. D., & Wittenbaum, G. M. (1995). Expert roles and information exchange during discussion: The importance of knowing who knows what. *Journal of Experimental Social Psychology, 31*, 245–265. doi:10.1006/jesp.1995.1012

Steiner, I. D. (1966). Models for inferring relationships between group size and potential group productivity. *Behavioral Science, 11*, 273–283.

Thorndyke, P. W., & Hayes-Roth, B. (1982). Differences in spatial knowledge acquired from maps and navigation. *Cognitive Psychology, 14*, 560–589.

Tversky, B. (1993, September). Cognitive maps, cognitive collages, and spatial mental models. In *European conference on spatial information theory* (pp. 14–24). Berlin: Springer.

Tversky, B., Bauer Morrison, J., Franklin, N., & Bryant, D. J. (1999). Three spaces of spatial cognition. *The Professional Geographer, 51*, 516–524.

van Ginkel, W. P., & van Knippenberg, D. (2008). Group information elaboration and group decision making: The role of shared task representations. *Organizational Behavior and Human Decision Processes, 105*, 82–97. doi:10.1016/j.obhdp.2007.08.005

Watson, G. B. (1928). Do groups think more efficiently than individuals? *Journal of Abnormal and Social Psychology, 23,* 328–336. doi:10.1037/h0072661

Wegner, D. M. (1986). Transactive memory: A contemporary analysis of the group mind. In B. Mullen & G. T. Gothals (Eds.), *Theories of group behavior* (pp. 185–208). New York: Springer-Verlag. doi:10.1007/978-1-4612-4634-3_9

Wegner, D. M., Erber, R., & Raymond, P. (1991). Transactive memory in close relationships. *Journal of Personality and Social Psychology, 61,* 923–929. doi:10.1037/0022-3514.61.6.923

Weisberg, S. M., & Newcombe, N. S. (2018). Cognitive maps: Some people make them, some people struggle. *Current Directions in Psychological Science, 27*, 220–226. doi:10.1177/0963721417744521.

Weisberg, S. M., Schinazi, V. R., Newcombe, N. S., Shipley, T. F., & Epstein, R. A. (2014). Variations in cognitive maps: Understanding individual differences in navigation.

Journal of Experimental Psychology: Learning, Memory, and Cognition, 40(3), 669–682. doi:10.1037/a0035261

Wiener, J. M., Büchner, S. J., & Hölscher, C. (2009). Taxonomy of human wayfinding tasks: A knowledge-based approach. *Spatial Cognition & Computation, 9*, 152–165 doi:10.1080/13875860902906496

Wijnmaalen, J., Voordijk, H., Rietjens, S., & Dewulf, G. (2018). Intergroup behavior in military multiteam systems. *Human Relations*. Online in advance of print. doi:10.1177/0018726718783828

Wolbers, T., & Hegarty, M. (2010). What determines our navigational abilities? *Trends in Cognitive Sciences, 14*(3), 138–146. doi:10.1016/j.tics.2010.01.001

Wright, P. (1974). The harassed decision maker: Time pressures, distractions, and the use of evidence. *Journal of Applied Psychology, 59,* 555–561. doi:10.1037/h0037186

Applications and Techniques

11 Wildfire Protective Actions and Collective Spatial Cognition

Thomas J. Cova and Frank A. Drews

Introduction

Wildfire is an increasing threat to communities in fire-prone regions.[1] This is primarily due to residential development in the wildland–urban interface (WUI) in conjunction with climate change (Syphard et al., 2019). Droughts and extreme wind events have resulted in many recent large, intense wildfires in fire-prone regions around the world. The United States, Australia, Greece, Canada, and Portugal are example countries that have experienced wildfire events in the last decade that many would consider unprecedented in fire behavior, structure loss, and affected populations.

As wildfires increase in size and frequency and communities expand into the WUI, protecting citizens is becoming more challenging. While protecting people from most wildfires is relatively routine, recent extreme fires have resulted in casualties. The chief causes of these losses generally include warning failures, residents that choose not to evacuate (or delay evacuation), low-mobility residents, and poor community egress (Church & Cova, 2000; Cova et al., 2013). The 2018 Camp Fire in Paradise, California, is a recent example where all of these factors came together to create a scenario that resulted in 85 civilian fatalities and more than 19,000 lost structures.

Wildfires are managed by large, diverse response teams that together face collective cognitive challenges in rapidly assessing changing risks, allocating personnel and resources, and protecting life and property. The goal of this chapter is to highlight wildfire protective-action decision-making as a promising application area for studying collective spatial decision-making. We review the area in terms of research and practice, provide examples of relevant cognitive psychology research, and discuss potential areas for research into collective spatial decision-making in the context of wildfires.

Spatial Protective-Action Decision-Making in Wildfires

When a wildfire is reported, officials must assess the risk and decide who should take action, what action to recommend, and when action should be initiated (Cova et al., 2017; Lindell & Perry, 2004). An initial spatial question arises in determining

DOI: 10.4324/9781003202738-16

those at greatest risk. While most fires offer substantial time for officials (and residents) to "wait and see" if any action is necessary, some offer much less time. A first step is delimiting a boundary around those most at risk to target them for warnings and recommended protective actions. In many cases, this boundary is delimited on-the-fly based on an assessment of the unfolding scenario, but in select cases, it may have been defined from prior planning, as in a staged evacuation conducted using predefined zones. It is also common for the resulting evacuation area to be expanded to include communities on the boundary as fire direction, speed, and behavior changes. In some cases, a designated evacuation zone may be canceled without issuing a warning if a fire is no longer viewed as a threat.

Wildfire protective options include the broad categories of evacuation and shelter-in-place, where the latter can be divided into in-home shelter (e.g., stay and defend), refuge shelter (i.e., a short trip to a designated refuge structure or zone) (Cova et al., 2009), or last-resort survival shelter opportunities such as a body of water or a large area cleared of vegetation (Whittaker et al., 2017); see Figure 11.1. While evacuation is the most common action, there are cases where this is not possible or it may place residents at greater risk than seeking protective shelter.

Two questions that arise in determining when a protective action should be initiated include (1) how much time is available and (2) how much time will it take for the recommended action to be completed? Figure 11.2 shows three different protective-action triggers (t_1, t_2, and t_3) that represent environmental thresholds that would result in issuing a warning to a given community (e.g., wildfire crossing a ridgeline toward a community). The application of protective-action triggers is akin to any situation requiring the detection of a signal

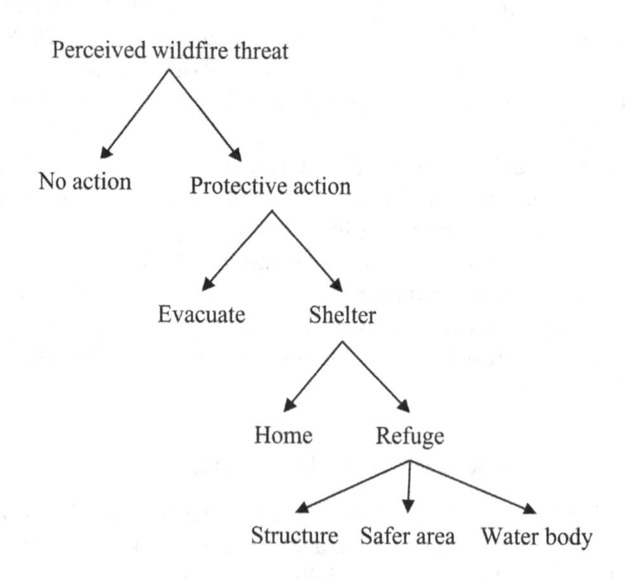

Figure 11.1 Protective-action options.

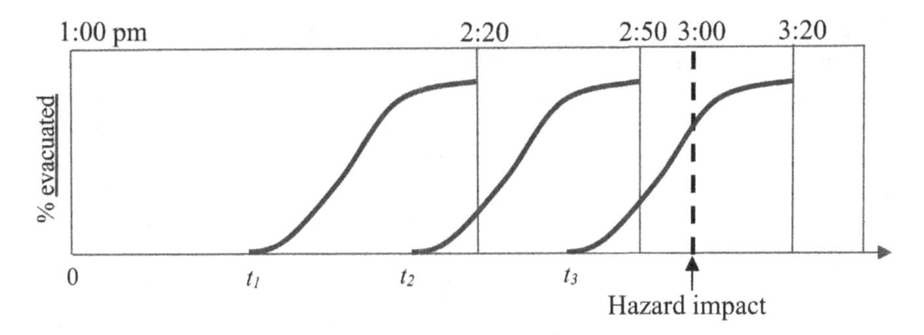

Figure 11.2 The importance of a well-timed trigger in evacuation planning and modeling where t_1, t_2, and t_3 represent three different trigger conditions being met.

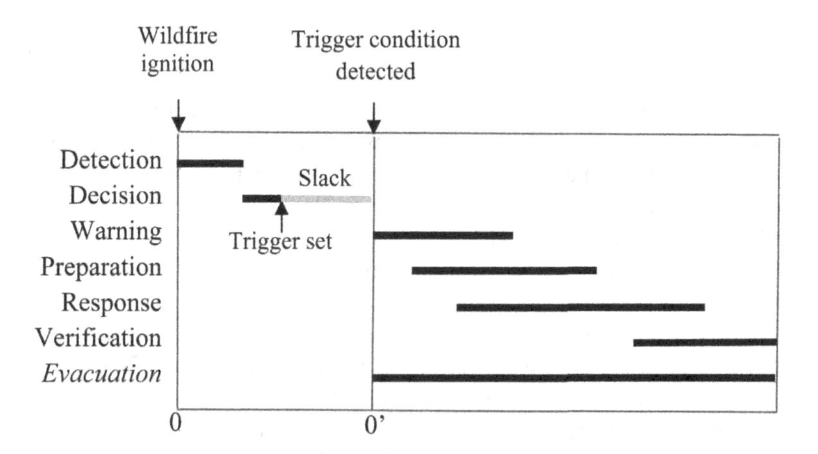

Figure 11.3 Protective-action time phases (detection through evacuation) with trigger duration (slack) shown in gray for a post-detection trigger point. A wildfire occurs at time 0, with 0' being the time at which the trigger condition is detected (e.g., wildfire crossing ridgeline toward community). The wildfire continues throughout the complete duration expressed on the abscissa.

in a noisy environment. Signal detection theory directly applies to this context (Green & Swets, 1966). While t_1 depicts a situation where the action may be recommended too early (i.e., unnecessary disruption), t_3 is a much more serious situation, as some residents were not able to carry out the recommended action prior to the wildfire's impact.

The cumulative evacuation curve in Figure 11.2 can be divided into decision time phases that range from the initial time to detect the wildfire ignition (detection time) through the time required to verify which residents have complied with the warning (verification time). Decision time is the time it takes officials to decide whether to recommend any actions (Figure 11.3). A trigger point is often set in this case where officials agree that if a given environmental condition occurs, a warning will have to be issued (e.g., wildfire crossing a ridgeline, road, or river). If the trigger condition is met, then the aggregate time to carry out the action is a composite of overlapping time phases that include the time to warn residents (warning time), the time for the public to prepare to leave (preparation time), and the time for residents to carry out the action (e.g., response or travel time). In some cases, the recommendations may be staged where different communities carry out the recommended action at different initial times, usually to get the most at-risk to safety first, as well as to minimize traffic delays.

In cases where it is not possible for everyone to evacuate, in-place shelter may be a viable alternative. This type of protection includes structures, safety zones, and water bodies (i.e., places of refuge). Structures can be separated into community shelters or private shelters. Community shelters generally require a short trip on the part of residents to reach the shelter, which may have been designated prior to the wildfire, or they can be selected on-the-fly during the event (e.g., temporary refuge area). Private shelters include those where a resident seeks shelter on his or her property. Private shelters can also be constructed prior to any wildfire or chosen on-the-fly during an event. In some contexts, particularly in Australia, residents may have prepared their house to survive a wildfire such that it can also serve as a place of refuge as a fire passes. A significant amount of research has been done on the "leave early or prepare, stay, and defend" policy in Australia regarding how to prepare and actively defend a home (McLennan et al., 2012). This decision must also be made under varying levels of uncertainty (Cheong et al., 2016).

Modern warning systems increasingly support more refined spatial decisions about whom to warn in a wildfire and at what rate (Li et al., 2015). Historically, the most effective options to warn the public were radio and television, but subscription-based reverse-911 systems, the wireless emergency alert (WEA) system (Doermann et al., 2021; Sutton & Kuligowski, 2019; Wood et al., 2018), and social media (Goodchild, 2009; Goodchild & Glennon, 2010; Li et al., 2021; Wang et al., 2016) offer three additional means of alerting and warning communities (i.e., officials to citizens, citizens to one another, and citizens to officials). The advantage of WEA is that it can be used to send a pop-up alert to cell-phones within range of a selected set of cellular towers. This allows an alert to be sent to all cell phones in a region at once that includes a vibration and tone to create strong cues for the cell phone owner to attend to the incoming notification (e.g., Amber Alert). One part of this spatial decision is the selection of towers to broadcast the alert. Another element involves the potential error associated with radio waves that may reach cell phones that are not part of the targeted at-risk population. Subscription-based reverse 911 systems (e.g., Code Red, Nixle, and Everbridge) call households that have signed up to receive a warning. The advantage of this approach is that officials have more precise spatial control over which households receive the warning to provide the detailed instructions, but the systems are bandwidth-limited in how fast they can call households, and past events have shown that subscription rates can be very low in some communities (e.g., <30%). Subscription-based systems also exclude tourists and transients, which can be a significant segment of the population in some wildfire areas and seasons.

In addition to text-based warning messages, there is growing potential for the application of visual and map-based displays in communicating with the public. This is facilitated by the increasing adoption of smartphones, which can be used to transmit map-based information directly to where people are standing. Cao et al. (2017) explored new map designs for communicating wildfire warnings to the public, and Liu et al. (2017) investigated how maps might affect public response to disaster information. MacPherson-Krutsky et al. (2020) studied the cognitive processes involved in hazard map comprehension of an interactive hazard map. While this area of research and application is just beginning, map-based visualizations are expected to play a larger role in public warnings as technologies to support this medium improve.

Advancing Cognitive Understanding of Collective Spatial Decision-Making in Wildfires

Situational Awareness

The ability of an incident commander (IC), or team, to identify the available protective options and time trajectory associated with each option is a function of how the dynamic and uncertain scenario is cognitively represented individually, but also as a team. Often, such representations are referred to as mental models that include existing knowledge about previous incidents and information that is specific to the situation at hand (Craig, 1943). At the group level, such a mental model would be referred to as a shared mental model (Cannon-Bowers et al., 1993), where each member of a team has a cognitive representation of aspects of the complete situation while also having specific technical expertise that goes beyond each team member's shared mental model. A complementary approach to conceptualize cognitive processes at the individual and the team level, especially in highly dynamic situations, is to conceptualize this as situational awareness (SA; Endsley, 1995; Gilson, 1995).

Endsley (1995) distinguished three levels of SA. Level 1 involves the detection of an event and includes information about the existence of an event and the associated changes in the environment. Level 2 SA involves a diagnosis of the information that was previously received. Here, an IC integrates all of the available information (e.g., spatial characteristics, environmental factors, resources available, and evacuation times) in conjunction with a set goal (e.g., minimizing resource use) and gaining an understanding of the meaning of this information. For an expert, this level of SA may result in the emergence of a gestalt or a pattern that allows (based on the level of expertise) the effortless integration of this information to develop hypotheses about the underlying causes of the events. The third level of SA is related to the prediction of the scenario into the future. Here, the IC predicts the future scenarios based on existing knowledge and the results of processes that are involved in generating the lower levels of situation awareness. At this level, an IC may develop plans and consider the effects of the implementation of such plans. It is important to emphasize that the three levels of SA are not static but are constantly updated in a dynamically changing environment.

Situational awareness also serves as a construct to understand team-related processes (Marks et al., 2001). However, team SA represents more complexity than individual SA because it results in more than the combination of the SA of the individual team members. One way to think about team SA is as a team's theory of the situation (Bolman, 1980), as proposed by the theory of practice, that is, team members examine their theories by sharing and assessing information, in combination with the team's ability to combine their individual skills to identify a mental model of the best theory to explain the emerging situation. Critical in this context are several aspects involved in the development, maintenance, and modification of a team's SA. Among these factors is the monitoring of position-specific information, confirming and validating this information among team members, and effective communication of this information to other team members or non-members. One of the particular challenges in the context of ICs and their teams includes the factors associated with monitoring of position-specific information, which is related to spatial cognition.

Early work by Mondschein (1994) identified the information elements that need to be included in community maps when developing emergency preparedness plans. Of special importance in this context are information elements that are related to the identification of smoke dispersion, population areas at potential risk, schools, nursing homes and special institutions, location of emergency resources that need to be deployed, and evacuation routes and alternates. Many other variables need to be considered in the context of wildfire decision-making, including the location, intensity, and direction of the fire, predicted weather conditions, and surrounding fuel conditions. It is of critical importance that the information described can be included in a cognitive representation of the geographical layout of the hazard area that the IC and the team develop. In addition, this information needs to be updated on a regular basis to help with maintaining IC and team situation awareness. Artifacts, such as maps or computer-based visualizations of this information, can support the decision processes but do not necessarily substitute well-developed cognitive representations that result in heightened spatial awareness. Wickens (2002) described some factors that are of importance when developing artifacts that promote spatial awareness. He identifies the frame of reference issue that specifies whether information should be presented in an egocentric or exocentric perspective. He also identifies the degree of integration issue that relates to the question of whether it is better to use an integrated information display that provides all of the information necessary on one display or a separated display that supports analytical reasoning but does not integrate the information. Finally, he points out the importance of predictive information that should be integrated in any information display to support the SA level. Another important consideration relates to the task that the IC or the team is performing. As a result, decision factors need to be outlined.

Drews et al. (2014) examined the critical factors that influence wildfire-protective-action recommendations and their relative importance for the decision derived. In this empirical study, 47 ICs with different levels of expertise (assessed on the basis of previously managed number of wildfires) participated to reveal the mean factor-importance scores of information. In addition, the authors performed additional analyses to identify the similarity of the factors to identify how close the factors were in terms of their mental representation by ICs. Table 11.1 lists the factor-importance ratings for the more and less experienced incident commanders.

One of the findings that emphasizes the difference between experienced and less experienced ICs was that the former group emphasized the importance of dynamic, fire-related factors, where the latter group emphasized static, community-related factors. A visualization of the relationship between the 15 factors identified as important for experienced ICs is shown in Figure 11.4.

Table 11.1 Mean Factor-Importance Ratings for Less and More Experienced Incident Commanders

More experienced (n = 23; managed more than 70 wild fires)			*Less experienced (n = 24; managed less than 70 wild fires)*		
Rank	*Factor*	*Mean (SD)*	*Rank*	*Factor*	*Mean (SD)*
1	Fire spread rate	6.71 (0.59)	1	Fuel in/near community	6.65 (0.49)
2	Wind direction	6.65 (0.49)	2	Shelter/safe zones	6.59 (0.87)
3	Wind speed	6.59 (0.51)		Distance fire to comm.	6.59 (0.71)
4	Spotting and branding	6.56 (0.73)		Fire intensity	6.59 (0.62)
5	Fire intensity	6.53 (0.51)		Forecast weather	6.59 (0.71)
6	Forecast weather	6.35 (1.06)		Fuel moisture	6.59 (0.71)
7	Shelter/safe zones	6.24 (1.09)	7	Fire direction	6.53 (0.72)
8	Defensible structures	6.18 (1.13)		Terrain charac-teristics	6.53 (0.87)
	Time to warn population	6.18 (1.01)		Wind speed	6.53 (0.72)
	Population at risk	6.18 (1.01)		Wind direction	6.53 (0.80)
	Resources available	6.18 (0.81)	11	Humidity	6.47 (0.94)
12	Fuel in/near community	6.18 (0.81)	12	Spotting and branding	6.41 (0.71)
13	Personnel available	6.12 (0.99)		Fuel along exiting roads	6.41 (0.94)
14	Distance fire to comm.	6.06 (0.90)	14	Population at risk	6.35 (0.93)
15	Terrain characteristics	6.06 (0.83)	15	Fire spread rate	6.35 (1.06)

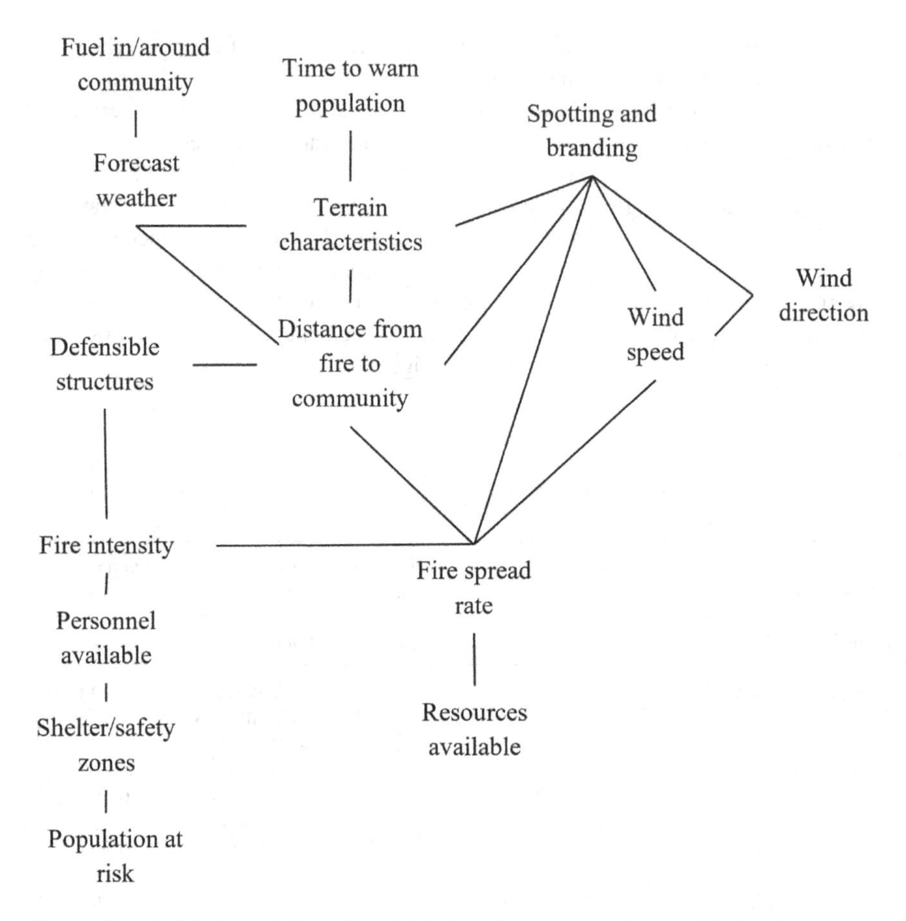

Figure 11.4 Pathfinder solution of top 15 factors for more experienced ICs ($n = 18$).

Drews et al. (2015) also explored the information search behavior and decision-making processes of ICs in computer-based wildfire simulations. In this chapter, the authors examine the effect of expertise (as measured by IC level) on IC information search and decision-making. In addition, they examined the process of information updating of ICs during the evolution of a complex simulated wildfire scenario. Of special interest was the frequency by which ICs of different levels of expertise accessed information during the simulation. Table 11.2 shows this information based on information item and level of expertise.

Overall, this work demonstrated that expertise (i.e., a well-developed mental model, typically found in IC with substantial experience in managing wildfires) leads to more targeted information search. Another important finding was that a specific subset of information contributed to the difference in information search behavior between groups of different expertise. Among these variables were fire intensity, fire spread rate, the number of structures, percentage of fire containment,

Table 11.2 Frequency of Access of Specific Information Items Across All Scenarios Between Groups for Selected Variables; IC1–2, $n = 10$; IC3 (high), $n = 9$; and IC3 (low), $n = 10$

Information Item	*Group*		
	IC1–2	*IC3$_{high}$*	*IC3$_{low}$*
Fire intensity	10.8	13.2	20.8
Fire spread	10.9	13.4	20.6
Number of structures	11.4	10.8	22.4
Percent contained	13.3	11.1	20.3
Time to warn	11.8	12.5	20.5

Note. IC1–2 = high expert; IC3$_{high}$ = medium expert; IC3$_{low}$ = low expert.

and time to warn the community. Remarkably, low expertise ICs examined this set of variables twice as often as higher expertise ICs. This suggests that an increase in expertise emphasizes a stronger focus on the dynamic properties of a scenario (see Table 11.1 top rakings of more experienced ICs). Integration and rapid updating of this information into a mental model allows for maintenance of high levels of situation awareness, resulting in faster decision-making in light of changes in the nature of a wild fire. ICs with less well-developed mental models do not consider these dynamically changing variables to the same extent, consistent with the idea that the mental model of the dynamic aspects is not as well developed and needs additional refinement, which can be acquired with more expertise in managing wildfires.

Operating Picture

Another framework related to team spatial cognition that originates in military operations or industrial emergency operations is the framework of the operating picture. The operating picture results from the information that is perceived and its integration into an understanding of the situation as it presents itself (Looney, 2001). Often the information is provided as a spatial visualization. For example, Looney (2001) described the need of an information architecture that involves high-level command information (providing the common operating picture) down to low-level command information for field commanders (providing the common tactical picture), with the latter less being concerned about the overall picture and more so with the local factors. What is important is that the common operating picture uses information from the local tactical picture as building blocks to integrate into the common operational picture. For the purpose of our discussion, ICs may operate sometimes at the level of the tactical picture, while sometimes when dealing with larger fires, they may operate more at the common operating level.

Additional information on the different organizational levels involved in developing an understanding of the emerging situation is provided by McLennan et al. (2006), who investigated decision-making effectiveness in wildfire incident management teams. As a result of their work, they identified three factors with direct impact on IC effectiveness for four different incident management team functions. The four functions identified relate to command, operations, planning, and logistics as part of the operations. The factors that were identified across all of the functions were "1. fast, efficient and judicious information sharing and management among the four functions, 2. effective matching of the activities of the four functions to the incident management team (IMT) goals, and 3. monitoring of IMT processes in order to detect and correct non-rational and other potentially task-disruptive effects" (p. 34). Clearly, these are important elements of a common operating picture and need to be integrated to provide optimal and consistent access to the common operating picture as well.

More recent work by Danielsson et al. (2014) discusses both situational awareness and the operating picture in their relationship, focusing on the transition process from the operating picture to situational awareness. In their paper, the authors emphasize the importance of information to support cooperation between different stakeholders either from within the organization or from outside of the organization. Important factors that influence the operating picture according to these authors are information from all available sources. This involves, for example, information about the geographical layout, information about potential victims and structure losses, and meteorological conditions, but also information from other collaborating organizations, information from all involved working in the same spatial location, and, finally, information from those who serve as a liaison between organizations ("boundary spanners").

The results of this work suggest that situational awareness is influenced by a number of other factors, which include functional situational awareness (defined as occupationally oriented situational awareness), situational awareness based on role, responsibility, and expertise, and others (e.g., personal networks, use of different terminology, and differences in organizational culture). In addition, their results suggest that the operational picture that consists of information about the event is a result of role-based situational awareness.

Overall, of importance in this context is that the formation of the common operating picture is facilitated by having the actors operate in a shared location, which allows the formation of a common operating picture, which includes a description of the evolving situation, such as its location and properties. Of critical importance for the development of a common operating picture is also that the actors jointly interpret the information to create the common operating picture. Boundary spanners who may facilitate information flow both vertically and horizontally inside and outside of the organization also contribute to the common operational picture.

Based on Danielsson et al. (2014), the operational picture and situational awareness focus on different aspects of the emergent situation, but both are needed to create a common operational picture that combines role- and expertise-specific knowledge with perspectives that are potentially organization specific into a broader understanding that can guide operations.

Clearly, more work is required to explore the relationship between situational awareness, operational picture, and common operational picture. Based on the initial work reported earlier, situational awareness is a function of the background of a specific individual, where the development of an operational picture is more a function of an interaction of the team in the same location to develop a common understanding or a common operating picture.

Synthesis for Collective Spatial Cognition and the Common Operating Picture

Wildfire protective-action decision-making involves processing and integrating substantial amounts of information that emergency managers use to determine protective actions. Thus, one problem is that individually operating emergency managers or their staff may formulate a wide range of actions for the same event, with such a situation resulting in a range of not only different assessments but also different interpretations about the necessary interventions. Because assessments, intervention, and action recommendations need to be articulated in concert with other stakeholders, it is important that a common operating picture emerges, which includes aspects of collective spatial cognition.

In addition, there is a need to improve our understanding of experts' mental models, processes affecting the development and maintenance of situational awareness, and potentially different cognitive styles in formulating triggers. Furthermore, note that while the use of artifacts, such as maps or computer-based visualizations of this information, can support the decision processes, they do not necessarily substitute well-developed cognitive representations that result in heightened spatial awareness. Finally, the specific processes that lead to the integration of situational awareness into an operational picture and how numerous operational pictures are integrated into a common operational picture still need to be better understood. Emerging hypotheses then need to subsequently be tested compared to existing protective-action decision theory and models (Lindell & Perry, 2012).

Some of these issues and research questions can be examined at the individual level of the IC as described earlier and illustrated by the research focus of Drews et al.'s (2014, 2015) work. While such an approach is helpful to identify the factors that drive decision-making in general and more specifically in the context of spatial cognition, such an approach does not take into account the group-based factors that play an important role in the development of a common operational picture. As a result, future research should focus on group processes, shared mental models, shared situation awareness, the operational picture, and the common operational picture to better understand how groups that involve individuals from different levels of the same organization and different organizations as well, make decisions, and how to improve such decision-making processes.

Future Directions

Based on the literature discussed earlier, there are a number of directions the work on collective spatial cognition should pursue. These directions involve issues related to the process of developing and updating situational awareness and the common operational picture, information integration in common platform, and display development to maintain common operational pictures, improving ways of intra- and inter-organizational information sharing and training interventions to advance ICs understanding about the development of situational awareness and the operational picture. There is also a need to develop new tools that can improve spatial collective decision-making in emergencies by assisting incident management teams in rapidly constructing and sharing situational awareness and a common operating picture. These tools will be enabled by many real-time data-collection technologies, including advanced satellites, airborne platforms, geosensors, and smartphone technologies.

Note

1 For a general overview of the increasing wildfire threat, see https://en.wikipedia.org/wiki/List_of_wildfires.

References

Bolman, L. (1980). *Aviation accidents and the theory of the situation.* Cambridge, MA: Graduate School of Education, Harvard.

Cannon-Bowers, J. A., Salas, E., & Converse, S. (1993). Shared mental models in expert team decision making. Individual and group decision making. In N. John Castellan (Ed.), *Individual and group decision making: Current issues* (pp. 221–245). Hillsdale, NJ: Lawrence Erlbaum.

Cao, Y., Boruff, B. J., & McNeill, I. M. (2017). The smoke is rising but where is the fire? Exploring effective online map design for wildfire warnings. *Natural Hazards, 88*(3), 1473–1501. https://doi.org/10.1007/s11069-017-2929-9

Cheong, L., Bleisch, S., Kealy, A., Tolhurst, K., Wilkening, T., & Duckham, M. (2016). Evaluating the impact of visualization of wildfire hazard upon decision-making under uncertainty. *International Journal of Geographical Information Science, 30*(7), 1377–1404. https://doi.org/10.1080/13658816.2015.1131829

Church, R. L., & Cova, T. J. (2000). Mapping evacuation risk on transportation networks with a spatial optimization model. *Transportation Research Part C: Emerging Technologies, 8*, 321–336.

Cova, T. J., Dennison, P. E., Li, D., Drews, F. A., Siebeneck, L. K., & Lindell, M. K. (2017). Warning triggers in environmental hazards: Who should be warned to do what and when? *Risk Analysis, 34*(4), 601–611.

Cova, T. J., Drews, F. A., Siebeneck, L. K., & Musters, A. (2009). Protective actions in wildfires: Evacuate or shelter-in-place? *Natural Hazards Review, 10*(4), 151–162.

Cova, T. J., Theobald, D. M, Norman, J., & Siebeneck, L. K. (2013). Mapping wildfire evacuation vulnerability in the western US: The limits of infrastructure. *Geojournal, 78*(2), 273–285.

Craig, K. J. W. (1943). *The nature of explanation* (Reprinted 1952). Cambridge: Cambridge University Press.

Danielsson, E., Alvinius, A., & Larsson, G. (2014). From common operating picture to situational awareness. *International Journal of Emergency Management, 10*(1), 28–47.

Doermann, J. L., Kuligowski, E. D., & Milke, J. (2021). From social science research to engineering practice: Development of a short message creation tool for wildfire emergencies. *Fire Technology, 57*, 815–837.

Drews, F. A., Musters, A., Siebeneck, L. K., & Cova, T. J. (2014). Environmental factors that influence wildfire protective-action recommendations. *International Journal of Emergency Management, 10*(2), 153–168.
Drews, F. A., Siebeneck, L. K., & Cova, T. J. (2015). Information search and decision making in computer based wildfire simulations. *Journal of Cognitive Engineering and Decision Making, 9*(3), 229–240.
Endsley, M. R. (1995). Toward a theory of situational awareness in dynamic systems. *Human Factors, 37*, 32–64.
Gilson, R. D. (1995). Introduction to the special issue on situation awareness. *Human Factors, 37*(1), 3–4.
Green, D. M., & Swets, J. A. (1966). *Signal detection theory and psychophysics* (Vol. 1). New York: Wiley.
Goodchild, M. F. (2009). NeoGeography and the nature of geographic expertise. *Journal of Location Based Services, 3*(2).
Goodchild, M. F., & Glennon, J. A. (2010). Crowdsourcing geographic information for disaster response: A research frontier. *International Journal of Digital Earth, 3*(3), 231–241.
Li, D., Cova, T. J., & Dennison, P. E. (2015). A household-level approach to staging wildfire evacuation warnings using trigger modeling. *Computers, Environment and Urban Systems, 54*, 56–67.
Li, L., Ma, Z., & Cao, T. (2021). Data-driven investigations of using social media to aid evacuations amid Western United States wildfire season. *Fire Safety Journal, 126*.
Lindell, M. K., & Perry, R. W. (2004). *Communicating environmental risk in multiethnic communities*. Thousand Oaks, CA: Sage Publications.
Lindell, M. K., & Perry, R. W. (2012). The protective action decision model: Theoretical modifications additional evidence. *Risk Analysis, 32*(4), 616–632.
Liu, B. F., Wood, M. M., Egnoto, M., Bean, H., Sutton, J., Mileti, D., & Madden, S. (2017). Is a picture worth a thousand words? The effects of maps and warning messages on how publics respond to disaster information. *Public Relations Review, 43*(3), 493–506.
Looney, C. G. (2001). Exploring fusion architecture for a common operational picture. *Information Fusion, 2*(4), 251–260.
MacPherson-Krutsky, C., Brand, B. D., & Lindell, M. K. (2020). Does updating natural hazard maps to reflect best practices increase user comprehension of risk? *International Journal of Disaster Risk Reduction, 46*, 101487. https://doi.org/10.1016/j.ijdrr.2020.101487.
Marks, M. A., Mathieu, J. E., & Zaccaro, S. J. (2001).A temporally based framework and taxonomy of team processes. *Academy of Management Review, 26*, 356–376.
McLennan, J., Elliott, G., & Omodei, M. (2012). Householder decision-making under imminent wildfire threat: Stay and defend or leave? *International Journal of Wildland Fire, 21*(7), 915–925.
McLennan, J., Holgate, A. M., Omodei, M. M., & Wearing, A. J. (2006). Decision making effectiveness in wildfire incident management teams. *Journal of Contingencies and Crisis Management, 14*(1), 27–37.
Mondschein, L. G. (1994). The role of spatial information systems in environmental emergency management. *Journal of the American Society for Information Science, 45*(9), 678–685.
Sutton, J., & Kuligowski, E. D. (2019). Alerts and warnings on short messaging channels: Guidance from an expert panel process. *Natural Hazards Review, 20*(2), 04019002.
Syphard, A., Rustigian-Romsosa, H., Mann, M., Conlisk, E., Moritz, M. A., & Ackerly, D. (2019). The relative influence of climate and housing development on current and

projected future fire patterns and structure loss across three California landscapes. *Global Environmental Change, 56*, 41–55.

Wang, Z., Ye, X., & Tsou, M.-H. (2016). Spatial, temporal, and content analysis of Twitter for wildfire hazards. *Natural Hazards, 83*, 523–540.

Whittaker, J., Blanchi, R., Haynes, K., Leonard, J., & Opie, K. (2017). Experiences of sheltering during the Black Saturday bushfires: Implications for policy and research. *International Journal of Disaster Risk Reduction, 23*, 119–127.

Wickens, C. D. (2002). Situation awareness and workload in aviation. *Current Directions in Psychological Science, 11*(4), 128–133.

Wood, M. M., Mileti, D. S., Bean, H., Liu, B. F., Sutton, J., & Madden, S. (2018). Milling and public warnings. *Environment and Behavior, 50*(5), 535–566.

12 Modeling and Simulating the Impact of Human Spatial and Social Behavior on Infection Spread in Hospitals

Dario Esposito, Davide Schaumann, Megan Rondinelli, Yehuda E. Kalay, Kevin M. Curtin, and Penelope Mitchell

Introduction

The World Health Organization recognizes the spread of hospital-acquired infections (HAIs) as a major risk to both individual patient and public health (Girard et al., 2002). HAIs are the third most common cause of death in the United States, killing more Americans than AIDS, breast cancer, or automobile accidents (Ulrich et al., 2004). It is estimated that United States hospitals experience almost two million HAI cases annually, causing over 100,000 deaths and an additional $30.5 billion in hospital costs each year (Barnes et al., 2010). The global pandemic associated with SARS-CoV-2 and the large number of HAIs that have been documented (Barranco et al., 2021) provide dramatic evidence of the serious consequences of the spread of disease in hospital settings.

In the hospital setting, the participants can be considered a "collective" with a common-goal—the health and well-being of the patients. Healthcare is provided to patients through the collective action of doctors, nurses, and other support staff, visitors—often family and friends—who aid the patients, and the patients themselves. This collective operates in a complex built environment where there are a variety of users and procedures (Schaumann, Pilosof et al., 2016). Each hospital ward can be represented as a complex sociotechnical system because its environment is heterogeneous and dynamic, and both technological and medical practices are simultaneously operating. Operations are non-linear and depend on several local factors, including patients' acuity, staff organization, and the physical configuration of the facilities. All subsystems (environment, personnel, and technology) share common goals; even so, their objectives may conflict, such as in cases when they are competing for the same space to conduct different activities at the same time (Jiménez et al., 2013). For these reasons, hospital operational efficiency and safety are heavily influenced by the relationship between the design of the built environment and its actual use since this constantly affects the human capacity to learn and make appropriate decisions at both individual and collective levels.

DOI: 10.4324/9781003202738-17

The goal of the research presented here is to instantiate what is known about the spread of HAIs, the spatial arrangements of hospitals and the activities within them, and the interactions among the collective actors in the hospital environment in order to determine whether there is emergent behavior that can be used to better understand how and why infections spread and eventually to mitigate against that spread. In order to reach this goal, this work modeled and simulated HAI propagation dynamics by a contact transmission route in a hospital ward. It considers exogenous cross-contamination followed by cross-infection, meaning that a hospital agent comes into contact with an infective pathogen, becomes contaminated, and may subsequently develop an infection. To this end, the model accounts for the role of human spatial and social behavior and the effects of intervention policies, environmental organization, and spatial design on the contamination spread. This is developed with the event-based modeling and simulation approach, a flexible system that can be calibrated with high sensitivity to the behavior and interaction of agents (Schaumann et al., 2015). This has never previously been used to model HAIs. Our approach is innovative in its interpretation of the HAI-environment relationship, considering both the physical aspects, that is, the built space's role as vectors for pathogens, as well as agents' perceptions, that is, the social and spatial impact on human behavior regarding hand hygiene. The produced framework aims to support decision-making of hospital design vis-à-vis HAIs to foster infection prevention and control. However, once developed into a simulator working as a decision support system (DSS), it could also be used for real-time management of infections and outbreaks, and the long-term design and planning of healthcare infrastructures (Esposito, Schaumann et al., 2020a).

In the following section, we review the literature in three areas pertinent to this research: the causes and consequences of the transmission of HAIs in hospital environments, the use of agent-based simulations in interior environments, and the areas of the literature on spatial cognition where the present work could contribute to or benefit from. This is followed by a description of the event-based visual simulation methods used to determine the influence of the built environment on human spatial behavior. The specific spatial and social elements (the setting, actors, activities, and events) are then outlined. Results of a case study of a hidden infected patient (HIP) are presented, and conclusions regarding the significance of this work and avenues for further research are discussed.

Background

The Causes and Consequences of HAIs

The contamination propagation phenomenon is derived from many sources and proceeds through a dynamic transmission mechanism, which may lead to outbreaks. It overlaps hospital processes, activities, and workflows, it is affected by infection prevention and control procedures, and it is influenced by spatial design-related aspects. Although the primary source of most hospital epidemics is infected patients, pathogenic microorganisms can enter a hospital ward through any colonized or infected person or contaminated object. Patients, health-care workers (HCWs), and visitors frequently interact, creating transmission opportunities. If a pathogen infects a member of the hospital collective, it can spread to many others within the hospital population (Barnes et al., 2010). One of the most common means of spread is by hand contact, and therefore, hand hygiene is a crucial element of many hospital processes, with well-documented guidelines (Boyce & Pittet, 2002). Transmission can occur between HCWs, between patients and HCWs, and between visitors and patients, or visitors and HCWs. It is also important to consider patient-to-patient routes when there is a non-negligible probability that two patients come into direct contact, such as in a pediatric ward or in the case of room or toilet sharing. Finally, pathogens can be released into the environment, remain on furniture and equipment in quantities that may exceed the minimal infective dose, and then contaminate others who subsequently enter this environment (Chartier et al., 2014).

Given the prevalence of this infection vector, there is substantial literature regarding hand hygiene compliance in the hospital setting. The main factors that affect whether a member of the hospital collective complies with hand hygiene prevention and control procedures are (1) the actor's perception or awareness of the risk of becoming infected or infecting others and (2) the availability of time to perform the hand hygiene procedure. More specifically, the first factor depends on two conditions. The first is the level of knowledge of the likelihood of contamination of other actors (objects or spaces). From this point of view, hand hygiene is methodical behavior carried out to ensure self-protection. However, self-protection is not always a response purely to a microbiological threat, that is, other infected people, but also to emotive sensations including feelings of unpleasantness, discomfort, or disgust. It may frequently occur that these sensations are associated with a patient who is regarded as "unhygienic" either through appearance, age, or demeanor, or after touching an "emotionally dirty" area such as the axillae, groin, or genitals (Whitby et al., 2006). In these cases, HCWs are more likely to wash their hands because they perceive a risk of infection (Boyce & Pittet, 2002). The second factor in the perception of risk is the activity type combined with contact duration (or intensity of patient care). Longer durations and closer or more vigorous contact contribute to the perception of higher risk level and, in turn, adherence to hand hygiene as reported by HCWs (Fuller et al., 2011). The frequency of hand hygiene practices rises with the contact duration, being lowest after brief patient encounters (less than two minutes), which accounts for a substantial proportion of all observed hand contamination opportunities (Dedrick et al., 2007).

The second factor—the availability of time—is associated with two conditions related to perceived barriers to compliance and is dependent on the contextual spatial configuration. The first condition relates to the quantity and arrangement of hand hygiene facilities. Providing more rapid and easy access to hand hygiene materials is fundamental in promoting compliance. If hand washing facilities (i.e., sinks, soap, detergent dispensers, and disposable towels) are inadequate, inaccessible, inconveniently located, or scarce, adherence to hand hygiene prescription is reduced. This is because the longer time required to leave a patient's bedside, walk to a sink, wash and dry hands, and return to patient care impedes incorporation of full hand hygiene procedure adherence within the workflow (Quan et al., 2015). Instead, visible and easily accessible basin locations should permit the total time spent engaged in hand hygiene to be no more than 60 seconds (Voss & Widmer, 1997). Increasing the availability of sufficient hand hygiene facilities at convenient locations, for example, providing antimicrobial hand-rub dispensers at the point of care, is a contributing factor to hand hygiene compliance (Stiller et al., 2016). This is achievable in most healthcare environments due to alcohol-based hand rubs (ABHRs), which save time; these have been welcomed by HCWs, who are more likely to use them than to wash their hands (Quan et al., 2015). If placed in a convenient location such as a patient's bedside, about one quarter of the time is needed to use an ABHR compared to hand washing (Basurrah & Madani, 2006). Several studies with baseline hand hygiene compliance below 50% showed a significant increase in hand hygiene compliance after the introduction of ABHR solutions; in contrast, studies with baseline compliance greater than 60% observed no significant increase. These findings suggest that high-profile hospital settings may require more comprehensive strategies to achieve further improvement. However, hand hygiene behavior will continue to require hand washing with water and soap when there is visible soiling on hands, especially because ABHR is ineffective for some pathogens (e.g., *Clostridium difficile*) the spores of which are unaffected by the chemical action of ABHR (Landelle et al., 2014).

The second condition influencing the availability of time for hand hygiene is workload level and multi-tasking demands from overcrowding, understaffing, and high patient bed occupancy rates. Higher levels of each of these elements lead to lower hand hygiene compliance. Outbreak investigations have shown positive correlations between these factors and infections; numerous cases have been reported in the literature (Borg, 2003; Fridkin et al., 1996; Harbarth et al., 1999; Pessoa-Silva et al., 2002). These studies not only demonstrate a correlation but also recognize the intermediate cause of contamination spreading as a consistent relaxation of attention to control procedures and poor adherence to hand hygiene.

The Use of Agent-Based Simulations in Interior Environments

The direct observation of human behavior in built environments is usually considered to be the best way to understand and evaluate how a building meets the needs of the spatial behavioral processes of its intended users. Consequently, the post-occupancy evaluation (POE) paradigm has provided several approaches and techniques to assess whether the goals of a building project have been met (Zimring, 2002). However, POE can be applied only after an infrastructure has been completed and occupied, at which point it is costly to correct critical errors and failures, such as inconsistencies with user safety requirements (Schaumann, Morad et al., 2016). Moreover, in a hospital setting, full-scale POE experimentation entails logistical and ethical concerns: Inpatients are an extremely heterogeneous population, often too debilitated to cooperate, and are sometimes available only for a short period of time, making them difficult subjects for study. Data availability and strict privacy regulations and laws require that great care be taken with patient information and that persons under study are willing and informed human subjects (Cooper et al., 2003). Moreover, the hospital setting is highly complex, with a diverse collective of actors and a wide range of activities underway simultaneously. In such an environment and with such restrictions, it is both difficult and costly to perform research regarding collective spatial behavior.

As an alternative, simulations of interior spaces, and modeling agent behavior within those simulations, provides a means of studying the spaces before they are built or studying changes to spaces to evaluate alternative arrangements. Moreover, using computer simulations, complex systems can be modeled with dynamic and transient effects (Pidd, 2004), and such simulations have been recognized as an efficient method for evaluating the performance of designed systems when the relationships among variables cannot be established analytically (Kalay, 2004). Moreover, simulation appears to be the best choice for investigating human behavior representation (HBR), also known as computer-based models, which imitate either the behavior of a single person or the collective actions of a team of people (Majid, 2011; Pew & Mavor, 1998; Esposito, Abbattista et al., 2020), where human spatial behavior is too complex to allow for an analytical computationally efficient solution (Borshchev & Filippov, 2004).

These agent-based simulations of interior spaces have been shown to be useful across a range of applications (e.g., Hajibabai et al., 2007; Pan et al., 2005; Simeone, Schaumann et al., 2013), including in healthcare settings. At the present time, models and simulations are frequently employed to support decision-making in two areas of healthcare management: optimization of the use of hospital resources and control of the spread of HAIs (Ferrer et al., 2013; Friesen & McLeod, 2014). Manipulating a simulated version of a healthcare environment is inexpensive, does not place patients at risk, and allows one to draw conclusions while controlling the premises and assumptions made in the environment's design.

The Potential for Agent-Based Simulations in Studying Collective Spatial Cognition

Given that agent-based simulations provide an opportunity to study behavior and the influence of the environment when the direct observation of human behavior is impractical or impossible, it is reasonable to ask how the cognition that drives spatial behavior can be studied with only agents in a simulated environment. Agents are not cognitive beings, so regardless of the advantages of using an agent-based model to describe an agent's state and location and its suitability for representing some aspects of human (spatial) behavior, they are still limited to representing activity patterns of interaction (e.g., agents-agents-space) rather than true cognition. Agents are designed to mimic or emulate cognition using simple decision rules and probabilistic decision-making. The sequential changes in those rules or the probabilities that drive the behavior in the simulation are the closest a simulation can come to modeling learning or cognition. But this emulation has value in and of itself. The simulation can assess whether or not the decision rules drive the many types of emergent behavior that arise from spatial cognition, such as imitation effects, competition for shared space and between different activities, co-operation, and the natural tendency to maintain the distance (Hall, 1966; Stokols, 1972). When decision rules faithfully reproduce collective spatial behavior, the simulated environment itself can be modified to determine its influence on outcomes such as gathering and crowding, queuing, and interruption of activity (Hajibabai et al., 2007; Hoogendoorn & Bovy, 2004; Pan et al., 2005; Shelby et al., 1989).

In our simulation-driven understanding of human spatial behavior, space comprises physical places of activities and interactions (Gärling & Golledge, 1987). However, it is not only the morphological configuration of space, which allows for collective interaction patterns, but also the function of space (which is codified in the semantics of the space and shared by actors), allowing for or impeding certain determined activities within it. Finally, based on the distribution of given features, for example, facilities availability to perform hand hygiene and the local perception of contamination risk level, the space becomes an active element, which adds to the probability that certain spatial behaviors, such as hand hygiene, are carried out, or not, by agents.

When we study human behavior from a cognitive point of view, analyzing the mechanisms that oversee behavior, we select a very specific form of behavior, for example, the cognitive process that attains an understanding of dimensions in space. The more we can advance its description while attempting to discover how it develops and why it leads to situations occurring, the more we can then try to generalize this for all types of different circumstances. While a variety of literature on the role of environment in human behavior and performance exists (Altman, 1975; Esposito, Santoro et al., 2020; Gärling & Golledge, 1993; Gehl, 2011; Rudofsky, 1981; Yan & Kalay, 2005), it is not straightforward to assess such a multifaceted relationship, especially in regard to qualitative aspects such as human satisfaction, productivity, comfort, and safety. Indeed, a combination of qualitative variables derived from social and cognitive sciences, environmental psychology, and behavioral geography are relevant to agents' spatial choices.

Given this background, our simulation framework offers the potential to build novel understanding of how a healthcare environment may operate; it aims to model spatial influences and social dynamics in order to observe the spatial spread of contaminants due to human spatial behavior and social interaction. Its ultimate purpose is to produce insights into the relative impacts of infection prevention and control measures. This might help us find an optimal plan to manage hospital resources with regard to the impacts of built space and of organizational measures regarding HAIs, which could reduce costs and risk of errors in implementing choices.

Methodology

Recently, a new agent-based modeling (ABM) approach has emerged on the basis of the notion of *events*, which add to the typical bottom-up structure of ABMs a system architecture designed to manage coordinated behavior of many agents in a top-down fashion (Schaumann et al., 2015). Using the event-based approach, we are able to simulate complex and realistic scenarios, and a flexible calibration system allows us to apply it to HAI spread. This allows us to coherently integrate an ecosystem of hospital facilities in a simulation in which the space is the hospital, the collective is its occupants, the interactions involve daily procedures, and spatial cognition is investigated with respect to prerequisites for crucial hand hygiene behavior. Although currently available data in the literature are not yet detailed enough to accurately quantify the weights of interrelations between all involved variables, our framework supports a fine-grained simulation of infection dissemination dynamics, and it adapts to diverse physical layouts and contingent situations (Coen, 2012).

More specifically, our research was carried out with the event-based method as a reference approach, which is designed to discover and represent how built space influences people through visual–spatial simulation emergent observable patterns. Established by Schaumann, Pilosof et al. (2016), this is a modeling and simulation technique for human use of buildings where spaces, actors, and activities are computationally modeled; it considers the users and the processes of use of the space in a hospital environment by modeling events caused by user behaviors in the space. The event is a computational entity that combines information concerning people (who?), the activity they perform (what?), and the spaces they inhabit (where?). Events are designed to coordinate temporal, goal-oriented routine activities performed by agents. They provide us with a variety of behavior and interactions during the simulation of hospital workflow and use. The event-based approach can represent user–space interactions, that is, activities as specific modeling entities on their own, distinct from—yet related to—the spaces they occur within. In addition to actors, activities, and spaces, the simulation presented here was extended to include pathogens and the contamination states of actors and spaces in order to evaluate HAI spread (Figure 12.1). Interaction of these elements during the

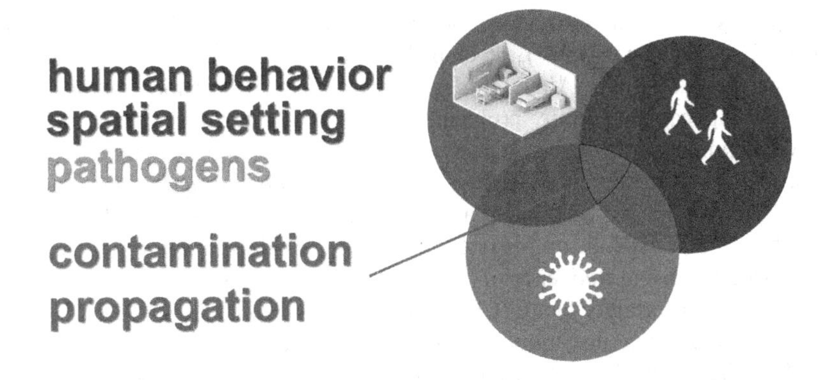

Figure 12.1 Component framework for contamination propagation.

simulation drives pathogen transmission and thus propagates contamination within the ward.

To develop our research, we selected as a setting the Cardiology Unit (CU) in the Sammy Ofer Heart Building, Sourasky Tel-Aviv Medical Center, designed by Sharon Architects & Ranni Ziss Architects 2005–2011, done in collaboration with the internal healthcare staff and management. This environment is appropriate because the CU does not share the extremely regimented situations and procedures of an intensive care unit (ICU) or surgery ward. References, guidelines, and sessions with experienced medical practitioners and healthcare workers supplemented our understanding of daily life workflows, established protocols, and best HAI management practices. These collaborations helped us to build a plausible model with which to test our hypotheses and within which we could manage and represent the interrelations between major elements driving infections spread.

While this work is based on the event-based approach, we also developed the method further. We extended the event-based approach to simulate contamination propagation and visualize HAI transmission on spaces and actors via a contact route in a spatially explicit, heterogeneous environment, through the use of a contagion risk map. A detailed explanation of the proposed Volterra integral equation that drives the flow of contamination during a contact between two agents can be found in Esposito, Schaumann et al. (2020c). This allows us to model cognitive awareness, perceived barriers, and other relevant local conditions affecting the collective's spatial and social behavior. The framework covers a multitude of aspects, including representation of user activities in a hospital environment and their reciprocal influences, as well as elements of spatial perception and cognition. Of particular interest are the impacts on human decision-making and behavior—especially related to hand hygiene—of the spatial cognition of the setting layout and the perception of infection risk due to other agents. The framework considers perception and awareness of risk factors, such as self-protection from infected others, objects perceived as potentially contaminated (Boyce & Pittet, 2002), and the threat of the activity and contact duration (Dedrick et al., 2007). Moreover, perceived local constraints include high workloads, overcrowding, understaffing, physical configuration of space, and distribution of facilities (Quan et al., 2015; Whitby et al., 2006).

Expert systems were introduced for the first time in the MYCIN Expert System (Shortliffe & Buchanan, 1975), developed at Stanford University to aid physicians in diagnosing infections. Although designed for medical decision-making, it has potential applications for any problem in which real-world knowledge and subjective judgments must be combined to inform an evaluation that can explain the consequences of observations or suggest a future course of action. For our system, we introduced for each agent a "Cleanliness" feature, denoted CL, which is a coefficient that depends on his or her compliance with prevention policies, that is, the occurrence and frequency of hand hygiene procedures. The CL variable was formalized by a certainty factor (CF) method (Hayes-Roth et al., 1983). CFs were developed to describe possibilities suggested by evidence to drive the expert system. Specifically, a CF combines degrees of certainty or uncertainty derived from different pieces of evidence into one number. It exploits the tendency of a piece of evidence (condition) to prove or disprove a given hypothesis (h). Our case study considered whether the conditions for the hand hygiene procedure to have been performed were met; the CF measured confidence that hand hygiene would be performed when such conditions were satisfied. Mathematically, the final evidence that supports our hypothesis is composed of two parts of evidence, $e+$ and $e-$, each in turn composed of occurring conditions (pieces of evidence). Total evidence $e = e+ + e-$, where $e+$ represents all confirming conditions and $e-$ represents all non-confirming conditions acquired to date. Therefore, the CF for hypothesis h with regard to evidence e is

$$\mathrm{CF}[h, e] = \mathrm{CF}[h, e+ \wedge e-] = \mathrm{MB}[h, e+] - \mathrm{MD}[h, e-].$$

The measure of belief (MB) and measure of disbelief (MD) measure how much the evidence validates the hypothesis or its negation. Shortliffe and Buchanan (1975) presented a function to combine different conditions to obtain MB[h, $e+$] and MD[h, $e-$], which represent how each item of evidence ($e+$ and $e-$) is incrementally acquired.

The notation is given as follows:

> MB[h, e] = x, ($0 < x < 1$) means "the measure of increased belief in the hypothesis h, based on the evidence e, is x"
>
> MD[h, e] = y, ($0 < y < 1$) means "the measure of increased disbelief in the hypothesis h, based on the evidence e, is y."

In our case, the CF was based on a number of key conditions such as awareness, sensations, perceived barriers, and local spatial context for each agent; these conditions influence an actor's predisposition to comply with hand hygiene prevention and control procedures. Moreover, these principles were derived from expert knowledge, acquired through questionnaires submitted to healthcare managers and practitioners. We used estimates provided to measure the influence of each condition identified from the literature and weighted the confidence that hand hygiene procedures could increase or decrease in order to verify the hypothesis of full compliance with hand hygiene procedures.

In event-based modeling, the simulation of building use is represented as a story path in which events work as milestones. A story path is represented as linearly connected entities depicting—step-by-step—what happens next (Simeone et al., 2012). To represent complex building use scenarios, events are combined into a process-based structured sequence called a "human behavior narrative" (HBN). This is constructed with coexisting building-use story paths containing discrete, simultaneous, or interrelated activities involving several users, performed in a specific space and in sequence. This is articulated as a graph that connects multiple activity paths in a network where each branch's orientation indicates its logical sequence. Events can be shared by different building-use paths and are connected to define the scenario operational flow (Figure 12.2) (Simeone, Schaumann et al.,

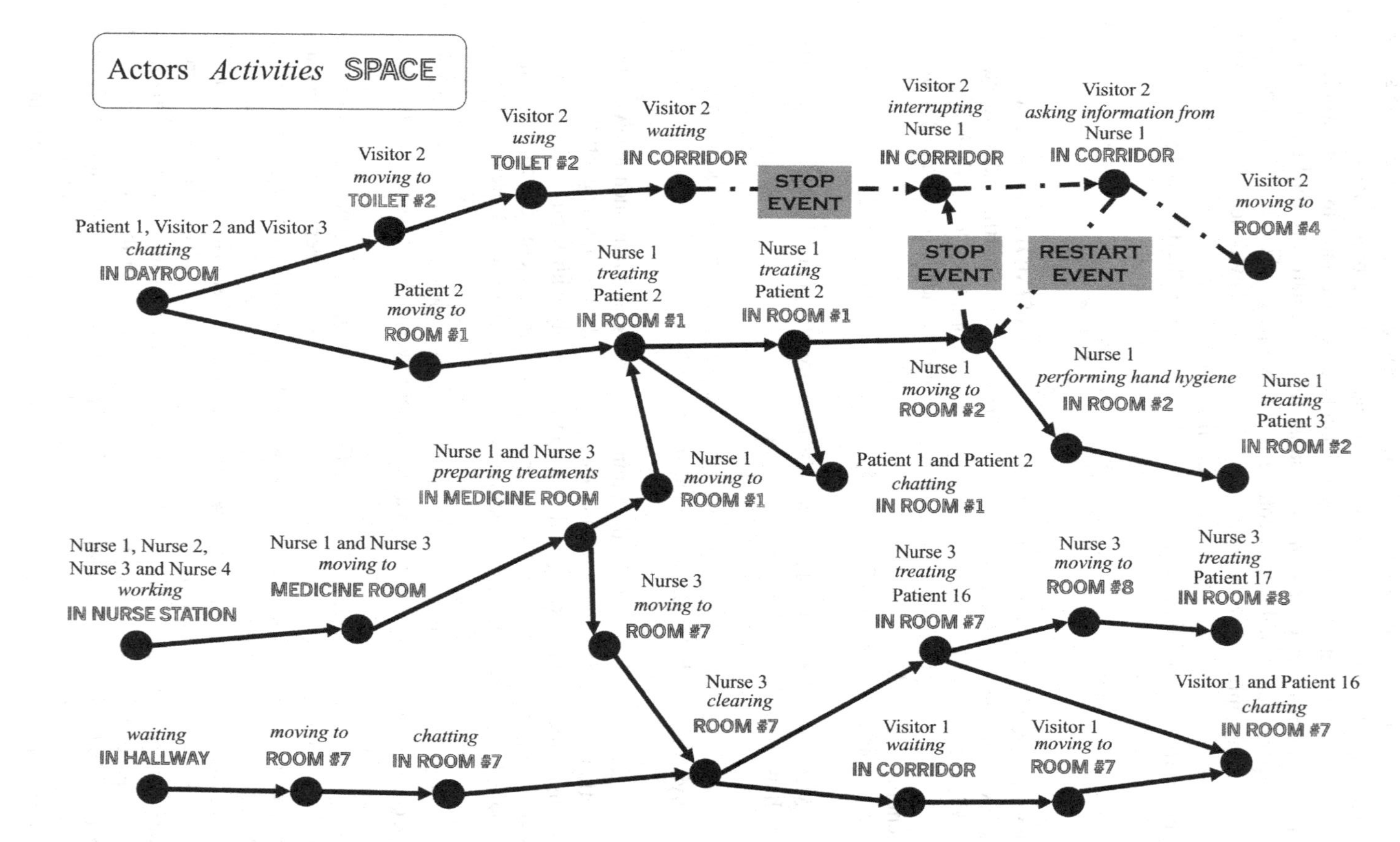

Figure 12.2 Human behavior narrative.

2013). According to structural programming rules, established by the Bohm–Jacopini theorem, HBN may contain hierarchical events, aggregate events, sub-events, sequences, or either parallel or selection connections (Böhm & Jacopini, 1966). A modeler's high-level expert knowledge can structure sequences of events to act as both the process and the engine for simulated social behavior in space.

It requires extensive work to build an HBN with as many branches of evolution as possible—the more detailed the narrative, the more convincing the simulation. Writing it involves a researcher's in-depth knowledge of the context and of the numerous potential situations, which may arise when changing the script; this knowledge can be sourced from real-life observations or interviews and discussions with experts. Once this architecture is built, the system can rearrange the script in numerous ways, mirroring the wide range of possible interactions that can arise. This permits the simulation of many possible scenarios, each with different combinations of input factors. Counterintuitive or emergent outputs can be both qualitatively and quantitatively assessed.

Our virtual-use scenario simulations include the building and its intended users, represented activities, and use of space in the built environment, and produce results and a data log for several HAI risk scenarios. The simulations are built with a Unity 3D simulation engine and coded in C#. The virtual model of the hospital ward is integrated with the HBN to simulate the use of space and model the associated contamination propagation. The simulation follows the sequence of events taking place in the environment—both those coded explicitly (such as treatments) and those that emerge from the collective's varied interactions (e.g., patient contacts). There are stochastic elements in the scenario development, both in opportunities for interactions among actors and in the random entrance of visitors. Specifically, the system simulates unplanned events generated by the local interactions of the actors in the social and spatial surrounding environment, enlarging the spectrum of options available for an agent's behavior. Indeed, if, for instance, the paths taken by two occupants bring them close enough, they may choose to stop and talk or to ignore each other and continue their scheduled tasks. Thus, each actor displays simplified decision-making to choose among different possible behavior completion options, correlated to their psychological and physiological status, motivations, desires, and contextual conditions (Simeone, Kalay et al., 2013). More specifically, the system uses the Bayesian conditional probability method to account for the agents' predisposition to comply with prevention and control procedures, depending on factors surrounding the collective (local conditions, perceived barriers, and risk awareness). Our purpose was to maintain the narrative's organized complexity and force the event-based HBN system architecture to incorporate more likely human spatial perception processes. This also allows for a simulated local low-level cognition of risk perception that requires a certain level of autonomy for each agent's spatial and social behavior. For any run, the way that system elements are combined is unpredictable before it ends, mirroring a real system evolution. To assess the hospital ward performances, we visualize a dynamic contagion risk map in real time as output.

The Spatial Setting in Which the Collective Operates

In the event-based model, the space entity answers the "where" question. It is composed of static geometry, and real-time semantic and dynamic simulation information in a single model, integrating all the information required to manage the simulation of realistic actors' spatial behaviors. In our case study, the nature of the space under consideration is an indoor workplace built environment, which is defined by two different data types. First, the architect's spatial design provides the data required to render scenes for a specific case study. Second, semantics, given by space functions and observation of current use of the space, determine the data that support the human simulation of complex patterns of spatial behaviors. Thus, the modeled setting provides the conceptual connection between the building design solutions proposed by the architect and the building use process based on the organization's operational dynamics.

The space can be derived from the real-world environment, representing floor plans as they are, or from a synthesized environment that can be generated by the modeler with simplifications (Demianyk, 2011). Moreover, the space can be semantically subdivided into a set of zones and subzones that define the function of and activities in the space (Brodeschi et al., 2015; Gibson, 1979). Our simulation utilizes a synthesized, slightly modified prototypical "Ichilov ward" layout (Friesen & McLeod, 2014). In this layout, a nurse station zone affords patient-record keeping, administration, consultation, and communication activities. Likewise, a section of a corridor affords passage, social encounters, patient treatments, and other medical activities, if needed (Sopher et al., 2016) (Figure 12.3).

The virtual setting comprises one five-patient, 14 double, and two single rooms, accommodating a total of 35 patients, as well as a medicine room, a

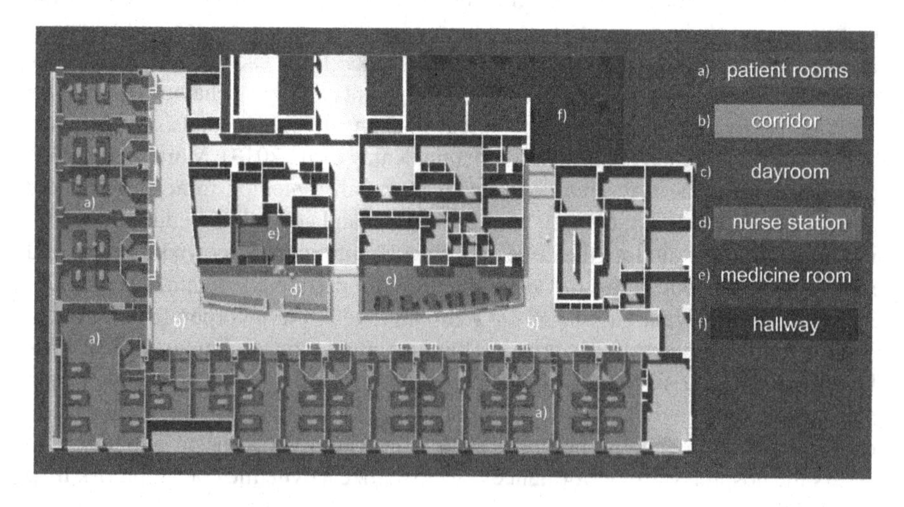

Figure 12.3 Simulated ward plan.

nursing station, and a dayroom. A corridor connects the rooms and houses the nursing station, which is adjacent to a central medicine room where medications can be prepared before distribution. The representation includes furniture and equipment needed for patient care. Space may become a vector for pathogens containing contaminated surfaces, furniture, and objects. Even if not simulated, we assume that actors operate in rooms so that the contamination level of the room is reciprocally affected by their presence. The contamination level for each space is visualized through a range of colors, which dynamically changes when the contamination level exceeds the preset thresholds. These data can be dynamically retrieved via control dashboard in the form of numerical values associated with the space.

Space plays a key role in the spread of infections, both as a direct vector and as a prerequisite for individual and collective spatial interaction between agents. The case under study refers to a well-defined and enclosed layout at high risk of infection spread, which is a paradigmatic example to aid understanding of the complex phenomenon of spatial and social behavior impact (Flyvbjerg, 2006). Specifically, the infection is driven by collective spatial behaviors carried out at close distance, about 1 or 2 m apart, in the range of proxemic personal and social distance.

The Hospital Collective

When designing an agent-based model, an agent's characteristics and rules of interactions (both physical and social) are chosen through careful consideration and definition of the agent's environment and are unique to the model's context and objectives. The model's value then depends on how environmental conditions were read and interpreted—in our case, patient activities, HCW workflow, and prevention and control behaviors. Such modeling considerations must match the technique chosen to simulate the system, which sets the extent of the phenomenon representation and, thus, the research breadth (Box & Draper, 1987). This means that ultimately, the collective nature is defined by the observer, who recognizes and classifies elements of different classes of objects. Categorizing is based on both individual and distinctive characteristics of actors as well as the dynamic factors of their spatial and social relationships.

In the modeling proposed here, different types of actors are represented. Specifically, these include patients, HCWs, and visitors; they share common classes of functions, behavior, and objectives, although each agent can still have individual features incorporated. In the development of daily workflows, these agents interact, forming temporary collectives for the duration of an event. This results in coordinated action of different types of agents in the same place for a certain amount of time. They form a collective that conducts a cooperative spatial behavior or task. However, if and how the activity is collectively performed and what other activities may arise changes depending on contextual circumstances. In the following paragraphs, actors and activities are described, as well as how they relate and unfold in space through the event entity.

Actors

Event-based modeling formally represents people as actors. Its purpose is to answer the "who" question. In the model, people have a fixed semantic role but changeable physiological and psychological attributes. In principle, actors are anthropomorphic agents with profiles and statuses that account for human heterogeneity in psychological, social, and cultural traits and abilities. Computationally, actors are entities with physical descriptions; they have abilities and can move and perform activities, represented by rules that describe the response of actors to their physical and social surroundings, based on each actor's individual state and features.

Actors do not incorporate autonomous strategic decision-making abilities because these are provided by the event—a process model—that controls them in real time, as described in the following. The actor is the recipient of an activity that needs to be performed, communicated to them by the event. The performance of the activity is modulated by the actor's current surroundings (physical and social), which is conveyed by the space semantics and is affected by the actor's internal state (e.g., tiredness). Together, they produce an individual unique reaction to the event's directive. In such a way, actors are provided with the ability to autonomously adapt their behavior within a predefined range based on the status of the environment. In turn, the simulated actor's action outputs influence the process model, providing information that influences the nature of the space (e.g., increasing the risk of infection in a room) and potentially changing the behavior of other members of the collective.

Each actor possesses basic capabilities of navigation and movement through space; these imply an origin and destination within the layout and the ability to avoid obstacles. Similarly, actors possess abilities of perception that allow them to detect the presence of other actors or objects if they are within a certain distance that permits local interaction. These components are intended as baseline autonomous operative aspects of an actor's spatial behavior within the actor entity itself.

We populated the simulation with three actor types: HCWs, patients, and visitors. Actors display properties that define their role, current status (e.g., the activity they are currently engaged in), and their relation to other actors (e.g., nurse medicates patient and visitor visits patient). To create actor profiles, we drew from works by Sopher et al. (2016), which include characteristics, abilities, preferences, knowledge, and states. As shown in Figure 12.4, the attributes are user accessible.

The collective of actors in the developed case study consists of four nurses, 35 patients, and nine visitors. Their starting level of contamination can be set at the start of the simulation within the actor contamination console for each one or randomly generated with the 'Setrand' option (Figure 12.5). As shown in the literature, any person arriving in a hospital ward has a probability of being colonized by pathogenic microorganisms. Otherwise, the quantities of non-colonized, colonized, and infected actors can be adjusted to reflect conditions, for example, the proportion of colonized patients that may be admitted to the hospital or transferred from other hospitals.

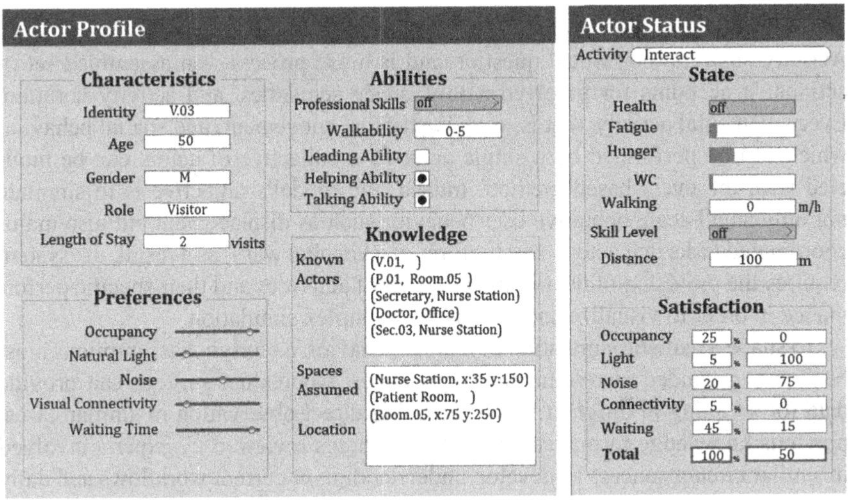

Figure 12.4 Actor profiling dashboard (Sopher et al., 2016).

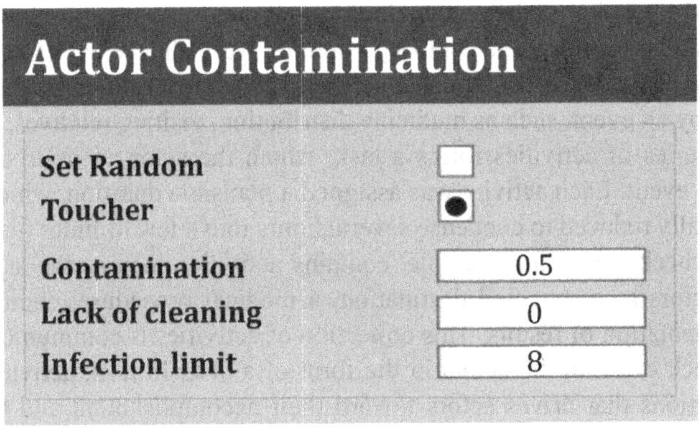

Figure 12.5 Actor contamination console.

Actors may carry pathogens on their skin, clothing, and equipment; their contamination status is visualized by a range of colors (or numeric values) that change dynamically when the level of contamination exceeds preset thresholds.

As this simulation condenses hours of activity into a few minutes, the dynamic contamination develops at the same rate. Therefore, when an actor's contamination level exceeds the infection level, he will not suddenly manifest sickness but, rather, is sufficiently contaminated to develop the disease over the following days.

Activities

Activity answers the "what" question and is a use process—an assembled set of actions. It accounts for involved actors, space semantics, and activity duration. Every elemental activity, that is, social activities, movement, and spatial behavior, which can be performed by a single actor or a collective of actors can be modeled with the event-based method. Indeed, the model's objective is to simulate not only small-scale operative user behavior such as displacement but also major coordinated tasks that actors may perform in a hospital ward; as a result, the system requires the provision of details about the set of activities and their specific performance in order to visualize and analyze the complex simulation.

To shape a realistic simulation of human behavior, underlying assumptions must be firmly grounded in real human experiences. Contextual surveys can provide data for scripting in different ways, including direct observation of similar cases, previous knowledge formalization, and hypotheses reviewed by experts involved in similar circumstances. To develop understanding of current workflows and daily life in the ward, we carried out direct observations; tracking of people; interviews with HCWs (doctors, nurses, and administrative staff), patients, and visitors; and extensive meetings with medical staff and hospital managers (Esposito & Abbattista, 2020). We established behavioral patterns from these data that related to real-world hospital situations, including treatments, nurse and physician commitments, and prevention and control guidelines. Accordingly, actors in the simulation behave in space to accomplish a set of activities determined by their role and requested by an event, such as medicine distribution, visiting relatives, and patient check. A series of activities forms a task, which the actor needs to complete to resolve the event. Each activity was assigned a plausible duration, which was then proportionally reduced to condense several hours into a few minutes of simulation. A patient-check event, for example, contains activities such as the arrival of the relevant actors at a specified destination, a medical procedure, communication, and documentation of results. This collection of activities is communicated by the patient-check event to the actors in the form of a task. In turn, activities provide a set of actions that drives actors toward their accomplishment and that are the main hospital processes that indirectly promote infection spread between actors and spaces.

The simulation environment modeled furniture and medical equipment during the activities. Although objects and spaces cannot become infected, they can be contaminated with pathogens and become sources of contagion. However, the trial case study does not explicitly visualize all objects and equipment present in a hospital ward; rather, it determines which objects belong to the space or to the actor by whether they are utilized by actors in a simulated activity. Examples of frequently utilized objects include doorknobs or taps. Because a contaminated object is part of the space, its contamination level informs the contamination level of the space. For example, if a medicine distribution activity involved a contaminated trolley, patients could be colonized by the trolley even if, in a physical space, the pathogen would have come into contact with a different part of the trolley than that which the patient touches. This assumption is useful in dealing with the representation of bacteria transported by common objects such as smartphones.

Event

Within the simulation, the combination of the space where actors operate, the actors themselves, and the activities they carry out interacting with each other and with the space is combined into a meta-level entity termed an "event" (Simeone et al., 2012). Events embed the knowledge to perform a task in virtual settings. They are meaningful interpretations of information, applying activities and space semantics to actors and envisioning the specific use process of that space for a certain time span.

Events are computational entities that manage the performance of a specific behavior pattern, querying the actors, space, and activities involved. They are useful in coordinating activities of a collective of agents and the variety of social interactions occurring simultaneously between actors during their behavior development. As a result, this combination becomes an emerging and dynamic system of reciprocal, collective, and spatial relationships and influences. Rather than describing collaborative behavior from the point of view of each actor, events allow for the description of behavior from the point of view of the procedures that need to be performed to achieve a task (Schaumann et al., 2017). An example is the patient-check event: Three actors (doctor, nurse, and patient) must be present at the same place and time to perform this medical activity. In the simulation, the event entity behaves like a movie director, managing and coordinating a single actor's behaviors during a scene, but leaving a low level of adaptation to each actor. When triggered, the event reduces the autonomy of the actors and tactically coordinates them through a series of actions, related to the task and contextual building function. The event unit for a certain space checks the list of possible space semantics and the actual one, to see whether it can instruct an activity to be performed or not. The novelty of this approach lies in the real-time adaptability of the action process execution to emerging circumstances. Indeed, an event does not directly predict how people will behave in a built environment but rather creates a knowledge base that can be modified and adapted by local, specific circumstances and actor states to describe how a task will occur. It is assigned preconditions, performance procedures, and post-conditions.

Preconditions are facts about the virtual world which can be true or false; they account for the decision-making ability of the system and represent the actors and space required for an event to be triggered. When pre-conditions are satisfied, the event is triggered. Agent goal-oriented behavior is defined as a procedure that runs if certain pre-conditions are recognized, and such procedures will produce some effects and post-conditions (Weiss, 2000). Performance procedures guide the event execution. They are provided by the activity entity and comprehend the duration of the activity, unless it concerns a displacement activity, which depends on the space conformation. At the end of each activity, the simulation engine collects information and updates the state and statistics of the space and actor entity, creating

Table 12.1 The Cohort Principle

Nurses	*No. of Rooms*	*No. of Patients*
Nurse 1	5	10
Nurse 2	1	5
Nurse 3	6	10
Nurse 4	5	10

post-conditions. However, activities, as functions, cannot be affected or modified by post-conditions; thus, in an event-based method, actors do not have learning abilities.

The case study of the building-use scenario simulates a number of event types in which the sequence of activities, the involvement of actors, and the location of the actions are known in advance. Specifically, in a treatment procedure, nurses begin at their station, move to the central medicine room to prepare medicines or medicaments, and subsequently move through the patients' rooms to look after them one by one. Patients do not leave their rooms. Each nurse is assigned a cohort of patients and operates only in some patients' rooms on the ward, as shown in Table 12.1.

After visiting some patients in randomly initialized cases, nurses return to the medicine room to prepare additional medication and/or pick up new equipment before returning to their workflow.

In the visiting relative's event, a random number of visitors enter the hospital to meet their relatives undergoing treatment, each one visiting a single patient in the ward. They walk through the corridor to the patient's room, where social interaction takes place. Afterward, visitors leave the ward by the same entrance. Besides these two, emerging and unplanned events could be triggered when their spatial and social preconditions arise. For example, if a nurse arrives to check the patients in a room, the visitor leaves and waits in the corridor until the nurse finishes the check and moves to the next patient's room. Moreover, when a visitor encounters a nurse, proximity between the two causes the visitor to begin a social interaction (e.g., asking for information) at their current location and interrupt the nurse's scheduled activities, before both return to their planned events.

Results: Modeling Emergent Behavior to Inform Studies of Collective Spatial Cognition

As stated previously, the purpose of simulating collective spatial behavior is not to test cognitive function since agents are not beings with cognition. Rather the purpose is to allow agents to mimic learning based on decision rules and probabilistic behavior in order to determine whether the outcomes that emerge can inform research into collective spatial cognition. Three key factors shaped our case study development: visualizing contamination transmission to assess the effects of prevention and organizational measures (hand hygiene practice, contact precautions, and cohorts); investigating impacts of architectural design and space distribution (single room, double rooms, and multi-rooms) on pathogen propagation; and understanding better the potential of our event-based simulation approach to model collective human spatial behavior related to infection spread. Toward those ends, we took an event-based modeling approach.

Although many simulation runs could be generated, we present here the hidden infected patient (HIP) scenario as a demonstrative example. This scenario models the frequent condition where the primary source of transmission is uncertain. One or more HCWs, patients, or visitors may transmit pathogens to other members of the collective and may include asymptomatic carriers. One potential use of this scenario is to challenge the user to detect the initial cause(s) of the outbreak. Patients newly admitted or moved from other units are usually placed in the first available double room or, if an infection state has been detected, in a single room. However, because infection screening test results can take several days, in the HIP, two hidden infected patients occupy two rooms, a double and a single. Inpatients may either begin the simulation already infected or become colonized during their stay in the ward. In this scenario, the infection risk is the same for all patients involved (preset at "minimum risk") and is lower for all HCWs and visitors, reflecting their status as healthier than the patients. HCWs are initialized at a non-contaminated state. In this run, neither objects nor spaces can be carriers. However, if one seeks to investigate how infection propagates from the healthcare environment, as in the case of exogenous environmental infections (de Oliveira & Damasceno, 2010), the initial cause of infection spread can be a contaminated space.

In our simulation, the nurse caring for the first cohort is compliant with hand hygiene measures, whereas the nurse caring for the second cohort is not. Infected patients start contaminating their surroundings at the same rate, and they receive the same invasive treatment—an injection. The infected patient sharing the room contaminates his roommate through the environment, while in the case of the patient in the single room, this does not occur. Even if both nurses become equally contaminated after touching the infected patient, the one compliant with hygiene procedures prevents transmitting contamination further, namely to other rooms, whereas for the other nurse, the contamination propagates to subsequent rooms. The spread proceeds through one single and numerous double rooms. Doubles reveal higher contamination than singles where the patient has a lower chance of contagion from environmental contamination (Figure 12.6).

Figure 12.6 shows both rooms and their patients at different levels of contamination as the contagion spreads through the space. The contagion risk map allows visualizing the contamination dissemination dynamic, for example, clusters of infected patients and patterns of occurrence. By varying the initial conditions, the behavioral propensities of the members of the collective, organizational process, and the nature of the space itself, this output can illuminate how pathogen circulation varies on the basis of these factors. In this case, the simulation illustrates the effectiveness of hand hygiene practice to prevent and control contamination diffusion. It reflects how nurses, via cohort guidelines, may transmit pathogens to patients in their cohort but not to any other patients in the ward, unless interruptions

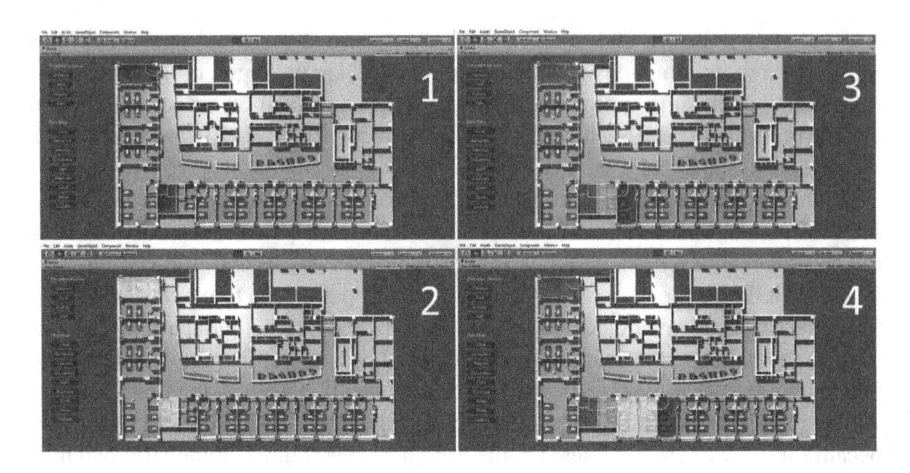

Figure 12.6 Contagion risk map development from hidden infected patient scenario simulation.

occur in their workflow pattern. This also proves the effectiveness of isolating patients suspected or recognized as infected so that infected patients are prevented from transferring the bacteria to others directly or, if hand hygiene procedures are followed, through HCWs' hands.

Qualitative results help us understand the effectiveness of implementing different prevention strategies, namely procedures such as hygiene behavior and contact precautions, and the effect of patient distribution, ward design, and appropriate HCW-patient ratios on pathogen transmission; they also help shed light on possible protocol breaches in infection outbreak management. Indeed, the modeled scenarios and initial analysis provide a certain degree of confidence about the validity of the results that are consistent with the literature, demonstrating varying degrees of reduction of the HAI risk within the range of prevention strategies (Health Act, 2006; Fuller et al., 2011; Hornbeck et al., 2012; Pittet et al., 2006; Sehulster & Chinn, 2003; Temime et al., 2009; van Kleef et al., 2013). As a matter of fact, only through validation and integration with complete and real-time data is it possible to support decision-making for the evaluation of interventions (Crooks et al., 2014; Esposito, Schaumann et al., 2020b; Rubin et al., 2013). Although it requires further calibration through data collection to retrieve all the information needed to thoroughly feed the system, a key insight is that if social and spatial behaviors do not fully comply with safety practices, a different spatial design could help prevent and control infections. This is because designs influencing human spatial behavior may compensate for some behavioral deficiencies that impact contamination propagation.

Multiple runs of the HIP scenario provide qualitative insights into the progression of an infectious disease, given various initial conditions and settings. By representing many interactions among the members of the collective, the system reveals real-time contamination transmission that depends on agent profiles and spatial choices, agents' activities, pathogen characteristics, and the roles of spaces and the inanimate objects contained within them.

Conclusions and Future Research

As stated earlier, the primary goal of the research presented here is to instantiate what is known about the spread of HAIs, what is known about the spatial arrangements of hospitals and the activities within them, and what is known about the interactions among the collective actors in the hospital environment in order to determine whether there is emergent behavior that can be used to better understand how and why infections spread and eventually to mitigate against that spread. The example results above demonstrate that the system can provide a testing environment whereby experts can implement "what–if" scenario analyses to compare the behavior of the hospital collective under varying conditions and behavioral preferences. This process can thereby provide practical support for decision-making processes in hospital risk management, including active interventions to present the spread of disease. The simulation can facilitate managers' instructions to healthcare staff members and act as a knowledge support tool for training. It can help to reduce non-compliance with policies while optimizing hospital resources.

Beyond this fundamentally practical goal, we believe that this work provides avenues for contributions to the larger research area of collective spatial cognition. Since the complex hospital system, its dynamic collective, and the spread of infections are extremely difficult (if not impossible) to recreate in an experimental setting, the emergent behaviors and outcomes provided by the simulation give insight into specific spatial or behavioral changes that could be designed as experiments. For example, if hub and spoke layouts in ward designs that appear to increase efficiency also increase pathogen spread in the hospital simulation, then alternative spatial layouts that lead to less spread could be tested, and the loss of efficiency documented. The tradeoff may well ultimately prove worthwhile. In this way, our study aims to bridge the gap between theoretical research on human spatial behavior and applied design. If the simulation complements the design process for building healthcare environments, this will help architects to foresee the consequences of their conceived design choices.

We see potential theoretical, practical, and methodological future avenues for this research. Regarding the first of these, there is a computational tradeoff with the thoroughness of the depiction of the cognitive processes that underlie spatial behavior in the simulation. That is, increasing the diversity and complexities of the behaviors attributed to agents in order to mimic their cognitive abilities decreases the heterogeneity of agents, activities, and behaviors, which may more realistically model the collective. This added complexity can create computational barriers for the system, but perhaps more importantly, create additional variables which must be managed, and which make it more difficult to determine which variable is driving any emergent behavior. The determination of the proper cognitive architecture that is both tractable and most clearly gives insight into the collective's behavior will be an ongoing question.

With regard to practical advances, further development of the system to create a practical forecasting tool would permit users to understand and assess the impact of infrastructural design on collectives and, consequently, support public and private decision-makers in improving sustainable spatial planning and design of human-centered built environments. This advance should be complemented by methodological advances that produce more formal quantitative results. The spread of infection can be measured with regard to speed, direction, and acceleration. The level of clustering of infections can be measured in either space or time, or simultaneously in space–time. These and other quantitative metrics of spatial presence and spread can add objective measures to the qualitative and interpretive infection risk map.

We do not aim to draw final, detailed conclusions of the impact of spatial and social behavior on infection spread, but to demonstrate the validity and potential of the developed simulation framework to properly work as a knowledge support tool in a case-based analysis, in order to envision how social interaction and spatial influences can affect the spread of HAIs. This should assist decision-makers in forecasting outcomes based on informed speculation acting as a decision support system (DSS) (Jit & Brisson, 2011). Therefore, unlike a prognostic model focused on accurately predicting the future, our approach is diagnostic—it is exploited to understand and explore the system and phenomena under study (Saltelli et al., 2008).

Acknowledgments

We wish to thank Prof. Jacob Yahav and Prof. Dino Borri for their methodological assistance and the following research group members for their useful comments: K. Date, E. Eizenberg, M. Gath Morad, L. Morhayim, N. Pilosof, and E. Zinger.

References

Altman, I. (1975). *The environment and social behavior: Privacy, personal space, territory, crowding.* Pacific Grove: Brooks/Cole Publishing Company.

Barnes, S., Golden, B., & Wasil, E. (2010). MRSA Transmission reduction using agent-based modeling and simulation. *INFORMS Journal on Computing, 22*(4), 635–646. https://doi.org/10.1287/ijoc.1100.0386

Barranco, R., Vallega Bernucci Du Tremoul, L., & Ventura, F. (2021). Hospital-acquired SARS-CoV-2 infections in patients: Inevitable conditions or medical malpractice? *International Journal of Environmental Research and Public Health, 18*(2). https://doi.org/10.3390/ijerph18020489

Basurrah, M. M., & Madani, T. A. (2006). Handwashing and gloving practice among health care workers in medical and surgical wards in a tertiary care centre in Riyadh, Saudi Arabia. *Scandinavian Journal of Infectious Diseases, 38*(8), 620–624. https://doi.org/10.1080/00365540600617025

Böhm, C., & Jacopini, G. (1966). Flow diagrams, Turing machines and languages with only two formation rules. *Communications of the ACM, 9*(5), 366–371. https://doi.org/10.1145/355592.365646

Borg, M. A. (2003). Bed occupancy and overcrowding as determinant factors in the incidence of MRSA infections within general ward settings. *Journal of Hospital Infection, 54*(4), 316–318. https://doi.org/10.1016/S0195-6701(03)00153-1

Borshchev, A., & Filippov, A. (2004). From system dynamics and discrete event to practical agent based modeling: Reasons, techniques, tools. In *Proceedings of the 22nd international conference of the system dynamics society* (p. 23). Oxford. http://www.systemdynamics.org/publications.htm#ConfProc

Box, G. E. P., & Draper, N. R. (1987). *Empirical model-building and response surfaces* (pp. xiv, 669). Chichester: John Wiley & Sons.

Boyce, J. M., & Pittet, D. (2002). *Guideline for hand hygiene in health-care settings* (RR-16; Recommendataions and Reports, p. 56). Atlanta, GA: Centers for Disease Control and Prevention.

Brodeschi, M., Putievsky Pilosof, N., & Kalay, Y. (2015). *The definition of semantic of spaces in virtual built environments oriented to BIM implementation. The next city* (pp. 331–346). Sao Paulo: Biblioteca Central Cesar Lattes.

Chartier, Y., Emmanuel, J., Pieper, U., Pruss, A., Rushbrook, P., Stringer, R., Townend, W., Wilburn, S., & Zghondi, R. (2014). *Safe management of wastes from health-care activities* (p. 329). World Health Organization. http://www.who.int/water_sanitation_health/publications/wastemanag/en/

Coen, P. (2012). Models of hospital acquired infection. *Infection Control - Updates, Section, 2*, 39–64.

Cooper, B. S., Stone, S. P., Kibbler, C. C., Cookson, B. D., Roberts, J. A., Medley, G. F., Duckworth, G. J., Lai, R., & Ebrahim, S. (2003). Systematic review of isolation policies in the hospital management of methicillin-resistant *Staphylococcus aureus*: A review of the literature with epidemiological and economic modelling. *Health Technology Assessment (Winchester, England), 7*(39), 1–194.

Crooks, A. T., Patel, A., & Wise, S. (2014). Multi-agent systems for urban planning. *Technologies for Urban and Spatial Planning: Virtual Cities and Territories*, 29–56. https://doi.org/10.4018/978-1-4666-4349-9.ch003

de Oliveira, A. C., & Damasceno, Q. S. (2010). [Surfaces of the hospital environment as possible deposits of resistant bacteria: A review]. *Revista Da Escola De Enfermagem Da U S P, 44*(4), 1118–1123. https://doi.org/10.1590/s0080-62342010000400038

Dedrick, R. E., Sinkowitz-Cochran, R. L., Cunningham, C., Muder, R. R., Perreiah, P., Cardo, D. M., & Jernigan, J. A. (2007). Hand hygiene practices after brief encounters with patients: An important opportunity for prevention. *Infection Control & Hospital Epidemiology, 28*(03), 341–345. https://doi.org/10.1086/510789

Demianyk, B. C. P. (2011). *Development of agent-based models for healthcare: Applications and critique* (p. 139). Master's thesis. IEEE.

Esposito, D., & Abbattista, I. (2020). *Dynamic network visualization of space use patterns to support agent-based modelling for spatial design* (pp. 260–269). https://doi.org/10.1007/978-3-030-60816-3_29

Esposito, D., Abbattista, I., & Camarda, D. (2020). A conceptual framework for agent-based modeling of human behavior in spatial design. In *Agents and multi-agent systems: Technologies and applications 2020* (pp. 187–198). Singapore: Springer. https://doi.org/10.1007/978-981-15-5764-4_17

Esposito, D., Santoro, S., & Camarda, D. (2020). Agent-based analysis of urban spaces using space syntax and spatial cognition approaches: A case study in Bari, Italy. *Sustainability, 12*(11), Article 11. https://doi.org/10.3390/su12114625

Esposito, D., Schaumann, D., Camarda, D., & Kalay, Y. E. (2020a). A multi-agent simulator for infection spread in a healthcare environment. In Y. Demazeau, T. Holvoet, J. M. Corchado, & S. Costantini (Eds.), *Advances in practical applications of agents, multi-agent systems, and trustworthiness. The PAAMS collection* (pp. 408–411). Springer International Publishing. https://doi.org/10.1007/978-3-030-49778-1_36

Esposito, D., Schaumann, D., Camarda, D., & Kalay, Y. E. (2020b). Decision support systems based on multi-agent simulation for spatial design and management of a built environment: The case study of hospitals. In O. Gervasi, B. Murgante, S. Misra, C. Garau, I. Blečić, D. Taniar, B. O. Apduhan, A. M. A. C. Rocha, E. Tarantino, C. M. Torre, & Y. Karaca (Eds.), *Computational science and its applications—ICCSA 2020* (pp. 340–351). Springer International Publishing. https://doi.org/10.1007/978-3-030-58808-3_25

Esposito, D., Schaumann, D., Camarda, D., & Kalay, Y. E. (2020c). Multi-agent modelling and simulation of hospital acquired infection propagation dynamics by contact transmission in hospital wards. In *Advances in Practical Applications of Agents, Multi-Agent Systems, and Trustworthiness. The PAAMS Collection: 18th International Conference, PAAMS 2020, L'Aquila, Italy, October 7–9, 2020, Proceedings 18* (pp. 118–133). Springer International Publishing. New York, https://doi.org/10.1007/978-3-030-48778-1_10

Ferrer, J., Salmon, M., & Temime, L. (2013). Nosolink: An agent-based approach to link patient flows and staff organization with the circulation of nosocomial pathogens in an intensive care unit. *Procedia Computer Science*, *18*, 1485–1494. https://doi.org/10.1016/j.procs.2013.05.316

Flyvbjerg, B. (2006). Five misunderstandings about case-study research. *Qualitative Inquiry*, *12*(2), 219–245. https://doi.org/10.1177/1077800405284363

Fridkin, S. K., Pear, S. M., Williamson, T. H., Galgiani, J. N., & Jarvis, W. R. (1996). The role of understaffing in central venous catheter-associated bloodstream infections. *Infection Control and Hospital Epidemiology*, *17*(3), 150–158.

Friesen, M. R., & McLeod, R. D. (2014, January). A Survey of Agent-Based Modeling of Hospital Environmentss [Environmentss read Environments]. *Hda*, *2*, 227–233. https://doi.org/10.1109/ACCESS.2014.2313957

Fuller, C., Savage, J., Besser, S., Hayward, A., Cookson, B., Cooper, B., & Stone, S. (2011). "The dirty hand in the latex glove": A study of hand hygiene compliance when gloves are worn. *Infection Control & Hospital Epidemiology*, *32*(12), 1194–1199. https://doi.org/10.1086/662619

Gärling, T., & Golledge, R. G. (1987). Environmental perception and cognition. In E. H. Zube & G. T. Moore (Eds.), *Advances in environment, behavior, and design* (Vol. 2, pp. 203–236). New York: Plenum Press.

Gärling, T., & Golledge, R. G. (Eds.). (1993). *Behavior and environment: Psychological and geographical approaches*. Amsterdam: North-Holland, Advances in Psychology 96.

Gehl, J. (2011). *Life between buildings: Using public space* (6th ed.). Washington, DC: Island Press.

Gibson, J. J. (1979). *The ecological approach to visual perception*. Boston: Houghton Mifflin.

Girard, R., Perraud, M., Herriot, H. E., Prüss, A., Savey, A., Tikhomirov, E., Thuriaux, M., Vanhems, P., & Bernard, U. C. (2002). *Prevention of hospital-acquired infections: A practical guide* (p. 72). Geneva: World Health Organization.

Hajibabai, L., Delavar, M. R., Malek, M. R., & Frank, A. U. (2007). Agent-based simulation of spatial cognition and wayfinding in building fire emergency evacuation. *Geomatics Solutions for Disaster Management*, 255–270. https://doi.org/10.1007/978-3-540-72108-6_17

Hall, E. T. (1966). *The hidden dimension*. Garden City, NY: Doubleday.

Harbarth, S., Sudre, P., Dharan, S., Cadenas, M., & Pittet, D. (1999). Outbreak of *Enterobacter cloacae* related to understaffing, overcrowding, and poor hygiene practices. *Infection Control & Hospital Epidemiology*, *20*(09), 598–603. https://doi.org/10.1086/501677

Hayes-Roth, F., Waterman, D. A., Donald, A., & Lenat, D. B. (1983). *Building expert systems*. Boston, MA: Addison-Wesley Pub. Co.

Health Act. (2006). *Part 2, c. 14–16*. https://www.legislation.gov.uk/ukpga/2006/28/contents

Hoogendoorn, S. P., & Bovy, P. H. L. (2004). Pedestrian route-choice and activity scheduling theory and models. *Transportation Research Part B: Methodological*, *38*(2), 169–190. https://doi.org/10.1016/S0191-2615(03)00007-9

Hornbeck, T., Naylor, D., Segre, A. M., Thomas, G., Herman, T., & Polgreen, P. M. (2012). Using sensor networks to study the effect of peripatetic healthcare workers on the spread of hospital-associated infections. *The Journal of Infectious Diseases*, *206*(10), 1549–1557. https://doi.org/10.1093/infdis/jis542

Jiménez, J. M., Lewis, B., & Eubank, S. (2013). *Hospitals as complex social systems: Agent-based simulations of hospital-acquired infections* (pp. 165–178). https://doi.org/10.1007/978-3-319-03473-7_15

Jit, M., & Brisson, M. (2011). Modelling the epidemiology of infectious diseases for decision analysis. *Pharmaco Economics*, *29*(5), 371–386. https://doi.org/10.2165/11539960-000000000-00000

Kalay, Y. E. (2004). *Architecture's new media: Principles, theories and methods of computer-aided design*. Boston, MA: MIT Press.

Landelle, C., Verachten, M., Legrand, P., Girou, E., Barbut, F., Brun-Buisson, C., & Buisson, C. B. (2014). Contamination of healthcare workers' hands with Clostridium difficile spores after caring for patients with C. difficile infection. *Infection Control and Hospital Epidemiology*, *35*(1), 10–15. https://doi.org/10.1086/674396

Majid, M. A. (2011). *Human behavior modeling: An investigation using traditional discrete event and combined discrete event and agent-based simulation* (PhD thesis). University of Nottingham.

Pan, X., Han, C. S., & Law, K. H. (2005). *A multi-agent based simulation framework for the study of human and social behavior in egress analysis* (pp. 1–12). https://doi.org/10.1061/40794(179)92

Pessoa-Silva, C. L., Toscano, C. M., Moreira, B. M., Santos, A. L., Frota, A. C. C., Solari, C. A., Amorim, E. L. t, Carvalho, M. D. G. S., Teixeira, L. M., & Jarvis, W. R. (2002). Infection due to extended-spectrum beta-lactamase-producing Salmonella enterica subsp. Enterica serotype infants in a neonatal unit. *The Journal of Pediatrics*, *141*(3), 381–387.

Pew, R. W., & Mavor, A. S. (Eds.). (1998). *Modeling human and organizational behavior: Application to military simulations*. The National Academies Press. https://doi.org/10.17226/6173

Pidd, M. (2004). *Computer simulation in management science*. Hoboken, NJ: Wiley.

Pittet, D., Allegranzi, B., Sax, H., Dharan, S., Pessoa-Silva, C. L., Donaldson, L., Boyce, J. M., & WHO Global Patient Safety Challenge, World Alliance for Patient Safety. (2006). Evidence-based model for hand transmission during patient care and the role of improved practices. *The Lancet. Infectious Diseases*, *6*(10), 641–652. https://doi.org/10.1016/S1473-3099(06)70600-4

Quan, X., Taylor, E., & Zborowsky, T. (2015). *Clean hands saves lives: A systems approach to improving hand hygiene* (pp. 1–17). Concord, NH: The Center for Health Design.

Rubin, M. A., Jones, M., Leecaster, M., Khader, K., Ray, W., Huttner, A., Huttner, B., Toth, D., Sablay, T., Borotkanics, R. J., Gerding, D. N., & Samore, M. H. (2013). A simulation-based assessment of strategies to control clostridium difficile transmission and infection. *PLoS One*, *8*(11), e80671. https://doi.org/10.1371/journal.pone.0080671

Rudofsky, B. (1981). *Strade Per La Gente. Architettura e Ambiente Umano*. Laterza. Washington, DC: Van Nostrand Reinhold.

Saltelli, A., Ratto, M., Andres, T., Campolongo, F., Cariboni, J., Gatelli, D., . . . Tarantola, S. (2008). *Global sensitivity analysis: The primer*. Chichester: John Wiley & Sons.

Schaumann, D., Date, K., & Kalay, Y. E. (2017). An event modeling language (EML) to simulate use patterns in built environments. In *Proceedings of the symposium on simulation for architecture and urban design* (pp. 1–8).

Schaumann, D., Kalay, Y. E., Hong, S. W., & Simeone, D. (2015). Simulating human behavior in not-yet built environments by means of event-based narratives. In *Proceedings of the symposium on simulation for architecture & urban design* (pp. 5–12).

Schaumann, D., Morad, M. G., Zinger, E., Pilosof, N. P., Sopher, H., Brodeschi, M., Date, K., & Kalay, Y. E. (2016). A computational framework to simulate human spatial behavior in built environments. In *Proceedings of the Symposium on Simulation for Architecture & Urban Design* (pp. 121–128).

Schaumann, D., Pilosof, N. P., Date, K., & Kalay, Y. E. (2016). A study of human behavior simulation in architectural design for healthcare facilities. *Annali dell'Istituto Superiore di Sanità*, *52*(1), 24–32. https://doi.org/10.4415/ANN_16_01_07

Sehulster, L., & Chinn, R. Y. W. (2003). Guidelines for environmental infection control in health-care facilities. Recommendations of CDC and the healthcare infection control practices advisory committee (HICPAC). *MMWR. Recommendations and Reports: Morbidity and Mortality Weekly Report. Recommendations and Reports/Centers for Disease Control*, *52*(RR-10), 1–42. https://doi.org/DX

Shelby, B., Vaske, J. J., & Heberlein, T. A. (1989). Comparative analysis of crowding in multiple locations: Results from fifteen years of research. *Leisure Sciences*, *11*(4), 269–291. https://doi.org/10.1080/01490408909512227

Shortliffe, E. H., & Buchanan, B. G. (1975). A model of inexact reasoning in Medicine. *Mathematical Biosciences*, *9*(1), 45–74. https://doi.org/10.1016/0025-5564(75)90047-4

Simeone, D., Kalay, Y. E., Achten, H., Pavlicek, J., Hulin, J., & Matejovska, D. (2012, September). An event-based model to simulate human behaviour in built environments. *Ecaade 2012*, *1*(1), 525–532.

Simeone, D., Kalay, Y. E., Schaumann, D., & Hong, S. (2013). Modelling and simulating use processes in buildings. *Proceedings of ECAADe*, *31*(2), 59–68.

Simeone, D., Schaumann, D., Kalay, Y. E., & Carrara, G. (2013). Adding users' dimension to BIM. *EAEA-11 conference (Track 3) Conceptual Representation: Exploring the layout of the built environment* (pp. 483–490).

Sopher, H., Schaumann, D., & Kalay, Y. (2016). Simulating human behavior in (Un)built environments: Using an actor profiling method. *World Academy of Science, Engineering and Technology International Journal of Computer, Electrical, Automation, Control and Information Engineering*, *10*(12), 2030–2039.

Stiller, A., Salm, F., Bischoff, P., & Gastmeier, P. (2016). Relationship between hospital ward design and healthcare-associated infection rates: A systematic review and meta-analysis. *Antimicrobial Resistance & Infection Control*, *5*(1), 51. https://doi.org/10.1186/s13756-016-0152-1

Stokols, D. (1972). On the distinction between density and crowding: {Some} implications for future research. *Psychological Review*, *79*(3), 275–277. https://doi.org/10.1037/h0032706

Temime, L., Opatowski, L., Pannet, Y., Brun-Buisson, C., Boëlle, P. Y., & Guillemot, D. (2009). Peripatetic health-care workers as potential superspreaders. *Proceedings of the National Academy of Sciences of the United States of America*, *106*(43), 18420–18425. https://doi.org/10.1073/pnas.0900974106

Ulrich, R., Quan, X., Systems, H., Architecture, C., & Texas, A. (2004, September). The role of the physical environment in the hospital of the 21st century: A once-in-a-lifetime opportunity. *Environment*, *439*, 69.

van Kleef, E., Robotham, J. V., Jit, M., Deeny, S. R., & Edmunds, W. J. (2013). Modelling the transmission of healthcare associated infections: A systematic review. *BMC Infectious Diseases*, *13*, 294. https://doi.org/10.1186/1471-2334-13-294

Voss, A., & Widmer, A. F. (1997). No time for handwashing!? Handwashing versus alcoholic rub: Can we afford 100% compliance? *Infection Control and Hospital Epidemiology*, *18*(3), 205–208.

Weiss, G. (2000). *Multiagent systems: A modern approach to distributed artificial intelligence*. Boston, MA: MIT Press.

Whitby, M., McLaws, M.-L., & Ross, M. W. (2006). Why healthcare workers don't wash their hands: A behavioral explanation. *Infection Control & Hospital Epidemiology*, *27*(05), 484–492. https://doi.org/10.1086/503335

Yan, W., & Kalay, Y. (2005). Simulating human behaviour in built environments. In B. Martens & A. Brown (Eds.), *Computer Aided Architectural Design Futures 2005*. Dordrecht: Springer. https://doi.org/10.1007/1-4020-3698-1_28

Zimring, C. (2002). *Post-occupancy evaluation: Issues and implementation. Handbook of environmental psychology* (pp. 306–319). Chichester: John Wiley & Sons.

13 Spatial Analytic Tools and Techniques to Inform Research on Collective Spatial Cognition

Kevin M. Curtin, Penelope Mitchell, and Megan Rondinelli

Introduction

The primary goal of the collective spatial cognition research effort is to lay the foundation for a new research paradigm that integrates elements of cognitive science, geography and spatial analysis, and team performance and cognition. Reviews of the literature (including Baumann, Kretz, and He in this volume) demonstrate that while there are pairwise integrations of these elements (e.g., spatial cognition in individuals or team cognition without a spatial focus), there are relatively few that have to date integrated all three. The focus of this chapter is the potential for further integration of spatial analytic techniques with the other two pillars of collective spatial cognition: collectives and spatial cognition. The impetus for this chapter was a review of the position papers that were presented at the Collective Spatial Cognition Specialist Meeting in Santa Barbara, CA, in April 2019. While there were many spatially informed—and even more spatially interested—participants among the specialists at that meeting, the position papers themselves were notable for the nearly complete absence of the spatial analytic techniques that have been developed in the field of geographic information science over the past half century, as well as other canonical geographic methods such as cartographic techniques. Moreover, a review of the broader literature in spatial cognition and team cognition reveals that even when the spatial aspect of cognition or team performance is an element of a research study, the extent to which spatial analyses or spatial metrics are employed is very limited.

Given this heretofore highly limited integration of spatial analytic techniques with research regarding collectives and cognition, this chapter seeks not to present what has been accomplished, but instead to consider the realm of the possible if the broad range of spatial analytic techniques were more thoroughly integrated into studies regarding collective spatial cognition. There is extensive literature surrounding spatial analytic methods and metrics. These range from simple descriptive measures (e.g., distance and area) to sophisticated spatial statistical analyses. The spatial domains on which these methods can operate (e.g., in the plane, at point sets, and on networks) are varied. Given the richness of spatial analytic techniques that are available, this chapter posits that there are many opportunities to advance the body of knowledge that is being developed

DOI: 10.4324/9781003202738-18

in the collective spatial cognition research effort. Toward that end, this chapter reviews—in significant detail—the spatial analytic methods that are available for use by researchers undertaking studies of collective spatial cognition. In some cases, we suggest ways in which these techniques could be incorporated into existing studies of team cognition or extend those studies by including a spatial aspect. In each of the subsequent sections, a family of spatial analytic techniques is described, specifically cartography and geovisualization, spatial statistics, network analysis, spatial optimization, spatial data query, 3D analysis, raster analysis, and spatio-temporal analysis. For each of these, the typical uses of these techniques are detailed and then possible uses of the methods in research studies of collective spatial cognition are suggested. The examples we provide are intended to prompt researchers to investigate these tools and integrate them when appropriate into their own research efforts.

Cartography and Geovisualization

Maps are a way to communicate and acquire knowledge of spatial components of the environment. Cartographic techniques have historically been associated with geographic analyses, although their broad appeal and usefulness creates widespread interest in their use. Cartography of one form or another has certainly found its way into studies of spatial cognition, team cognition, and even nascent studies of collective spatial cognition; more thoroughly, in fact, than most of the other techniques described in this chapter.

Of course, there is a long history regarding map design and the perception of spatial information, and a comprehensive review of 20th-century cognitive map-design research can be found in Montello (2002). Even that work, however, concluded that an open question remained as to why there has not been greater interaction among those who study "map-design, map-psychology, and map-education research," a sentiment that foreshadowed the collective spatial cognition effort. Modern geovisualization interfaces present a myriad of design options for representation and interaction requiring educated decisions to promote usability and spatial understanding. The goal here is to describe how those techniques for implementing cartographic elements can be further brought to bear on studies of collective spatial cognition.

Color

Among the topics where cartographic representations are examined for their influence on perception, color is perhaps the most widely studied. The influence of color schemes—particularly in choropleth mapping (the use of color to indicate different variable values or classes of values)—is an area of long-standing interest (e.g., Breslow et al., 2009; Brewer et al., 1997; Brewer & Pickle, 2002; MacEachren et al., 1998). Imagine how the influence of color can be extended to collective spatial cognition. Do assignments of color to team members allow them to understand particular map elements assigned to them, focusing their attention on particular tasks and improving overall team performance? As another example, consider the collective spatial cognition challenge problem of the army squad moving through an unknown urban environment under changing conditions. Imagine that the squad is equipped with a headset that allows them not only to see their environment but also to see overlain graphics that indicate the likelihood of enemy combatants in the building around them. A "stoplight" choropleth scheme (or alternatively a "hot and cold" scheme) could be used to indicate the level of danger associated with different features. Differences in color could indicate where and how individual team members behave, and those color choices may influence team performance.

Orientation

The orientation (which way is up) of a map or diagram is a cartographic choice. Tests of map alignment in individuals have appeared in the literature (e.g., Liben & Downs, 1993), and unsurprisingly, there are noted differences based on characteristics such as age or gender. Map use requires visuospatial perspective taking, or in other words, imagining the visual perspective of the map. Cognitive psychologists have described perspective taking as an effortful spatial transformation that updates the self-representation after movements such as the rotation of one's body (Gunalp et al., 2019; May, 2004; Presson & Montello, 1994; Rieser, 1989). Consider extending map alignment work to test, for example, if providing maps of the same area, but with different orientations, helps or hinders the overall ability for a collective to learn the space or complete a navigation task in that space. Given the large individual differences in perspective taking, research should build upon the insights of current individual perspective-taking insights to guide investigation of perspective taking and other spatial orientation abilities in teams.

Scale

Once again, the issue of the influence of map scale on spatial cognition has appeared in the literature, particularly as it applies to individual learning (e.g., Hegarty et al., 2006), but our concern here is how this type of work could be extended to collectives or teams. Can individual differences in the ability to learn about or navigate spaces at different scales be leveraged to improve team performance? Does the generalization of features (simplifying geospatial data by omitting certain geometric or topological properties and retaining others) that is required as map scales become smaller suggest that for a holistic understanding of a complex space, some team members should focus on small-scale overviews of the space, while others should learn about portions of the space viewed at larger scales? Can these cross-scale interpretations be combined and communicated to generate a superior team understanding of the entire space? What level of detail leads to the most effective team leader performance? Can the level of generalization be optimally associated with team members based on their task assignment?

Symbology

The state-of-the-art cartographic technology allows for greater choice regarding graphic devices and greater ease in implementing those choices than ever before. The choices for thematic method (e.g., choropleth, dot density, charts, proportional symbols, and graduated symbols), for the number and kind of available symbols (often related to particular industries or cartographic traditions), and for the styles for features and supplementary text are numerous and growing. But the complexity that this multitude of symbolic opportunities can add to a map is not always desirable. In fact, existing studies of individual spatial cognition frequently employ highly simplified diagrams without the map elements that are typically expected of general-use maps (e.g., scale, orientation, legend, and spatial reference); as an example, see Richardson et al. (1999). This simplification is typically necessary, given the common practice of focusing subjects' attention on the spatial task and study area under experiment and eliminating factors that could confound the assessment of the cognitive or spatial performance. Moreover, it has been shown that with regard to the complexity of map, a simpler map may be more efficient in communicating information but is not necessarily the preferred map from the perspective of the user (Hegarty, 2013).

As above, one can imagine how this broad range of symbology tools and techniques can be brought to bear on studies of collective spatial cognition. Certainly, the extant experiments could be expanded to determine whether more complex symbology creates greater or lesser map communication for a team as a whole or for team members with specific tasks. Consider just the idea of changing symbol size, including the use of cartograms—a cartographic method that distorts the representation of space by replacing area with another variable value. Cartograms were historically difficult to draw by hand but can be generated in a highly automated way today. For an experiment into collective spatial cognition such as that presented in this volume by Andrews-Todd and Rapp (2024), where participants were asked to identify states, the results of correct and incorrect identifications could be presented as a cartogram where the size of the state indicated the number of times the participants correctly identified the state. In studies of collective wayfinding, symbol size could be used to indicate the importance of navigation aids as judged by the team as a whole.

In summary, while cartographic technique and map design has long been a subject of research in spatial cognition, there is ample room to extend this work into studies of collective spatial cognition. There is much evidence that task performance outcomes can dramatically differ with different visual displays of the same information (Armstrong & Densham, 1995; Breslow et al., 2009; Hegarty, 2011; Novick & Catley, 2007; Simkin & Hastie, 1987), given that informationally equivalent spatial representations are not cognitively computationally equivalent (Larkin & Simon, 1987). How to best use new information technologies to reveal patterns in complex spatial data is a geovisualization research challenge with implications for collective spatial cognition. A crosscutting research challenge identified by MacEachren and Kraak (2001) regarding the need to develop geovisualization methods and tools to support group work still holds true today. The breadth of available methods and their increasing ease of use makes implementing such research more tractable.

Spatial Statistics

One of the most dramatic advancements in spatial analytic techniques over the recent past has been the flourishing of new, yet robust, spatial statistics. While there have certainly been spatial statistics created in the more distant past (e.g., Moran, 1950), this subdiscipline within geography flourished in the 1990s, and innovative development continues to the present. Entire textbooks exist to give broad and/or in-depth coverage of this topic, so a comprehensive review is not possible here; the intent here is to roughly describe the main categories of spatial statistics and suggest where they could be useful in studies of collective spatial cognition.

Centrography

Centrographic measures are measures of spatial central tendency. These straightforward descriptive statistics are often overlooked yet provide robust means of performing comparative analyses. Very briefly, measures of the spatial center include the mean center (mean in the *x*- and *y*-directions), the median center (intersection of the lines through the medians in the *x*- and *y*-directions), and the 1-median point (or Weber point), which is the location that minimizes the total distance to all observation points, among others. Consider that the mean center itself could be a useful comparative measure of team performance. Imagine that team members are asked to identify a target location on a map based on some external information. The mean center of the locations they choose is a summary of team performance and permits an objective measure (distance) from the actual location. The mean center (and associated offset distance) can be used to compare the spatial performances among different teams, or for the same team with different information inputs, or before and after training. The direction of the offset between mean center of the team observations and the actual target location can indicate whether there is some factor pushing the team to err in one way or another. If the mean center of the team members' observations is found to move closer to the true value over time or during training, then the speed and acceleration of that improvement can be calculated.

Centrographic measures of spatial spread (typically spread around a measure of the spatial center) include the standard distance (or standard radius), which is the two-dimensional equivalent of the standard deviation, and the standard deviational ellipse, which describes both spread and directional trend. Once again, these could be used as comparative measures of team spatial performance. Consider again the example where team members are identifying target locations on a map. If both mean center accuracy and spatial spread of team observations are considered, there are four possible categories of team spatial performance: (1) mean center near to true location, but observations spread out; (2) mean center near to true location and not spread out; (3) mean center far from the true location and spread out around the mean center; and (4) mean center far from the true location but located close to the mean center. Each of these potential categories of observation says something about the accuracy of the team observations taken as a whole and the internal consistency of the observations among team members.

Clustering in Space

The earliest measures of clustering in space were generated in the area of ecological modeling (Clark & Evans, 1954; Pielou, 1960). Among these methods are quadrat analysis, mean nearest neighbor analysis, and Ripley's K. Generally speaking, these are inferential statistics that determine whether or not a spatial point pattern differs significantly from the pattern to be expected (given the number of observations and the size of the study area) under the assumption of complete spatial randomness. There are tests for significance, which describe whether or not the spatial point pattern is significantly clustered, significantly dispersed, or indistinguishable from a random pattern. If a statistically significant deviation from random is found, this indicates that there may be a process that is driving that deviation. In the context of ecological modeling, such a process may be a population exploiting a resource and thus clustering around it or a population seeking unrivaled territory and thereby dispersing through the space.

Consider then how this determination of clustering or dispersion can be used to inform studies of collective spatial cognition using the challenge question of the teams of firefighters battling a blaze and evacuating residents in a complex and dynamic interior environment. The performance of these teams may be tested with different equipment and different maps of the environment. The measures of clustering or dispersion can describe the layout into which each team organizes itself during the operation. If high-performing teams were found to have greater dispersion in their operations, then this can become an element of operational training.

Clustering of Variable Values

There is a set of spatial statistics that are concerned not with clustering or dispersion with regard to pure spatial location but rather are concerned with whether or not similar variable values associated with observations are spatially associated. These measures of spatial autocorrelation include Moran's I (Moran, 1950)—which determines whether or not similar variable values tend to be neighbors—and the Getis-Ord General G (Getis & Ord, 1992)—which determines specifically if high variable values or low variable values are spatially associated (Figure 13.1). There are both local and global versions of these statistics.

Consider the case of a multiteam system such as those reviewed by Baumann et al. (2024) in this volume, perhaps a set of search and rescue teams who need to cover the territory of an earthquake disaster zone. Each member of each team may be asked to rate locations that have been searched with a confidence score

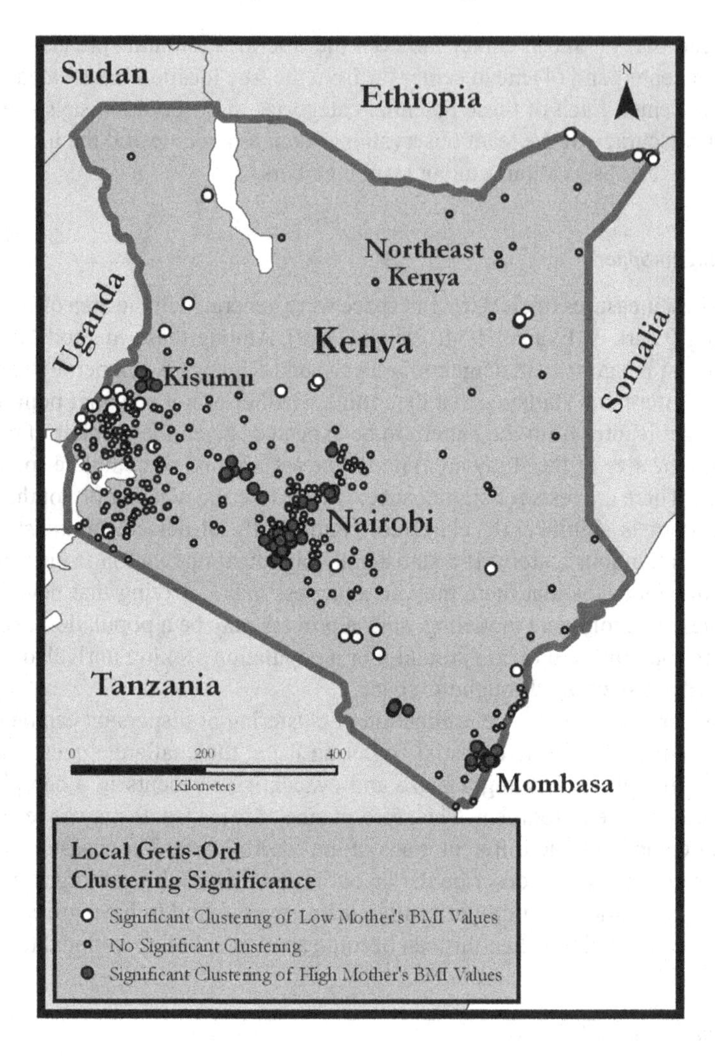

Figure 13.1 Test of clustering with the local Getis-Ord Gi* statistic (Pawloski et al., 2012).

that represents their level of belief that no further survivors are likely to be found in the searched location. Using tests of spatial autocorrelation, one can determine where there are clusters of high and low values of confidence, and these levels can later be back-checked against actual survivor/casualty numbers and locations. This would allow determinations of what types of locations are prone to more or less accurate confidence scores, tests of which teams are more accurate in their confidence scores, and ultimately an examination of the different ways in which successful/accurate teams operate and learn about their search environment.

Geographically Weighted Regression

Although means of incorporating spatial information into regression models have existed for many decades, a minor revolution in this area of methodology was created by the development of geographically weighted regression (GWR) (Fotheringham et al., 2002). Briefly stated, GWR produces a family of local regression models (rather than a single global model) that apply to a subset of the observations in a spatial dataset in order to more accurately capture spatial variation in the predictor variables. Other spatial regression models are available to explore other types of potential spatial dependence; for a review of performance between models, see Rüttenauer (2022).

Consider again the case above of multiple search teams in a disaster zone. A training officer may be able to assign performance scores to teams, and team members can develop a database of location-specific attributes. This scenario is ideal for GWR, given that the nature of the disaster may be distinct in different areas (local rather than global), the teams may perform in isolation in different local areas of the operation, and it would be useful to associate the local characteristics of the landscape with the variation in the performance scores.

Network Spatial Statistics

The analysis of stochastic point processes on Euclidean planes is well established, but for many applications, a network geographical space is the more appropriate and less biased context (Yamada & Thill, 2004). Consider, for example, traffic accidents which essentially always occur on the street network itself, roadside improvised explosive devices which by definition only exist in road network space or interactions among pedestrians who are confined to the network of sidewalks. Although the following section delves more deeply into the nature of network analysis for collective spatial cognition, the topic of network spatial statistics deserves mention here. Since the 1990s, there has been ongoing work to generate versions of spatial statistics that were originally designed for phenomena that operate in the plane, which will work appropriately for phenomena operating on defined networks (Okabe et al., 1995). These statistics have examined things such as the density of observations on a network (Okabe et al., 2009), measures of clustering on a network (Okabe & Yamada, 2001), and how to transform non-uniform networks for analysis (Okabe & Satoh, 2006). Toolsets have been developed for integration into GIS software in order to implement these statistics (Okabe et al., 2006). Given the presence of these network spatial statistics, the same speculation above regarding how collectives operating in space could be analyzed pertains to collectives restricted to network space.

In summary, there are many more spatial statistical methods than those described here, but this brief overview and speculation about potential uses demonstrate that there is potential for more prodigious use of the host of spatial statistical techniques in studies of collective spatial cognition.

Network Analysis

The network structure has proven to be one of the most long-standing and useful data structures in spatial analysis (Curtin, 2007). The analysis of networks precedes computerized geographic analyses, with studies bridging geography with sociology, urban planning, and other disciplines (Haggett & Chorley, 1969). However, the network structure has become a dominant spatial platform, due in part to the vector (point, line, and polygon) database systems that were at the heart of early geographic information systems (GISs) and persist in virtually all GIS software. Moreover, the network structure is widely adopted across disciplines, given its natural means of identifying connectivity between elements of a system. There is a large and growing set of quantitative methods in network analysis that can be brought to bear on studies of collective spatial cognition. Some potential means of integrating these methods are explored here.

The Network as a Modeling Structure

Simply organizing entities and their connections can be a valuable tool, particularly for teams. Storing information in the form of a network structure translates data into the fundamental components of a graph, namely edges, also known as links or arcs, and the connecting vertices, also termed junctions, points, or nodes. In the geographic sense, network structures include transportation networks, pipelines, and hydrographic networks to name only a few, and these structures have been the subject of study for decades (Garrison & Marble, 1962). The network perspective is also utilized conceptually across countless scientific disciplines (Boccaletti et al., 2006). The quantitative methods of the network structure can be used to analyze social network characteristics, communication network hierarchies, or the neural networks of the nervous system. Network science provides the framework to model the dynamics of a variety of physical and conceptual relationships at varying temporal and geographic scales, enabling the extraction of complex interactions and emergent behaviors to promote an enriched understanding of the system of interest. Team spatial cognition would benefit from the quantitative analysis of the team communication networks and hierarchies with respect to improving team performance and enhancing resilience against disturbance.

The structural attributes of transportation networks include stock aggregates such as miles of road or kilometers of canal which describe the extent of the network (Garrison & Marble, 1962). Structural layout can be characterized as centralized, decentralized, or distributed. A centralized network contains one central node of high accessibility, and this is a common characteristic of hub and spoke networks such as with airline networks. Decentralized structures include multiple central points of accessibility, creating subcenters, or neighborhoods with significant levels of accessibility. Finally, distributed networks contain no central node of accessibility significantly different from other nodes, as connectivity is high throughout this network; network structures such as this contain many redundant connections.

Static network structure has been studied extensively in the social sciences—by the 1930s, sociologists recognized the importance of connectivity patterns between people in gaining insights into the functioning of human society (Moreno, 1934; Newman, 2003). Typical social network studies address issues of centrality—which individuals are best connected to others or have the most influence and connectivity—or whether and how individuals are connected to one another through the network (Newman, 2003).

The vantage of the network can be at the microscale, looking at the properties of individual nodes and edges, at the mesoscale, looking at subsets of nodes and connections as modules or neighborhoods, or at the macroscale, which encompasses the entire network structure (Siew et al., 2019). With moderate-sized networks, connectivity can be ascertained from viewing a network diagram, but as a network size and complexity increases, more advanced descriptive and comparative network analysis is required to fully grasp network characteristics and functionality. This hierarchical structure would seem ideal for examining the connections between members of a collective as they learn or behave in space and the higher level connections between collectives in multiteam systems.

Descriptive and Comparative Analysis With Networks

The quantitative foundation for network analysis lies in the mathematical subdiscipline of graph theory, and within graph theory, there are metrics that can be used to describe and compare networks (Curtin, 2008). These metrics range from the most straightforward such as counts of the number of edges and nodes in the network as a measure of network size to more derivative measures such as those that measure levels of connectivity (see Curtin (2009) for descriptions of the alpha, beta, and gamma indices of connectivity). Other characteristics such as the network diameter (the length of the shortest possible path from the two most distant points), the average degree of the nodes, and the number of cycles in the network can offer useful insights into the scope and ability to traverse a network. There are also measures of the shape of a network (the Pi index) and measures of traffic or flow through nodes (Theta index), among others (DuCruet & Rodrigue, 2017).

In terms of usefulness for studies of collective spatial cognition, consider that the communications among team members engaged in a spatial learning exercise can be structured as a network with team members as the nodes and communications between members as links. The measures of network size and connectivity (how many communications and to what extent members are linked) can be used to compare across teams or for the same team before and after training. These measures of network size and scope can be correlated with better or worse team performance. How does information cycle through the team or teams of teams? Does a larger diameter (team members "further" apart in the communication network) hinder group outcomes due to poor communication? Where might there be an information sink that hinders information passage or potential leak that may threaten information security? Mapping the communication network promotes transparency in the flow of information and facilitates the encoding of transactive memory systems (TMSs). The concept of TMS processes was initially developed in reference to the implicit division of cognitive labor observed in intimate couples (Lewis & Herndon, 2011; Wegner et al., 1985). It was then acknowledged as a method of storing and retrieving information in groups, allocating knowledge and responsibilities throughout a team of people, and creating a form of social cognition (Moreland, 2008). Research has shown that a well-developed TMS can improve group performance (Lewis, 2004; Moreland, 2008) but has also found that organizations may have different TMS characteristics than pairs or small groups (Moreland, 2008). Applying network structure analysis to TMS directories could shed light on the efficacy of the nature and frequency of interactions, as well as distill the nature of the distribution of information to determine whether the allocation is appropriate for the task. One of the first uses of TMS in the context of collective spatial cognition is described in Chapter 8 of this volume (Maupin, et. al., 2024).

Processes on Networks

There is a fundamental relationship between network structure and function, in which the key determinant of movement or flow on a network is the spatial configuration of the network itself (Hillier & Stonor, 2017). Although the idea of process on networks and other spaces is explored further in the section of this chapter dealing with spatial optimization, this idea is a key premise in space syntax theory. Space syntax can be described as the configurational analysis of space (Porta et al., 2010), focusing on the relationship between human societies and space. The structure-function theory seeks to describe the spatial structure, capturing spatial form and functional patterns to develop a link between form and function. This methodology was developed with respect to cities but is applicable to other types of networks. How does the network structure, be it a conceptual network such as organizational relationships, multiteam systems, or a traditional spatial network, influence the potential process and flow of people, information, or goods on that network?

The syntactic structure-function theory, which is the ethos of space syntax, can be broken down into five key concepts, as described by Hillier and Stoner (2017). First, it provides a way to describe space quantitatively, by discretizing the chaos and complexity on network links into elements that can be subject to computational analysis. For example, a street can be quantified in several ways such as by length, angle, connections, the number of buildings, or capacity, among other spatial attributes. Then, the theory of natural movement demonstrates that connected elements in a network are more often used and that most people most of the time tend to take the cognitively simplest journey from their origin to destination as opposed to the shortest path. This is related to the syntactic concept of movement economy; the processes on the network take advantage of the natural movement pattern. For example, in the context of a city, shops will locate on highly connected networks, whereas residential areas congregate in the back and side streets or remote suburbs. The simultaneous multi-scale concept addresses global and local movement. Major arteries will likely have global and local movement, while local, less connected links will carry mostly local traffic. The multi-scale properties of a link have implications regarding transition probability, in other words, the ability or potential to move from one area to another, which influences what type of people, information, or goods you find where. Finally, the dual grid notion further breaks down the multi-scale concept into foreground and background grid components, with foreground network structures being major arteries of flow versus background structures used for localized movements.

Consider the following questions that might rely on syntactic structure-function theory in the context of collective spatial cognition. What spatial syntax characterizations are needed for complex vertical environments where a multiteam system is responding to a disaster in a multi-use skyscraper? How can leaders best communicate connectivity of paths through floors as well as the degree of connectivity between floors. How do team members inquire and communicate about dead ends or perceived dead ends? How does the spatial syntax influence the flow of natural movement of the multiple teams? What communication should be implemented to improve that flow?

In short, it is informative to understand both the network structure and the dynamic processes that operate within—and are influenced by—that structure. In the best-case scenario, this combined structure/process approach leads to the understanding of emergent behavior among a collective operating in that environment. The cognition of the space is once such emergent behavior.

Linear Referencing

One of the least used, yet surprisingly useful, sets of tools in state-of-the-art GIS are tools for linear referencing. Linear referencing is a term that encompasses a family of concepts and techniques for associating features with a spatial location along a network, rather than referencing those locations to a traditional spherical or planar coordinate system (Curtin & Turner, 2019). Linear referencing systems (LRSs) are regularly used in transportation networks to denote the distance from an origin (e.g., mile markers on U.S. Interstates). Routes in an LRS are the foundational features on which the referencing occurs. The measures are the values along the routes that are used to specify linear referenced locations. The measure values may be in standard metrics or could represent cost or may indicate a percentage of the total route traveled. Events in an LRS are the objects that are referenced along the route(s). Events could be anything that warrants recording along the referenced route, including singular objects such as landmarks or explosive devices, or the start or end point of a continuous phenomenon such as a high crime zone (Figure 13.2).

Implementing linear referencing requires a series of steps that successively lay the foundation for the referencing system, define how the references will be captured, and finally locate events on the linear features so that they may be visualized and analyzed (Curtin et al., 2007). When generating an LRS, the application should be understood in advance so that the nature of the network and the topological rules that apply to the network are known. With the application established, the linear

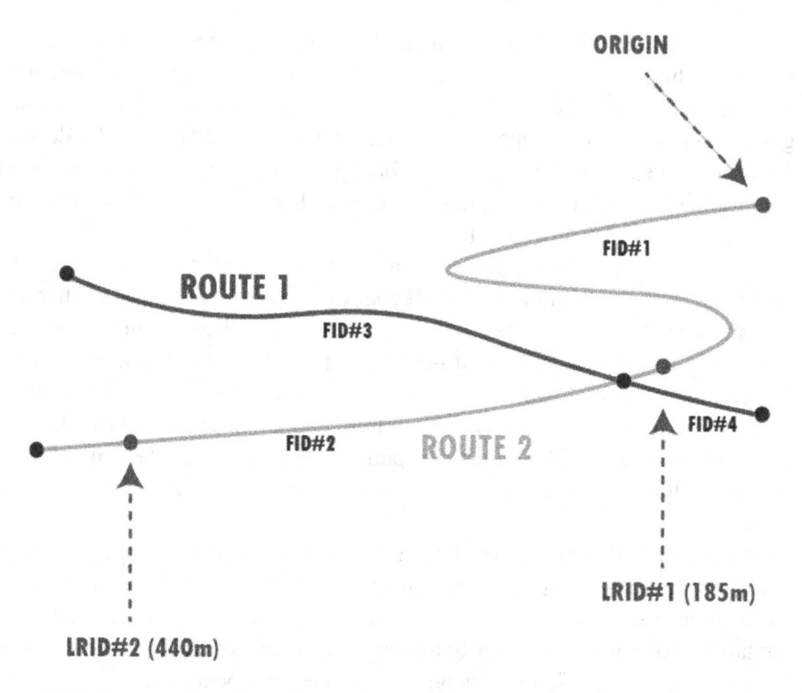

Figure 13.2 An example of how linear referencing is used to define the network space and locate events in that space (Curtin & Turner, 2019).

entities on which the measurements will be made should be identified, with each route having an associated origin. The set of routes provides the foundation on which to measure events and distill actionable information.

There are several advantages to using a linear referencing system over traditional spherical or planar projected referencing systems, including ease of use and interpretation of locations, database simplification (removal of duplicative features and attributes, and concomitant removal of potential for error), and a heightened level of accuracy, given that ground truth locations and distances can be associated with the network in lieu of more crudely calculated measurements generated by the GIS (e.g., not including elevation changes when calculating planar distances). The first of these advantages, ease of use and interpretation, is perhaps most pertinent to potential studies of collective spatial cognition. Natural language communication of distances and locations (e.g., "20 meters past the stop sign" and "halfway down the block") is much more likely to be the tool for communication than latitude/longitude pairs or state plane coordinates. The tools of linear referencing can be used to store the measures of important locations along the network paths (such as in a test of team navigation), to display stretches of a path that are more or less used by team members, or to communicate locations between team members in intuitive ways.

Spatial Optimization

The subdiscipline of geography known as location science—or, more broadly, spatial optimization—provides a number of opportunities for informing collective spatial cognition. The problems addressed with these methods are closely related to those of operations research and systems engineering—essentially, these are operations problems that have an explicitly spatial component. These techniques are primarily quantitative in nature and continue to develop closely alongside GIS.

Fundamentally, spatial optimization problems require definitions of location/attribute decision variables, an objective function (that serves as the goal of optimization), and a set of constraints (which limit the optimization of that objective function). The sense of optimization is to either minimize or maximize the objective function, which may consist of a single variable or a vector of decision variables. The many applications of spatial optimization are far too extensive to review thoroughly here, but some classic examples include minimizing distance to switching centers (Hakimi, 1964), maximizing the coverage of potential emergencies for ambulance location (Church & ReVelle, 1974), or finding the best bus routing pattern (Curtin & Biba, 2011).

Spatial Optimization in Individual Spatial Cognition

There are a number of spatial optimization methods that have been integrated into research regarding individual spatial cognition, and it is likely that they will appear in studies of collective spatial cognition as well. Perhaps the most common is the shortest path. When navigating through an environment, it is typical (although not universal) to choose the shortest—or least cost—path through the environment. The minimum cost path between a single source and destination on a network can be efficiently computed with the Dijkstra algorithm (Dijkstra, 1959), and this path can be compared against the path taken by participants to determine their ability to efficiently navigate the space. The measure of cost may be distance, time, amount of risk (real or perceived), or any other pertinent measure of impedance in traversing the network. Recent work has examined the use of landmark weights as an alternative to distances (Nuhn & Timpf, 2022).

When the navigation task extends to multiple stops between a source and the ultimate destination (such as in the challenge problem of a trio navigating a new city), the optimal path is the solution to the traveling salesman problem (Flood, 1956; Hoffman & Wolfe, 1985). This problem is much more difficult to solve optimally, and while there are purpose-built algorithms designed to extend the bounds of optimality (Applegate et al., 1998), many will instead use heuristics to determine a near-optimal solution, although in that case, the precise deviation from optimal cannot be determined (Curtin et al., 2014).

Spatial Optimization Solution Procedures as Tests of Collective Spatial Cognition

Optimization methods, when applied to all but the simplest problems, are often highly combinatorially complex—that is, there are an astronomical number of possible arrangements to examine in order to ensure that the optimal solution has been found. Because of this, a significant element of spatial optimization is concerned with the generation or improvement of solution procedures. There are optimal solution procedures that are used for a single problem, for example, the Dijkstra algorithm for the shortest path problem (Dijkstra, 1959) or the Ford and Fulkerson algorithm for maximizing flow across a network (Ford & Fulkerson, 1956); these are generally highly efficient but pertain only to that kind of problem. There are optimal procedures for general classes of optimization problems (e.g., the simplex method (Dantzig, 1957)), and while these methods extend the bounds of tractability, they may still fail to reach the optimal solution due to the size of the problem and limited time or available computing resources.

In those cases, heuristic procedures may be used to quickly determine solutions that have, in the past, been shown to generate near-optimal solutions. These heuristics typically are comprised of an iterative series of steps that search for a solution or for the next step toward a solution. In some cases, there is a fundamentally spatial component to the heuristics such as choosing the nearest neighbor in a "greedy"-type heuristic for solving a routing problem or a partitioning of the space such as in the Maranzana heuristic for the P-median problem (Maranzana, 1964). It is these solution procedures that suggest an avenue for insight into collective spatial cognition. Does the collective strategy resemble any optimization algorithms or heuristic techniques? If so, at what points in the sequence of steps does the strategy differ? To what extent do these differences affect the quality of the solution (nearness to optimality)? How do strategies used/steps taken by a trained team (military, search, and rescue) compare to those of an untrained team when compared to known heuristics?

It is actually possible that the observation of the heuristic that teams use to solve spatial optimization problems could generate new—and perhaps better performing—formal heuristic procedures that can be applied generally or to a class of problems. If this happened, decision-makers may be able to better predict a likely outcome of a collective acting in space. In the cases where the collective is attempting illicit or violent activities, this improved ability to predict their location or pattern could lead to more effective interdiction.

Broadening the Use of Spatial Optimization for Collective Spatial Cognition

Knowledge of the optimal location solution to a particular collective spatial task can serve as a baseline for evaluating team performance. For example, how closely do collective-chosen locations for aid stations during a natural disaster resemble the optimal locations for those stations? By varying the size of a team, dispersing decision-making power among a team, or varying the number of teams in an MTS, will outcomes converge on the optimal, diverge from it, or shift in some way? But, while some classic optimization tasks have been employed, the richness of problems that have been formulated for spatial optimization holds much more to offer for the field of collective spatial cognition.

For example, what if the task is to cover a space with teams that form an MTS, such as in a search and rescue operation? The maximal covering location problem (Church & ReVelle, 1974) provides the optimal solution to that problem for comparison with the collective's performance. Backup covering models may be of particular interest, given that they assume a collective where one member is able to provide backup to another when necessary (Curtin et al., 2010) (Figure 13.3). What if the team is charged with moving as much material as possible through a space and their capacity to carry the material is the constraint on performance? The maximal flow problem and its associated optimal algorithm can provide the optimal solution (Ford & Fulkerson, 1956). This same algorithm can be used to also

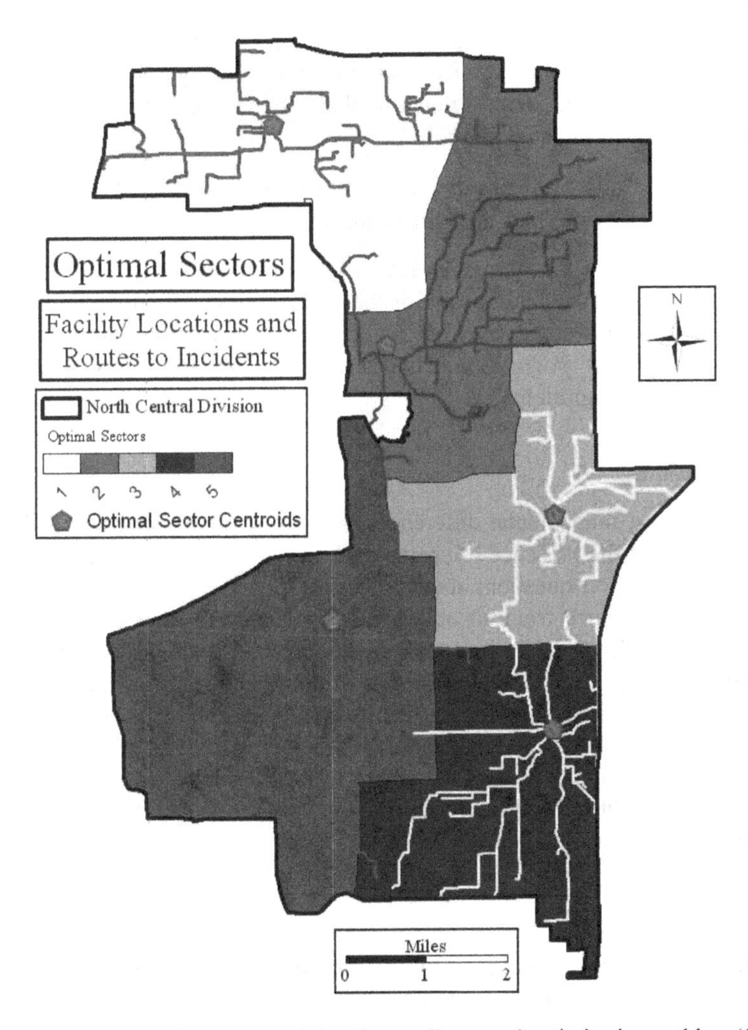

Figure 13.3 The optimal covering solution for a police patrol optimization problem (Curtin et al., 2010).

determine the minimum cut—the smallest number of edges or nodes that would need to be removed to disrupt the flow on the network. This would be useful if the goal of the team is to disrupt movement of an opposing collective. Should the team most efficiently reach all points in a network? Look at the minimum spanning tree problem (Kruskal, 1956). Does the team need to most efficiently divide the space into service areas and visit locations in those areas (e.g., delivering relief supplies) and return to the supply depots? The vehicle routing problem provides the solution (Dantzig & Ramser, 1959).

Many more examples could be provided here, but the point is that collectives may wish to accomplish tasks spanning a variety of spatial objectives (optimization) as well as multi-objective problems. Spatial optimization techniques provide a vast library of problem formulations that may very closely mimic those spatial goals and thereby provide a means of testing the collective's ability to achieve those goals. Although the extant literature contains uncounted numbers of these formulations, a bespoke new formulation that matches a collective goal and constraints could be formulated and solved.

Additional Classes of Tools and Techniques

It is not practical in this format to review even a small subset of the hundreds of analytical tools available in state-of-the-art GIS in any great depth. That said, in order to approach some level of completeness, this section highlights several additional classes of tools and techniques with potential application to studies of collective spatial cognition.

Spatial Database Query

One of the most fundamental differences between standard databases and spatial databases is that the presence of spatial data allows for explicitly spatial queries. These queries answer questions about properties such as intersection, containment, adjacency, or distance from an object. Studies along the lines of Fotheringham and Curtis (1999) or Andrews-Todd and Rapp (2024), which test the ability to identify states and capitals, could benefit from the ability to test mathematically whether or not a capital location as identified by a test subject is contained within the correct state boundary. If a spatial task involves team members identifying locations for search and rescue, the ability to identify those team members' (and under which conditions) chosen locations are within a critical distance of the true location would be useful.

Three-Dimensional Analysis

The transition from the traditional two-dimensional cartography with the inclusion of tools for three-dimensional visualization in modern GIS was a major shift in geovisualization. These tools have provided users with the ability to easily view surface topography including sub-surface features, to visualize movement through spaces (e.g., fly-throughs), and to more accurately model and display the spaces as they are used in real activities. These tools may be critical for studies of collective spatial cognition such as the challenge problem presented here where teams are fighting fires in a multi-story building. The movements of team members between floors of the building may be far more effectively communicated if they are visualized in three dimensions rather than as a series of two-dimensional maps of the individual floors. For studies in outdoor settings, three-dimensional tools permit test subjects to perceive and characterize elements of the natural landscape such as ridges and valleys, hills and mountains, and the height of woodlands.

Raster Analysis

A raster is a matrix of cells that each contains the values representing an attribute or observation. A raster can be used to represent discrete features as a collection of cells, but more commonly is used to represent continuous surfaces where the cells hold values associated with a continuous variable. Raster analysis is often used to analyze land use and cover, hydrologic flow and groundwater, elevation, viewsheds, and density mapping. It can also be used as the spatial basis for cellular automata.

Geographers frequently rely on remote sensing tools to both gather and visualize data about an environment. When evaluating the performance of a team in space, information about both their environment and their perception of the environment from various locations is valuable information. There are raster tools for determining viewsheds (the area a subject can see from a given location), computing hillshading for communicating changes in elevation, and calculating and representing slope and aspect, all of which are important pieces of information for a team traversing a landscape. Applying the concept of topology as separation to understand how topology impacts a team's navigation, a researcher could use raster-based measures of terrain traversal difficulty across the experimental environment to determine separation between individual members of a team (or teams within a multiteam system). There are raster-based tools for finding the least cost distance (or least difficulty, least risk, etc.) to every cell and the least cost path to get to that cell across the raster surface. These values could be compared with team performance in navigation in order to judge team performance or measure the change in performance after training or experience.

Spatio-Temporal Analysis

Spatio-temporal analysis attempts to answer questions and create information/knowledge about how phenomena (e.g., populations, land cover, individuals, and groups) change in both space and time. Geographers use a variety of tools to analyze spatial and temporal change for many purposes, including the tracking/analysis of vehicles, traffic, storms, criminal pathways, animal migration, human migration, diasporas, and more. While interest in spatio-temporal analysis is present among geographers (e.g., Eckley & Curtin, 2013), analytical tools have been slow to become integrated into off-the-shelf GIS software packages. The extent of spatio-temporal toolsets is largely limited to the ability to generate a temporal reference frame (equivalent to the spatial reference that is critical to spatial analysis) and tools for visualizing change or movement through time (e.g., hurricane tracks and migration pathways).

Outside of GIS, geographers share methods, models, and computational tools with ecologists and physicists to represent movement. Two extensively studied ecological phenomena are fish schooling and bird flocking (e.g., Ballerini et al., 2008). These studies not only offer applicable mathematical methods for modeling, tracking, analyzing, and predicting movement toward a common location with respect to space and time but also have contributed to experiments that offer insight into factors in collective spatial movement. For example, Berdahl et al. (2016) offered insight into the way that grouping enhances sensing capabilities in animal groups. Although cognition may be less significant in animal studies than in human studies, studies on cooperative (or competitive) foraging consider sharing of spatial information as well as decisions to share or not to share (Flack et al., 2018); this precedent of considering motivation in animals could be redressed by experiments related to team spatial cognition in humans.

It is straightforward to demonstrate that a spatial cluster of events may not be related to the same underlying process unless the events are also related to one another in time. Traffic accidents occurring at the same location but only many years apart are not likely to be caused by the same process. Similarly, traffic accidents that occur simultaneously but occur on different continents are not truly significant clusters. In addition to traffic accident analysis, spatio-temporal clusters have been studied in such varied contexts as spatial epidemiology, seismology, and animal behavior (Ansari et al., 2020). These methods have been applied to studies of collective spatial behavior in the context of goal-directed team movement (Barker, 2015) and could continue to inform experiments where the level of clustering or dispersion of team members is indicative of better or worse performance in learning about or achieving goals across space and time.

Conclusions

As stated previously, this chapter is concerned not with the ways in which spatial analytic techniques have been used in studies of collective spatial cognition, but rather what might come to pass if these methods are more well-known across the many disciplines that engage in this research. Although this approach is highly speculative, as one of the three pillars of this emerging research area, the full spectrum of spatial analytic techniques should be considered when seeking greater insight in group understanding of space and group behaviors that are a consequence of that understanding. It is hoped that this overview of the available methods will inform this growing research area. Of course, the set of spatial analytic tools and techniques—both inside GIS and out—is under continual development, and it is likely that this will lead to an even broader set of methods applicable to collective spatial cognition in the future.

References

Andrews-Todd, J., & Rapp, D. N. (2024). Adverse consequences of collaboration on spatial problem solving. In *Collective spatial cognition*. Milton Park: Routledge.

Ansari, M. Y., Ahmad, A., Khan, S. S., Bhushan, G., & Mainuddin. (2020). Spatiotemporal clustering: A review. *Artificial Intelligence Review*, *53*(4), 2381–2423. https://doi.org/10.1007/s10462-019-09736-1

Applegate, D., Bixby, R., Chvatal, V., & Cook, W. (1998). On the solution of traveling salesman problems. *Documenta Mathematica*, *3*, 645–656.

Armstrong, M. P., & Densham, P. J. (1995). Cartographic support for collaborative spatial decision-making. In *Proceedings of the International Symposium on Computer-Assisted Cartography*. Charlotte, NC: AutoCarto XII.

Ballerini, M., Cabibbo, N., Candelier, R., Cavagna, A., Cisbani, E., Giardina, I., Lecomte, V., Orlandi, A., Parisi, G., Procaccini, A., Viale, M., & Zdravkovic, V. (2008). Interaction ruling animal collective behavior depends on topological rather than metric distance: Evidence from a field study. *Proceedings of the National Academy of Sciences of the United States of America*, *105*(4), 1232–1237.

Barker, A. D. (2015). *Understanding the geographic dynamics of goal-directed social behaviors*. http://mars.gmu.edu/handle/1920/9809

Baumann, M. R., Kretz, D. R., & He, Q. (2024). A review of multiteam systems with an eye toward applications for collective spatial reasoning. In *Collective spatial cognition*. Milton Park: Routledge.

Berdahl, A., Westley, P. A. H., Levin, S. A., Couzin, I. D., & Quinn, T. P. (2016). A collective navigation hypothesis for homeward migration in anadromous salmonids. *Fish & Fisheries*, *17*(2), 525–542. https://doi.org/10.1111/faf.12084

Boccaletti, S., Latora, V., Moreno, Y., Chavez, M., & Hwang, D.-U. (2006). Complex networks: Structure and dynamics. *Physics Reports*, *424*(4), 175–308. https://doi.org/10.1016/j.physrep.2005.10.009

Breslow, L. A., Trafton, J. G., & Ratwani, R. M. (2009). A perceptual process approach to selecting color scales for complex visualizations. *Journal of Experimental Psychology: Applied*, *15*(1), 25–34. https://doi.org/10.1037/a0015085

Brewer, C. A., MacEachren, A. M., Pickle, L. W., & Herrmann, D. (1997). Mapping mortality: Evaluating color schemes for choropleth maps. *Annals of the Association of American Geographers*, *87*(3), 411–438.

Brewer, C. A., & Pickle, L. (2002). Evaluation of methods for classifying epidemiological data on choropleth maps in series. *Annals of the Association of American Geographers*, *92*(4), 662–681. https://doi.org/10.1111/1467-8306.00310

Church, R. L., & ReVelle, C. S. (1974). The maximal covering location problem. *Papers of the Regional Science Association*, *32*, 101–118.

Clark, P. J., & Evans, F. C. (1954). Distance to nearest neighbor as a measure of spatial relationships in populations. *Ecology*, *35*(4), 445–453. https://doi.org/10.2307/1931034

Curtin, K. M. (2007). Network analysis in geographic information science: Review, assessment, and projections. *Cartography and Geographic Information Science*, *34*(2), 103–111. https://doi.org/10.1559/152304007781002163

Curtin, K. M. (2008). Network analysis. In K. Kemp (Ed.), *The encyclopedia of geographic information science* (pp. 310–313). Los Angeles, CA: Sage Publications.

Curtin, K. M. (2009). *Network modeling [Chapter]. Handbook of research on geoinformatics*. IGI Global. https://doi.org/10.4018/978-1-59140-995-3.ch015

Curtin, K. M., & Biba, S. (2011). The transit route arc-node service maximization problem. *European Journal of Operations Research*, *208*(1), 46–56.

Curtin, K. M., Hayslett-McCall, K., & Qiu, F. (2010). Determining optimal police patrol areas with maximal covering and backup covering location models. *Networks and Spatial Economics*, *10*(1), 125–145. https://doi.org/10.1007/s11067-007-9035-6

Curtin, K. M., Nicoara, G., & Arifin, R. R. (2007). A comprehensive process for linear referencing. *URISA Journal*, *19*(2), 41–50.

Curtin, K. M., & Turner, D. (2019). Linear referencing. *Geographic Information Science & Technology Body of Knowledge*, *2019*(Q4). https://doi.org/10.22224/gistbok/2019.4.3

Curtin, K. M., Voicu, G., Rice, M. T., & Stefanidis, A. (2014). A comparative analysis of traveling salesman solutions from geographic information systems. *Transactions in GIS*, *18*(2), 286–301. https://doi.org/10.1111/tgis.12045

Dantzig, G. B. (1957). Discrete-variable extremum problems. *Operations Research*, *5*, 266–277.

Dantzig, G. B., & Ramser, J. H. (1959). The truck dispatching problem. *Management Science*, *6*(1), 80–91. https://doi.org/10.1287/mnsc.6.1.80

Dijkstra, E. W. (1959). A note on two problems in connexion with graphs. *Numerische Mathematik*, *1*, 269–271.

DuCruet, C., & Rodrigue, J. P. (2017, December 10). Graph theory: Measures and indices. *The Geography of Transport Systems*. https://transportgeography.org/?page_id=5981

Eckley, D., & Curtin, K. M. (2013). Evaluating the spatiotemporal clustering of traffic incidents. *Computers, Environment and Urban Systems*, *37*(1), 70–81. https://doi.org/10.1016/j.compenvurbsys.2012.06.004

Flack, A., Nagy, M., Fiedler, W., Couzin, I. D., & Wikelski, M. (2018). From local collective behavior to global migratory patterns in white storks. *Science*, *360*(6391), 911–914. https://doi.org/10.1126/science.aap7781

Flood, M. (1956). The traveling-salesman problem. *Operations Research*, *4*, 61–75.

Ford, L. R., & Fulkerson, D. R. (1956). Maximal flow through a network. *Canadian Journal of Mathematics*, *8*(0), 399–404. https://doi.org/10.4153/CJM-1956-045-5

Fotheringham, A. S., Brunsdon, C., & Charlton, M. (2002). *Geographically weighted regression: The analysis of spatially varying relationships*. Chichester: Wiley.

Fotheringham, A. S., & Curtis, A. (1999). Regularities in spatial information processing: Implications for modeling destination choice. *Professional Geographer*, *51*(2), 227. https://doi.org/10.1111/0033-0124.00159

Garrison, W., & Marble, D. (1962). The structure of transportation networks. *U.S. Army Transportation Command, Technical Report*, *62–II*, 100.

Getis, A., & Ord, J. K. (1992). The analysis of spatial association by use of distance statistics. *Geographical Analysis*, *24*(3), 189–206. https://doi.org/10.1111/j.1538-4632.1992.tb00261.x

Gunalp, P., Moossaian, T., & Hegarty, M. (2019). Spatial perspective taking: Effects of social, directional, and interactive cues. *Memory & Cognition*, *47*(5), 1031–1043. https://doi.org/10.3758/s13421-019-00910-y

Haggett, P., & Chorley, R. J. (1969). *Network analysis in geography*. London: Edward Arnold. https://trove.nla.gov.au/version/45731032

Hakimi, S. L. (1964). Optimum locations of switching centers and the absolute centers and medians of a graph. *Operations Research*, *12*(3), 450–459.

Hegarty, M. (2011). The cognitive science of visual-spatial displays: Implications for design. *Topics in Cognitive Science*, *3*(3), 446–474. https://doi.org/10.1111/j.1756-8765.2011.01150.x

Hegarty, M. (2013). Cognition, metacognition, and the design of maps. *Current Directions in Psychological Science*, *22*(1), 3–9. https://doi.org/10.1177/0963721412469395

Hegarty, M., Montello, D. R., Richardson, A. E., Ishikawa, T., & Lovelace, K. (2006). Spatial abilities at different scales: Individual differences in aptitude-test performance and spatial-layout learning. *Intelligence*, *34*(2), 151–176. https://doi.org/10.1016/j.intell.2005.09.005

Hillier, B., & Stonor, T. (2017, December 12). *Space syntax: Past, present and future*. London: Space Syntax: Urban Design Group, Kevin Lynch Memorial Lecture. https://www.youtube.com/watch?v=85BmaTMPQSA

Hoffman, A., & Wolfe, P. (1985). History. In *The traveling salesman problem—a guided tour of combinatorial optimization* (pp. 1–17). Chichester: John Wiley & Sons.

Kruskal, J. B. (1956). On the shortest spanning subtree and the traveling salesman problem. *Proceedings of the American Mathematical Society*, *7*, 48–50.

Larkin, J. H., & Simon, H. A. (1987). Why a diagram is (sometimes) worth Ten Thousand words. *Cognitive Science*, *11*(1), 65–100. https://doi.org/10.1111/j.1551-6708.1987.tb00863.x

Lewis, K. (2004). Knowledge and performance in knowledge-worker teams: A longitudinal study of transactive memory systems. *Management Science*, *50*(11), 1519–1533. https://doi.org/10.1287/mnsc.1040.0257

Lewis, K., & Herndon, B. (2011). Transactive memory systems: Current issues and future research directions. *Organization Science*, *22*(5), 1254–1265. https://doi.org/10.1287/orsc.1110.0647

Liben, L. S., & Downs, R. M. (1993). Understanding person-space-map relations: Cartographic and developmental perspectives. *Developmental Psychology*, *29*(4), 739–752. https://doi.org/10.1037/0012-1649.29.4.739

MacEachren, A. M., Brewer, C. A., & Pickle, L. W. (1998). Visualizing georeferenced data: Representing reliability of health statistics. *Environment and Planning A: Economy and Space*, *30*(9), 1547–1561. https://doi.org/10.1068/a301547

MacEachren, A. M., & Kraak, M.-J. (2001). Research challenges in geovisualization. *Cartography and Geographic Information Science*, *28*(1), 3–12. https://doi.org/10.1559/152304001782173970

Maranzana, F. E. (1964). On the location of supply points to minimize transport costs. *OR*, *15*(3), 261–270. https://doi.org/10.2307/3007214

Maupin, C. K., MacLaren, N. G., Goodwin, G. F., & Carter, D. R. (2024) Improving Wayfinding through Transactive Memory Systems. In *Collective Spatial Cognition*. Milton Park: Routledge.

May, M. (2004). Imaginal perspective switches in remembered environments: Transformation versus interference accounts. *Cognitive Psychology*, *48*(2), 163–206. https://doi.org/10.1016/s0010-0285(03)00127-0

Montello, D. R. (2002). Cognitive map-design research in the twentieth century: Theoretical and empirical approaches. *Cartography and Geographic Information Science*, *29*(3), 283–305.

Moran, P. A. P. (1950). Notes on continuous stochastic phenomena. *Biometrika*, *37*(1/2), 17–23. https://doi.org/10.2307/2332142

Moreland, R. L. (2008). Transactive memory systems in organizations: Implications for knowledge directories. In P. Jackson & J. Klobas (Eds.), *Decision support systems* (Vol. 44, pp. 409–424). http://www.sciencedirect.com/science/article/pii/S0167923607000784

Moreno, J. L. (1934). *Who shall survive: A new approach to the problem of human interrelations*. Beacon House. https://reflexus.org/wp-content/uploads/whoshallsurvive.pdf

Newman, M. E. J. (2003). The structure and function of complex networks. *SIAM Review, 45*(2), 167–256. https://doi.org/10.1137/S003614450342480

Novick, L. R., & Catley, K. M. (2007). Understanding phylogenies in biology: The influence of a gestalt perceptual principle. *Journal of Experimental Psychology. Applied, 13*(4), 197–223. https://doi.org/10.1037/1076-898X.13.4.197

Nuhn, E., & Timpf, S. (2022). Landmark weights—An alternative to spatial distances in shortest route algorithms. *Spatial Cognition & Computation*, 1–27. https://doi.org/10.1080/13875868.2022.2130330

Okabe, A., Okunuki, K., & Shiode, S. (2006). SANET: A toolbox for spatial analysis on a network. *Geographical Analysis, 38*, 57–66.

Okabe, A., & Satoh, T. (2006). Uniform network transformation for point pattern analysis on a non-uniform network. *Journal of Geographical Systems, 8*, 25–37.

Okabe, A., Satoh, T., & Sugihara, K. (2009). A kernel density estimation method for networks, its computational method and a GIS-based tool. *International Journal of Geographical Information Science, 23*(1), 7–32. https://doi.org/10.1080/13658810802475491

Okabe, A., & Yamada, I. (2001). The K-function method on a network and its computational implementation. *Geographical Analysis, 33*(3), 271–290. https://doi.org/10.1111/j.1538-4632.2001.tb00448.x

Okabe, A., Yomono, H., & Kitamura, M. (1995). Statistical analysis of the distribution of points on a network. *Geographical Analysis, 27*(2), 152–175.

Pawloski, L. R., Curtin, K. M., Gewa, C., & Attaway, D. (2012). Maternal–child overweight/obesity and undernutrition in Kenya: A geographic analysis. *Public Health Nutrition, 15*(11), 2140–2147. https://doi.org/10.1017/S1368980012000110

Pielou, E. C. (1960). A single mechanism to account for regular, random and aggregated populations. *Journal of Ecology, 48*(3), 575–584. https://doi.org/10.2307/2257334

Porta, S., Latora, V., & Strano, E. (2010). Networks in urban design. Six years of research in multiple centrality assessment. In E. Estrada, M. Fox, D. J. Higham, & G.-L. Oppo (Eds.), *Network science: Complexity in nature and technology* (pp. 107–129). London: Springer. https://doi.org/10.1007/978-1-84996-396-1_6

Presson, C. C., & Montello, D. R. (1994). Updating after rotational and translational body movements: Coordinate structure of perspective space. *Perception, 23*(12), 1447–1455. https://doi.org/10.1068/p231447

Richardson, A. E., Montello, D. R., & Hegarty, M. (1999). Spatial knowledge acquisition from maps and from navigation in real and virtual environments. *Memory & Cognition, 27*(4), 741–750. https://doi.org/10.3758/BF03211566

Rieser, J. J. (1989). Access to knowledge of spatial structure at novel points of observation. *Journal of Experimental Psychology. Learning, Memory, and Cognition, 15*(6), 1157–1165. https://doi.org/10.1037//0278-7393.15.6.1157

Rüttenauer, T. (2022). Spatial regression models: A systematic comparison of different model specifications using Monte Carlo experiments. *Sociological Methods and Research, 51*(2), 728–759. https://doi.org/10.1177/0049124119882467

Siew, C. S. Q., Wulff, D. U., Beckage, N. M., & Kenett, Y. N. (2019). Cognitive network science: A review of research on cognition through the lens of network representations, processes, and dynamics. *Complexity, 2019*, 2108423. https://doi.org/10.1155/2019/2108423

Simkin, D., & Hastie, R. (1987). An information-processing analysis of graph perception. *Journal of the American Statistical Association, 82*(398), 454–465. https://doi.org/10.1080/01621459.1987.10478448

Wegner, D. M., Giuliano, T., & Hertel, P. T. (1985). Cognitive interdependence in close relationships. In W. Ickes (Ed.), *Compatible and Incompatible Relationships* (pp. 253–276). Springer New York. https://doi.org/10.1007/978-1-4612-5044-9_12

Yamada, I., & Thill, J.-C. (2004). Comparison of planar and network K-functions in traffic accident analysis. *Journal of Transport Geography*, *12*(2), 149–158. https://doi.org/10.1016/j.jtrangeo.2003.10.006

Index

Printed in the United States
by Baker & Taylor Publisher Services